Electromagnetics and Transmission Lines

Electromagnetics and Transmission Lines

Essentials for Electrical Engineering

Second Edition

Robert A. Strangeway
Milwaukee School of Engineering
Milwaukee, WI, USA

Steven S. Holland
Milwaukee School of Engineering
Milwaukee, WI, USA

James E. Richie
Marquette University
Milwaukee, WI, USA

This edition first published 2023

© 2023 John Wiley & Sons, Inc.

Edition History

"Preliminary edition", RacademicS Publishing LLC, 2019

1 edition, RacademicS Publishing LLC, 2020

Registered Office

John Wiley & Sons, Inc., 111 River Street, Hoboken, NJ 07030, USA

Editorial Office

111 River Street, Hoboken, NJ 07030, USA

For details of our global editorial offices, customer services, and more information about Wiley products visit us at www.wiley.com.

Wiley also publishes its books in a variety of electronic formats and by print-on-demand. Some content that appears in standard print versions of this book may not be available in other formats.

Library of Congress Cataloging-in-Publication Data Applied for:

Hardback ISBN: 9781119881902

Cover Design: Wiley

Cover Image: © PitukTV/Shutterstock

Set in 9.5/12.5pt STIXTwoText by Straive, Pondicherry, India

This textbook is dedicated to our spouses, Debby, Melissa, and Jean, whose enduring love and unselfish sacrifices made this work possible, and to the late Dr. Thomas K. Ishii, our mentor and colleague.

Contents

Preface

It is the responsibility of every author to answer the proverbial question "Why is there another text on such an established subject?" There is a plentitude of quality texts that thoroughly cover electromagnetics and transmission lines, usually in two semesters. We deviate from the norm in this text with an explicit mission and goal:

Mission: What every electrical engineer, not just the electromagnetics specialist, should know about electromagnetic fields and transmission lines.

Goal: A textbook that enables completion of static electric and magnetic fields, dynamic electromagnetic fields, transmission lines, antennas, and signal integrity within a single course.

With this mission and goal in mind, we have designed a text that conveys rigor to the depth consistent with the mission and goal of the text, yet focuses on the essential concepts of electromagnetics and transmission lines and excludes those that are generally considered specialized. Admittedly this approach involved experiential judgments and strategic trade-offs. We mitigated some of the agony of topic exclusion by placing select topics into chapter appendices. For example, the differential forms of electromagnetics laws are not utilized in the main body of the text, but Maxwell's equations in differential form are introduced in a chapter appendix.

It is intended that the reader progresses through the sections in the main body of the text without the chapter appendices. The page count of the entire text was intentionally constrained for use in a single semester course. Overall, this focused approach on electromagnetic fundamentals provides the essentials for all electrical engineers yet also provides a solid electromagnetics foundation for students who wish to continue studies in elective or graduate-level electromagnetics courses.

Chapter appendices can be included or excluded without compromising the pedagogic and topical flow within the body of the work. If an appendix depends on a previous appendix, this association is declared initially within the appendix. The availability of chapter appendices adds a degree of content flexibility for those programs that have covered content from the earlier chapters in other courses (such as the first chapter), have multiple courses, have more credit hours, and so forth, to accommodate additional instructional content per specific course requirements.

A primary feature of the text is the inclusion of the underlying concepts of the vector network analyzer (VNA) as well as a brief but focused introduction to the instrument. Modern VNAs are cost-effective and utilized significantly in a wide range of electronic industries. We hope that this text serves as a primer for effective understanding of the VNA.

End-of-text appendices are included upon the recommendation of multiple reviewers. A valuable suggestion in this regard was to include a symbols listing (Appendices A and B). Four pages of symbols resulted! Students who are new to the subject matter often mention the difficulty of learning so many new symbols. This exercise confirmed the multitude of symbols that are used in electromagnetics (so instructors, be patient when you hear this comment). Tables with approximate values of physical constants and material properties (Appendices C and D) are given to communicate a "feel for the numbers" and for consistent usage in homework problems, not to be a reference of precise numbers which are readily available online. The equation summaries at the end of each chapter are collected in Appendix E.

The student background assumptions are multivariable calculus, DC and AC electric circuits, electromagnetic physics, and elementary differential equations. Multivariable integration is used extensively in several chapters. Cartesian vectors, the solution of two-dimensional point charge – static electric field problems using Coulomb's law, and basic concepts of static magnetic fields are normally a subset of electromagnetic physics courses. Additional topics in the physics of electromagnetics are helpful but not necessary. Elementary differential equations, namely a second-order linear differential equation with constant coefficients, are required for the modeling and solution of the phasor transmission line wave equation.

Organization

The latter part of this text covers applications with which any electrical engineer should be familiar: transmission lines, antennas, and basic concepts behind signal integrity. The pedagogy in the earlier chapters builds the essential concepts of electromagnetics and transmission lines that are prerequisite to these applications. The chapter descriptions follow.

- **Chapter 1**: Initially, vector algebra and coordinate systems are developed and examined carefully. The authors recommend that this material be mastered – do not assume it is automatically known. We have found in decades of teaching electromagnetic fields courses that students who have mastered vector algebra and coordinate systems perform and retain concepts significantly better than their counterparts who have not done so. The time spent in Chapter 1 is more than made up for in subsequent chapters because students learn subsequent material with a stronger background in "the vector tools," which results in more efficient and effective learning.
- **Chapter 2**: Coulomb's law and the Biot–Savart law, the superposition laws, are examined. We have grouped these two laws to reinforce the commonality of the superposition approach (we call this general approach "conceptual grouping"). The effective integration of a unit vector via implementation of symmetry is utilized to simplify resultant field expressions and, more importantly, to develop intuition into field behavior from canonical source geometries.
- **Chapter 3**: Gauss's law and solenoidal magnetic flux are similarly grouped to leverage the commonality of the flux viewpoint of fields. The visualization of an entire field pattern from canonical source distributions is the next natural step from the superposition laws of the previous chapter.
- **Chapter 4**: Electric potential, which we use interchangeably with the term "voltage" for static electric fields, and Ampere's circuital law are grouped to reinforce the commonality of the path law approach. Several fundamental electrical engineering applications are examined from the electromagnetic fields viewpoint, such as electric circuit laws, capacitance, self and mutual inductances, and basic magnetic circuit principles (the latter as a chapter appendix). They illustrate and reinforce the path law thinking as well as previously covered electromagnetic field concepts. Dielectric and magnetic material concepts are covered briefly. This static fields background sets the stage for time-changing electromagnetic fields.
- **Chapter 5**: Faraday's law, displacement current, and the integral form of Maxwell's equations are approached from a conceptual viewpoint to relate the mathematical expressions to physical understanding. Several fundamental electrical engineering applications are again utilized to reinforce the electromagnetic fields – circuit relationships. The differential forms of Maxwell's equations, including explanations of the del operator, divergence, and curl, are covered in a chapter appendix to provide exposure to the forms often encountered in the literature.
- **Chapter 6**: The concept of DC transient and AC steady-state waves and reflections on a transmission line (T-line) are examined initially. This approach leads into measures of reflection, Smith charts, and scattering parameters. The T-line model and electrical distance considerations are delayed until the next chapter.

We have found that separating waves and reflections from electrical distance concepts improves the pedagogy because students are learning fewer concepts simultaneously. The VNA is introduced at the end of this chapter.

- **Chapter 7**: The concepts from Chapter 6 are expanded by modeling the T-line and solving for the AC steady-state voltage, current, and impedance expressions. AC standing waves and the Smith chart are consequently expounded. The visualization of T-line behavior via the Smith chart is emphasized. Many of the T-line relations that are utilized in practice result from this chapter. Some of the detailed mathematical developments that are not essential to the pedagogic flow are separated into the chapter appendices.
- **Chapter 8**: An intuitive explanation of electromagnetic radiation is used to introduce antennas. Uniform plane wave concepts are introduced with analogies to T-lines, and the Poynting vector is used to determine the power received from a uniform plane wave by an antenna. Basic antenna parameters such as antenna gain and beamwidths are examined. Finally, link loss is developed and applied via the Friis transmission equation to determine the power received in a link.
- **Chapter 9**: Signal integrity concepts are introduced. T-line effects are considered in the context of signal integrity. Crosstalk is related to Faraday's law and displacement current from Chapter 5. Electromagnetic interference (EMI), especially the measurements, is explained to a somewhat greater extent given that most students will not be familiar with this topic. Ground bounce is only mentioned. Electromagnetic safety/human exposure and electrostatic discharge are not addressed in this text.

Each chapter has an equation summary and a variety of homework problems: quantitative, explanatory, and development, and some applications. Select answers are given at the end of the text so that students can check the results of their own work.

Ancillaries

Adopters of the text have the following ancillaries available to them:

- Homework solution manual
- Lecture notes (not slides) – blank and answered versions, to promote interactivity
- VNA experiments

Acknowledgments

The authors would like to acknowledge the following reviewers for their evaluation of the manuscript during its development: Dr. Donovan Brocker, Assistant Professor at Milwaukee School of Engineering; Dr. Justin Creticos, Principal Sensor Systems Engineer at the MITRE Corporation; Dr. David Haas, Visiting Assistant Professor (Physics) at Marquette University; Dr. Jovan Jevtic, Chief Scientist at Radom Corporation and Adjunct Associate Professor at Milwaukee School of Engineering; Dr. K.C. Kerby-Patel, Associate Professor at University of Massachusetts Boston; Dr. Owe Petersen, Professor Emeritus at Milwaukee School of Engineering; Professor Brian Petted, Technology Leader at Laird Connectivity and Adjunct Associate Professor at Milwaukee School of Engineering; Professor David J. Schmocker, P.E., Instructor at Milwaukee Area Technical College, call sign KJ9I; Mr. Eric Strangeway, P.E., Principal Field Applications Engineer at Wisconsin Public Service Corporation; Dr. Travis Thul, P.E., Senior Fellow and Director of Operations at Technological Leadership Institute, University of Minnesota, and Chair of ANSI C63.30 Working Group for Wireless Power Transfer; Dr. Mark Wolski, Senior Antenna and RF Engineer at Laird Connectivity; and Dr. Jie Xu, Professor at Loyola Marymount University. We thank them for their generous efforts, comments, and critique, all of which have improved the quality of this work. The authors request that any errors, corrections, or suggestions be forwarded to Dr. Robert Alan Strangeway (strangew@msoe.edu).

Robert A. Strangeway, PhD
Steven S. Holland, PhD
James E. Richie, PhD

About the Authors

Robert A. Strangeway, PhD, is a professor of electrical engineering and a transfer track coordinator in the Department of Electrical Engineering and Computer Science at Milwaukee School of Engineering (MSOE) where he has taught electromagnetic fields and transmission lines, electric circuits, and signals and systems courses for 40 years. He has received two teaching awards at MSOE. He is also an adjunct professor in the Department of Biophysics at Medical College of Wisconsin (MCW) where he was involved in research, development, and implementation of low phase noise microwave sources and multi-arm bridge configurations for electron paramagnetic resonance experiments for 35 years. Before this, he was a millimeter-wave engineer at TRW for three years. He is the lead author of two educational textbooks, *Electric Circuits* and the current text. He is a coauthor of several publications associated with his MCW work and in engineering education. Dr. Strangeway is a member of IEEE, ASEE, Eta Kappa Nu, and Tau Alpha Pi.

Steven S. Holland, PhD, is an associate professor and the electrical engineering program director in the Department of Electrical Engineering and Computer Science at Milwaukee School of Engineering (MSOE). He earned his BSEE degree from MSOE and his MSECE and PhD from the University of Massachusetts Amherst. Prior to joining the faculty at MSOE in 2013, he was a senior sensors engineer with the MITRE Corporation. He is active in both engineering education and technical research, regularly publishing papers and presenting workshops with interests in active learning methods, mobile measurement platform curriculum integration, antenna miniaturization, UWB antenna arrays, and HF antennas. He primarily teaches courses in analog electronics, electromagnetics, and antennas, and advises undergraduate research students. Dr. Holland is a senior member of the IEEE and is also a member of ASEE and Tau Beta Pi. He holds an Extra Class amateur radio license, call sign AC9UX.

James E. Richie, PhD, has been in the Department of Electrical and Computer Engineering at Marquette University, Milwaukee, WI, since 1988, where he is currently an associate professor and associate department chair. His research interests are in antennas, electromagnetic scattering, and inverse scattering. He teaches both undergraduate and graduate classes in microwaves, antennas, and electromagnetic theory. He has received several department- and college-level teaching awards. Dr. Richie is a member of IEEE, Eta Kappa Nu, Tau Beta Pi, and Sigma Xi.

About the Companion Website

This book is accompanied by a companion website.

www.wiley.com/go/Strangeway/ElectromagneticsandTransmissionLines

1

Vectors, Vector Algebra, and Coordinate Systems

CHAPTER MENU

Motivation *Why spend a chapter on vectors?* The degree to which you master the vector concepts and techniques in this chapter will generally determine the degree to which you understand and can apply the essential concepts and techniques of electromagnetic fields. Vectors are not just important tools in the calculation of electromagnetic field quantities. They are essential in the *visualization* of electromagnetic fields in an organized, dependable manner. In short, a mastery of vectors instills the "thought infrastructure" that allows one to visualize and apply electromagnetic fields in electrical engineering applications. *Enjoy*!

1.1 Vectors

This chapter develops the tools necessary to define and manipulate various types of vectors in three different coordinate systems. We begin with how vectors can be used to describe location and displacement.

What is a **vector**? It is a quantity with magnitude (scalar) and direction. What are examples of vectors? Velocity, force, acceleration, and so forth. How is direction in three-dimensions expressed? Start with the following vector definition:

> A **position vector** $\bar{r}$ locates a position in space with respect to the origin, that is, it is a vector that starts at the origin and ends at the point of the designated position in space (notation **r** is also used in some textbooks and literature).

Electromagnetics and Transmission Lines: Essentials for Electrical Engineering, Second Edition.
Robert A. Strangeway, Steven S. Holland, and James E. Richie.
© 2023 John Wiley & Sons, Inc. Published 2023 by John Wiley & Sons, Inc.
Companion website: www.wiley.com/go/Strangeway/ElectromagneticsandTransmissionLines

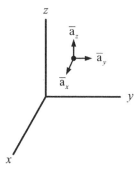

Figure 1.1 Cartesian (or rectangular) unit vectors.

Example 1.1.1

How is the position vector to point $P(2,3)$ expressed?

Strategy

$$\bar{r} = (\text{magnitude of } \bar{r})\,(\text{direction of } \bar{r})$$

Solution

What is the magnitude? Try the Pythagorean Theorem:

$$|\bar{r}| = r = \sqrt{x^2 + y^2} = \sqrt{2^2 + 3^2} = \sqrt{13} = 3.6056 \text{ (see footnote}^1)$$

Note: no bar over a vector variable means it is the *magnitude* of the vector, that is, $|\bar{r}| = r$

What is the direction of $\bar{r}$? Express it in terms of *known* directions.

Let's interrupt Example 1.1.1 with some discussion on vector direction. How are *known* directions expressed? By a **unit vector**: a vector with unit (one) magnitude. Thus, when a unit vector is multiplied by a scalar, it creates a vector in the desired direction with a magnitude equal to the scalar.

> **Unit vector notation**: $\bar{a}_x$ (or $\mathbf{a}_x$), where: $\bar{a}$ (or $\mathbf{a}$) indicates the vector is a unit vector (a without a bar or not bold is something else, often a dimension or a radius), and subscript x indicates the direction of the unit vector.

See Figure 1.1 for the three unit vectors in the Cartesian coordinate system.

Example 1.1.1 resumed: direction of $\bar{r}$

How does one form a unit vector in an *arbitrary* direction? Again, use known directions:

- "Start at 0, go 2 units in the x direction and 3 units in the y direction to reach P"
- Thus, $\bar{r} = 2\bar{a}_x + 3\bar{a}_y$ is the position vector from 0 to P, as shown in Figure 1.2.
- How is $\bar{r}$ expressed as (magnitude of $\bar{r}$) (direction of $\bar{r}$)?
- The (magnitude of $\bar{r}$) $= |\bar{r}| = r = 3.6056$, as before.
- The (direction of $\bar{r}$) is $\bar{a}_r$. In this example, $\bar{a}_r$ is not yet determined, so we need to express it in terms of the known Cartesian unit vectors.

Multiply r by the unit vector $\bar{a}_r$ to express $\bar{r}$:

$$\boxed{\bar{r} = r\bar{a}_r} \qquad (1.1)$$

How is the unit vector determined?

$$\frac{\bar{r}}{r} = \frac{r\bar{a}_r}{r} \quad \rightarrow \quad \boxed{\bar{a}_r = \frac{\bar{r}}{r}} \qquad (1.2)$$

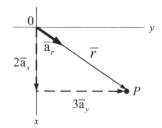

Figure 1.2 A position vector example.

In our example, $\bar{a}_r = \dfrac{\bar{r}}{r} = \dfrac{2\bar{a}_x + 3\bar{a}_y}{3.6056} = 0.55470\bar{a}_x + 0.83205\bar{a}_y$

Check: (magnitude of $\bar{a}_r$) $= |\bar{a}_r| = \sqrt{0.55470^2 + 0.83205^2} = 1$ (within rounding error)

Thus, $\bar{r} = 2\bar{a}_x + 3\bar{a}_y = r\bar{a}_r = \underbrace{(3.61)}_{\text{mag}} \underbrace{(0.555\bar{a}_x + 0.832\bar{a}_y)}_{\text{direction}}$

Note from the previous example that a general unit vector such as $\bar{a}_r$ is expressed in terms of known unit vectors. In vector terminology, $2\bar{a}_x$ and $3\bar{a}_y$ are the **vector components** of vector $\bar{r}$.

---

1 Often one uses five significant digits during manual calculations and rounds the final result to three significant digits. Whole numbers are often shown without the trailing significant zeros.

Example 1.1.2

Given $P(1,-2,4)$, determine:

a) the position vector to P, and
b) the unit vector in the direction from 0 to P.

Solution

See Figure 1.3.

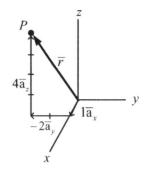

Figure 1.3 Position vector example.

$$\bar{r} = 1\bar{a}_x - 2\bar{a}_y + 4\bar{a}_z$$

$$\bar{a}_r = \frac{\bar{r}}{r} = \frac{+1\bar{a}_x - 2\bar{a}_y + 4\bar{a}_z}{\sqrt{1^2 + 2^2 + 4^2}} = \frac{+1\bar{a}_x - 2\bar{a}_y + 4\bar{a}_z}{4.5826} = +0.21822\bar{a}_x - 0.43644\bar{a}_y + 0.87287\bar{a}_z$$

$$\bar{r} = 1\bar{a}_x - 2\bar{a}_y + 4\bar{a}_z = \underbrace{(4.58)}_{|\bar{r}|}\underbrace{(+0.218\bar{a}_x - 0.436\bar{a}_y + 0.873\bar{a}_z)}_{\bar{a}_r}$$

Can the vector and unit vector expressions be generalized? Let $\bar{A}$ be a vector.

$$\bar{A} = A_x\bar{a}_x + A_y\bar{a}_y + A_z\bar{a}_z \tag{1.3}$$

where A_x is the magnitude of $\bar{A}$ in the x direction, and similarly for A_y and A_z. The magnitude of $\bar{A}$ is

$$A = |\bar{A}| = \sqrt{A_x^2 + A_y^2 + A_z^2} \tag{1.4}$$

and the unit vector in the direction of $\bar{A}$ is

$$\boxed{\bar{a}_A = \frac{\bar{A}}{A} = \frac{A_x\bar{a}_x + A_y\bar{a}_y + A_z\bar{a}_z}{\sqrt{A_x^2 + A_y^2 + A_z^2}} = \frac{A_x}{A}\bar{a}_x + \frac{A_y}{A}\bar{a}_y + \frac{A_z}{A}\bar{a}_z} \tag{1.5}$$

A position vector establishes displacement from the origin. What is a vector that establishes displacement between two points, neither of which must be the origin? Refer to Figure 1.4. This vector is named a distance vector:

> A **distance vector** $\bar{R}$ locates the position of one point with respect to another point (displacement), that is, it is a vector from one point in space to another point in space, neither of which are necessarily the origin (notation **R** is also used in some textbooks and literature).

How do the three vectors in Figure 1.4 relate to each other? By vector summation:

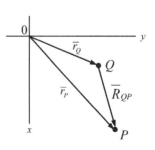

$$\bar{r}_P = \bar{r}_Q + \bar{R}_{QP} \quad \rightarrow \quad \bar{R}_{QP} = \bar{r}_P - \bar{r}_Q = \text{"end point" position vector}$$
$$- \text{"start point" position vector}$$

Thus, the distance vector from Q to P is:

$$\boxed{\bar{R}_{QP} = \bar{r}_P - \bar{r}_Q} \tag{1.6}$$

Note that P is often the "end point" and Q is often the "start point" in electromagnetics notation.

Figure 1.4 Position and distance vectors.

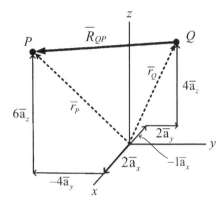

Figure 1.5 Sketch for distance vector example.

Example 1.1.3

Given $Q(-1, +2, +4)$ and $P(+2, -4, +6)$, determine:

(a) $\overline{R}_{QP}$, and (b) the unit vector in the direction of $\overline{R}_{QP}$.

Solution

Refer to Figure 1.5 to visualize $\overline{R}_{QP}$.

$$Q(-1, +2, +4) \rightarrow \overline{r}_Q = -1\overline{a}_x + 2\overline{a}_y + 4\overline{a}_z$$
$$P(+2, -4, +6) \rightarrow \overline{r}_P = +2\overline{a}_x - 4\overline{a}_y + 6\overline{a}_z$$
$$\overline{R}_{QP} = \overline{r}_P - \overline{r}_Q = \left(2\overline{a}_x - 4\overline{a}_y + 6\overline{a}_z\right) - \left(-1\overline{a}_x + 2\overline{a}_y + 4\overline{a}_z\right)$$

How does one subtract (or add) two vectors?
Subtract (or add) the corresponding vector components.

$$\overline{R}_{QP} = \overline{r}_P - \overline{r}_Q = \overline{\left(+2\overline{a}_x - 4\overline{a}_y + 6\overline{a}_z\right)}$$
$$-\left(-1\overline{a}_x + 2\overline{a}_y + 4\overline{a}_z\right)$$
$$\overline{R}_{QP} = +3\overline{a}_x - 6\overline{a}_y + 2\overline{a}_z$$

$$\overline{a}_R = \frac{\overline{R}_{QP}}{R_{QP}} = \frac{+3\overline{a}_x - 6\overline{a}_y + 2\overline{a}_z}{\sqrt{3^2 + 6^2 + 2^2}} = +0.42857\overline{a}_x - 0.85714\overline{a}_y + 0.28571\overline{a}_z$$
$$\overline{a}_R = +0.429\overline{a}_x - 0.857\overline{a}_y + 0.286\overline{a}_z$$

1.2   Vector Algebra

So far, the vector operations of addition and subtraction have been examined. Can we directly multiply or divide vectors? A better question is: what are some of the useful "relationships" between vectors?

- The extent to which two vectors are *parallel*.
- The extent to which two vectors are *perpendicular*.
- The projection of one vector onto another vector, that is, how much of one vector is in the direction of another vector. Two types of projections:

  - Scalar projection: magnitude only
  - Vector projection: magnitude and direction

How to determine these relationships between two vectors is examined in the following sections.

1.2.1   Dot Product

The dot product is a measure of the parallelness of two vectors. The definition of the vector **dot product** is

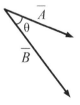

Figure 1.6 Angle between two vectors.

$$\boxed{\overline{A} \cdot \overline{B} = |\overline{A}||\overline{B}| \cos\theta = AB\cos\theta} \tag{1.7}$$

where θ is the angle between the two vectors, as shown in Figure 1.6. Note that the dot product of two vectors is a *scalar* result.

What is the dot product if : $\overline{A}$ and $\overline{B}$ are parallel? $\overline{A} \cdot \overline{B} = AB$ Why?

$\overline{A}$ and $\overline{B}$ are perpendicular? $\overline{A} \cdot \overline{B} = 0$ Why?

$\overline{A}$ and $\overline{B}$ are antiparallel? $\overline{A} \cdot \overline{B} = -AB$ Why?

The reason is the value of $\cos(\theta)$ when $\theta = 0°$, $\pm 90°$, and $\pm 180°$, respectively. Hence, the dot product result is somewhere within the range of:

$$-AB \leq \overline{A} \cdot \overline{B} \leq AB \tag{1.8}$$

As the dot product approaches $\pm AB$, the two vectors are more parallel (or antiparallel). Conversely, as the dot product approaches zero, the two vectors are more orthogonal (perpendicular).

While Eq. (1.7) defines the dot product, the angle between the two vectors must be known in order to determine the dot product. This angle is usually not known in advance. How is the dot product determined when the angle between the vectors is unknown? Start with vectors expressed in Cartesian form and apply the distributive property of the dot product:

$$\overline{A} \cdot \overline{B} = \left(A_x \overline{a}_x + A_y \overline{a}_y + A_z \overline{a}_z\right) \cdot \left(B_x \overline{a}_x + B_y \overline{a}_y + B_z \overline{a}_z\right) \tag{1.9}$$

$$\begin{aligned}
\overline{A} \cdot \overline{B} = {} & A_x \overline{a}_x \cdot B_x \overline{a}_x + A_x \overline{a}_x \cdot B_y \overline{a}_y + A_x \overline{a}_x \cdot B_z \overline{a}_z \\
& + A_y \overline{a}_y \cdot B_x \overline{a}_x + A_y \overline{a}_y \cdot B_y \overline{a}_y + A_y \overline{a}_y \cdot B_z \overline{a}_z \\
& + A_z \overline{a}_z \cdot B_x \overline{a}_x + A_z \overline{a}_z \cdot B_y \overline{a}_y + A_z \overline{a}_z \cdot B_z \overline{a}_z
\end{aligned} \tag{1.10}$$

Examine the first term:

$$A_x \overline{a}_x \cdot B_x \overline{a}_x = A_x B_x \cos\theta \quad \text{where } \theta \text{ is the angle between } \overline{a}_x \text{ and } \overline{a}_x \tag{1.11}$$

These two unit vectors are identical and therefore parallel; hence, the angle is $0°$ and $\cos(0°)$ is one. The key concept is that the *dot product between parallel unit vectors is one*. Thus, the first term is:

$$A_x \overline{a}_x \cdot B_x \overline{a}_x = A_x B_x \tag{1.12}$$

Similarly, the fifth and ninth terms are:

$$A_y \overline{a}_y \cdot B_y \overline{a}_y = A_y B_y \text{ and } A_z \overline{a}_z \cdot B_z \overline{a}_z = A_z B_z \tag{1.13}$$

What is the dot product between perpendicular unit vectors? Examine the second term:

$$A_x \overline{a}_x \cdot B_y \overline{a}_y = A_x B_y \cos\theta \quad \text{where } \theta \text{ is the angle between } \overline{a}_x \text{ and } \overline{a}_y \tag{1.14}$$

These two unit vectors are perpendicular; hence, the angle is $90°$ and $\cos(90°)$ is zero. Thus, the second term is:

$$A_x \overline{a}_x \cdot B_y \overline{a}_y = 0 \tag{1.15}$$

Similarly, all other terms where the unit vectors are perpendicular are zero. The key concept is that *the dot product between perpendicular unit vectors is zero*. Note that the dot product between two vectors with an unknown angle has been rearranged into dot products between unit vectors with known angles and known dot product results. Inserting these results into Eq. (1.10) gives:

$$\overline{A} \cdot \overline{B} = A_x \overline{a}_x \cdot B_x \overline{a}_x + A_y \overline{a}_y \cdot B_y \overline{a}_y + A_z \overline{a}_z \cdot B_z \overline{a}_z \tag{1.16}$$

$$\boxed{\overline{A} \cdot \overline{B} = A_x B_x + A_y B_y + A_z B_z \quad \text{(a scalar result)}} \tag{1.17}$$

This result is both powerful and convenient! When vectors are expressed in Cartesian form, the dot product is determined by summing the three products of corresponding vector components (those in the same direction). *We do not need to know the angle between the vectors in advance!*

Example 1.2.1

Given $\overline{A} = +2\,\overline{a}_x + 0\,\overline{a}_y - 3\,\overline{a}_z$ and $\overline{B} = +4\,\overline{a}_x - 6\,\overline{a}_y + 8\,\overline{a}_z$, determine

a) the dot product
b) the range of possible dot product values if the angle between the vectors were changed, and
c) the angle between the two original vectors.

Solution

a) $\overline{A} \cdot \overline{B} = \left(+2\,\overline{a}_x + 0\,\overline{a}_y - 3\,\overline{a}_z \right) \cdot \left(+4\,\overline{a}_x - 6\,\overline{a}_y + 8\,\overline{a}_z \right) = (2)(4) + (0)(-6) + (-3)(8) = -16$

b) $\left| \overline{A} \right| = A = \sqrt{2^2 + 0^2 + (-3)^2} = 3.6056$ $\quad \left| \overline{B} \right| = B = \sqrt{4^2 + (-6)^2 + 8^2} = 10.770$

   If the vectors are parallel,

$$AB = (3.6056)(10.770) = 38.8.$$

   If the vectors are perpendicular,

$$AB = 0.$$

   If the vectors are antiparallel,

$$AB = -(3.6056)(10.770) = -38.8.$$

   Thus, these two vectors are somewhere between perpendicular and antiparallel.

c) Can the dot product be used to determine the angle between two vectors in three-dimensional space?

$$\overline{A} \cdot \overline{B} = \left| \overline{A} \right| \left| \overline{B} \right| \cos\theta = AB\cos\theta \rightarrow \cos\theta = \frac{\overline{A} \cdot \overline{B}}{AB} \rightarrow \theta = \cos^{-1}\left(\frac{\overline{A} \cdot \overline{B}}{AB} \right) \tag{1.18}$$

   In this example:

$$\theta = \cos^{-1}\left(\frac{\overline{A} \cdot \overline{B}}{AB} \right) = \cos^{-1}\left(\frac{-16}{38.833} \right) = 114.33° \approx 114°$$

which confirms that the two vectors are somewhere between perpendicular and anti-parallel.

Another aspect of the relationship between two vectors is the length of one vector in the direction of another vector. One can apply trigonometry to Figure 1.7 to determine that the length of the vector component of $\overline{A}$ in the direction of $\overline{B}$ is $A\cos(\theta)$.

Can $A\cos(\theta)$ be determined from the dot product?

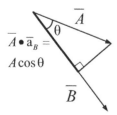

$$A\cos\theta = A\left(\frac{\overline{A} \cdot \overline{B}}{AB} \right) = \frac{\overline{A} \cdot \overline{B}}{B} = \overline{A} \cdot \overline{a}_B \tag{1.19}$$

Thus, the **scalar projection** of $\overline{A}$ onto $\overline{B}$ (that is, the length of vector $\overline{A}$ that is in the direction of vector $\overline{B}$) is

$$\overline{A} \cdot \overline{a}_B = \frac{\overline{A} \cdot \overline{B}}{B} \tag{1.20}$$

Figure 1.7 Scalar projection of one vector onto another vector.

How would the *vector component* of one vector in the direction of another vector be determined? Refer to Figure 1.8.

The **vector projection** of $\overline{A}$ onto $\overline{B}$ is found by multiplying the scalar projection with the direction of $\overline{B}$, that is, $\overline{a}_B$:

$$\left(\overline{A} \cdot \overline{a}_B\right)\overline{a}_B = \left(\overline{A} \cdot \frac{\overline{B}}{B}\right)\frac{\overline{B}}{B} = \left(\frac{\overline{A} \cdot \overline{B}}{B^2}\right)\overline{B} \tag{1.21}$$

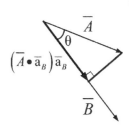

$$\left(\overline{A} \cdot \overline{a}_B\right)\overline{a}_B$$

Figure 1.8 Vector projection of one vector onto another vector.

Example 1.2.2

Given $\overline{A} = +12\,\overline{a}_x + 7\,\overline{a}_y - 2\,\overline{a}_z$ and $\overline{B} = +6\,\overline{a}_x - 5\,\overline{a}_y + 3\,\overline{a}_z$, determine:

a) $\overline{R}_{BA}$ (the vector from the point located by $\overline{B}$ to the point located by $\overline{A}$),
b) the length of $\overline{A}$ in the $\overline{B}$ direction (*scalar projection* of $\overline{A}$ onto $\overline{B}$), and
c) the vector component of $\overline{A}$ in the $\overline{B}$ direction (*vector projection* of $\overline{A}$ onto $\overline{B}$).

Solution

a) $\overline{R}_{BA} = \overline{A} - \overline{B} = \left(+12\,\overline{a}_x + 7\,\overline{a}_y - 2\,\overline{a}_z\right) - \left(+6\,\overline{a}_x - 5\,\overline{a}_y + 3\,\overline{a}_z\right) = +6\,\overline{a}_x + 12\,\overline{a}_y - 5\,\overline{a}_z$

b) $\overline{A} \cdot \overline{a}_B = \dfrac{\overline{A} \cdot \overline{B}}{B} = \dfrac{\left(+12\,\overline{a}_x + 7\,\overline{a}_y - 2\,\overline{a}_z\right) \cdot \left(+6\,\overline{a}_x - 5\,\overline{a}_y + 3\,\overline{a}_z\right)}{\sqrt{6^2 + 5^2 + 3^2}} = 3.7052$

c) $\left(\overline{A} \cdot \overline{a}_B\right)\overline{a}_B = \dfrac{\overline{A} \cdot \overline{B}}{B^2}\overline{B} = \dfrac{31}{(6^2 + 5^2 + 3^2)}\left(+6\,\overline{a}_x - 5\,\overline{a}_y + 3\,\overline{a}_z\right) = +2.66\,\overline{a}_x - 2.21\,\overline{a}_y + 1.33\,\overline{a}_z$

How is the dot product useful in physical problems? Recall that work is the component of force *in the direction* of motion multiplied by distance. The dot product can be used to determine this component of force: work $=\int \overline{F} \cdot \overline{dL}$. Relationships like this are common in electromagnetics. Another example of the usefulness of the dot product: parallel (tangential) and perpendicular (normal) components of a vector to a boundary.

Example 1.2.3

Determine the general expressions for the parallel and normal components of vector $\overline{A}$ to a straight boundary.

Solution

Define a vector $\overline{B}$ along the boundary. Refer to Figure 1.9.

$$\overline{A} = \overline{A}_t + \overline{A}_n$$
$$\overline{A}_t = \left(\overline{A} \cdot \overline{a}_B\right)\overline{a}_B$$
$$\overline{A} = \overline{A}_t + \overline{A}_n \rightarrow \overline{A}_n = \overline{A} - \overline{A}_t \rightarrow \overline{A}_n = \overline{A} - \left(\overline{A} \cdot \overline{a}_B\right)\overline{a}_B$$

Given specific numbers, how could you check this result? $(\overline{A}_t \cdot \overline{A}_n = 0)$

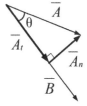

Figure 1.9 Tangential and normal components of vector $\overline{A}$ relative to a boundary, where $\overline{B}$ aligns with the boundary of concern.

1.2.2 Cross Product

Another "relationship" between vectors is the extent to which two vectors are perpendicular. The definition of the vector **cross product** is

$$\boxed{\overline{A} \times \overline{B} = |\overline{A}||\overline{B}| \sin\theta\,\overline{a}_n = AB\sin\theta\,\overline{a}_n} \tag{1.22}$$

where $\overline{a}_n$ is the unit vector *normal* (perpendicular) to the plane containing the two vectors (see Figure 1.10), as determined by the Right-Hand-Rule (RHR):

Use the curled fingers of your right hand to rotate the first vector ($\overline{A}$) into the second vector ($\overline{B}$) through the smallest angle, as illustrated in Figure 1.10. Extend the thumb of your right hand and it points in the $\overline{a}_n$ direction.

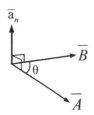
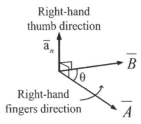

Figure 1.10 Unit vector normal to the plane containing vectors $\overline{A}$ and $\overline{B}$.

Note that the cross product of two vectors is a *vector* result.

What is the cross product if :	$\overline{A}$ and $\overline{B}$ are parallel?	$\overline{A} \times \overline{B} = 0$	Why?
	$\overline{A}$ and $\overline{B}$ are perpendicular?	$\overline{A} \times \overline{B} = AB\,\overline{a}_n$	Why?
	$\overline{A}$ and $\overline{B}$ are antiparallel?	$\overline{A} \times \overline{B} = 0$	Why?

The reason is the value of the $\sin(\theta)$ when $\theta = 0°$, $\pm 90°$, and $\pm 180°$, respectively. Hence, the magnitude of the cross product result is somewhere within the range of:

$$0 \le |\overline{A} \times \overline{B}| \le AB \tag{1.23}$$

As the magnitude of the cross product approaches AB, the two vectors are more orthogonal (perpendicular). Conversely, as the magnitude of the cross product approaches zero, the two vectors are more parallel or antiparallel.

Similar to what we initially encountered with the dot product, the cross product definition of Eq. (1.22) requires advance knowledge of the angle between the two vectors. How is the cross product determined when the angle between the vectors is unknown? Start with vectors expressed in Cartesian form and apply the distributive property of the cross product:

$$\overline{A} \times \overline{B} = \left(A_x\overline{a}_x + A_y\overline{a}_y + A_z\overline{a}_z\right) \times \left(B_x\overline{a}_x + B_y\overline{a}_y + B_z\overline{a}_z\right) \tag{1.24}$$

$$\begin{aligned}\overline{A} \times \overline{B} = {} & A_x\overline{a}_x \times B_x\overline{a}_x + A_x\overline{a}_x \times B_y\overline{a}_y + A_x\overline{a}_x \times B_z\overline{a}_z \\ & + A_y\overline{a}_y \times B_x\overline{a}_x + A_y\overline{a}_y \times B_y\overline{a}_y + A_y\overline{a}_y \times B_z\overline{a}_z \\ & + A_z\overline{a}_z \times B_x\overline{a}_x + A_z\overline{a}_z \times B_y\overline{a}_y + A_z\overline{a}_z \times B_z\overline{a}_z \end{aligned} \tag{1.25}$$

What is the cross product between parallel unit vectors? Examine the first term:

$$A_x\overline{a}_x \times B_x\overline{a}_x = A_xB_x \sin\theta\,\overline{a}_n \tag{1.26}$$

where θ is the angle between $\overline{a}_x$ and $\overline{a}_x$. These two unit vectors are identical and therefore parallel; hence, the angle is 0° and $\sin(0°)$ is zero. Thus, the first term is:

$$A_x\overline{a}_x \times B_x\overline{a}_x = 0 \tag{1.27}$$

Similarly, all other terms where the unit vectors are parallel are also zero. The key concept is that *the cross product between parallel unit vectors is zero*. Inserting these results into Eq. (1.25) gives:

$$\begin{aligned}\overline{A} \times \overline{B} = {} & + A_x\overline{a}_x \times B_y\overline{a}_y + A_x\overline{a}_x \times B_z\overline{a}_z + A_y\overline{a}_y \times B_x\overline{a}_x \\ & + A_y\overline{a}_y \times B_z\overline{a}_z + A_z\overline{a}_z \times B_x\overline{a}_x + A_z\overline{a}_z \times B_y\overline{a}_y \end{aligned} \tag{1.28}$$

What is the cross product between perpendicular unit vectors? Examine the first term of Eq. (1.28):

$$A_x\overline{a}_x \times B_y\overline{a}_y = A_xB_y \sin\theta\,\overline{a}_n \tag{1.29}$$

where θ is the angle between $\overline{a}_x$ and $\overline{a}_y$. These two unit vectors are perpendicular; hence, the angle is 90° and $\sin(90°)$ is unity (one). Perform the cross product between the unit vectors (use the RHR). Thus, the second term is:

$$A_x\overline{a}_x \times B_y\overline{a}_y = A_xB_y\overline{a}_z \tag{1.30}$$

Similarly evaluate all other terms where the unit vectors are perpendicular:

$$\overline{A} \times \overline{B} = + A_xB_y\overline{a}_z + A_xB_z\left(-\overline{a}_y\right) + A_yB_x\left(-\overline{a}_z\right) + A_yB_z\overline{a}_x + A_zB_x\overline{a}_y + A_zB_y\left(-\overline{a}_x\right) \tag{1.31}$$

Group the terms with the same unit vector directions together:

$$\boxed{\overline{A} \times \overline{B} = \left(+ A_yB_z - A_zB_y\right)\overline{a}_x + \left(+ A_zB_x - A_xB_z\right)\overline{a}_y + \left(+ A_xB_y - A_yB_x\right)\overline{a}_z} \tag{1.32}$$

This equation again illustrates that the cross product has a vector result. Note how the components and signs can be quickly determined by visualizing the RHR.

Example 1.2.4

Given: $\overline{A} = +12\,\overline{a}_x + 7\,\overline{a}_y - 2\,\overline{a}_z$ and $\overline{B} = +6\,\overline{a}_x - 5\,\overline{a}_y + 3\,\overline{a}_z$

Determine the unit vector perpendicular (*normal*) to both $\overline{A}$ and $\overline{B}$.

Solution

The key to solving this problem is to recognize that the cross product is a vector result and hence has a magnitude and a direction. The resultant direction is the unit vector perpendicular to both of the two original vectors.

$$\boxed{\overline{A} \times \overline{B} = |\overline{A} \times \overline{B}|\,\overline{a}_n}$$

$$\overline{a}_n = \frac{\overline{A} \times \overline{B}}{|\overline{A} \times \overline{B}|} = \frac{\left(+12\,\overline{a}_x + 7\,\overline{a}_y - 2\,\overline{a}_z\right) \times \left(+6\,\overline{a}_x - 5\,\overline{a}_y + 3\,\overline{a}_z\right)}{|\overline{A} \times \overline{B}|}$$

$$\overline{A} = +12\,\overline{a}_x + 7\,\overline{a}_y - 2\,\overline{a}_z \quad \rightarrow \quad A_x = +12 \quad A_y = +7 \quad A_z = -2$$
$$\overline{B} = +6\,\overline{a}_x - 5\,\overline{a}_y + 3\,\overline{a}_z \quad \rightarrow \quad B_x = +6 \quad B_y = -5 \quad B_z = +3$$

The x-component of the cross product is determined by $A_y B_z - A_z B_y = (7)(3) - (-2)(-5) = +11$, and similarly for the other vector components, resulting in

$$\overline{a}_n = \frac{+11\,\overline{a}_x - 48\,\overline{a}_y - 102\,\overline{a}_z}{\sqrt{11^2 + 48^2 + 102^2}} = +0.097\,\overline{a}_x - 0.424\,\overline{a}_y - 0.901\,\overline{a}_z$$

Hence, the cross product is useful in determining a direction normal to the plane that contains two vectors. Is the cross product useful in physical problems? Yes! For example, you may have heard about the Hall Effect, which is based on the magnetic force component in the Lorentz Force law $\left(\overline{F}_{\text{magnetic}} = Q\,\overline{v} \times \overline{B}\right)$. The cross product of the charge velocity and magnetic flux density vectors, multiplied by the charge Q, gives the magnitude and direction of the resultant force on a moving charge. The force on the moving charge carriers that constitute the current in a semiconductor results in a voltage that can be used to measure magnetic fields. Vector relationships like this are common in electromagnetics.

Example 1.2.5

An electron travels with a constant velocity of 5×10^6 m/s along the x-axis in the $+x$ direction. It is immersed in a uniform magnetic field of 4 T, which is directed in the $-y$ direction. Make a sketch. Determine the *force*, both magnitude and direction, on the electron at that instant.

Strategy

Lorentz Force law

$$\overline{F}_{\text{magnetic}} = Q\,\overline{v} \times \overline{B}$$

Solution

See Figure 1.11.

$$Q = e^- = -1.6022 \times 10^{-19}\,\text{C}$$
$$\overline{v} = 5 \times 10^6\,\overline{a}_x\,\text{m/s}$$
$$\overline{B} = 4\left(-\overline{a}_y\right)\,\text{T}$$

$$\begin{aligned}
\overline{F}_{\text{magnetic}} &= Q\,\overline{v} \times \overline{B} \\
&= -1.6022 \times 10^{-19}\left[5 \times 10^6\,\overline{a}_x \times 4\left(-\overline{a}_y\right)\right] \\
&= -3.2044 \times 10^{-12}\left[\overline{a}_x \times \left(-\overline{a}_y\right)\right] \\
&= -3.2044 \times 10^{-12}\left(-\overline{a}_z\right) \\
&\approx 3.20 \times 10^{-12}\left(+\overline{a}_z\right)\,\text{N}
\end{aligned}$$

Note that the negative charge of the electron reversed the direction of the cross product result.

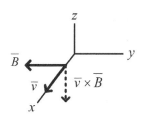

Figure 1.11 Sketch for Example 1.2.5.

Cross products also have a geometric interpretation that is both conceptually illuminating and practically useful in many applications. Given two vectors that lie on two adjacent sides of a parallelogram, the magnitude of the cross product of these two vectors is equal to the area of that parallelogram, as illustrated by the next example.

Example 1.2.6

Show that the area of a parallelogram equals $\left|\overline{A} \times \overline{B}\right|$.

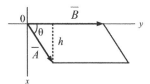

Solution

Refer to Figure 1.12.

$$h = A \sin \theta$$

Figure 1.12   Parallelogram area.

$$\text{area} = Bh = BA \sin \theta = AB \sin \theta = \left|\overline{A} \times \overline{B}\right|$$

The cross product is also used in the definition of a *right-handed coordinate system*. For example, the Cartesian coordinate system is a right-handed coordinate system:

$$\begin{aligned}
\overline{a}_x \times \overline{a}_y &= \overline{a}_z &(x \rightarrow y \rightarrow z) \\
\overline{a}_y \times \overline{a}_z &= \overline{a}_x &(y \rightarrow z \rightarrow x) \\
\overline{a}_z \times \overline{a}_x &= \overline{a}_y &(z \rightarrow x \rightarrow y)
\end{aligned} \tag{1.33}$$

Note the *x–y–z* order of the subscripts in the cross products of the unit vectors.

1.3   Field Vectors

Are there other types of vectors besides position and distance vectors? Yes! Examples include velocity, force, electric field, magnetic field, and so forth. The concept of a **field** is a quantity that varies in magnitude and possibly direction as a *function* of position. There are two general types of fields:

- *vector* fields (examples above)
- *scalar* fields (such as the magnitude of a vector field; other examples follow)

For a *scalar field*, functions or plots express the value of the scalar field as a function of position.

Example 1.3.1

Name examples of scalar fields from weather maps.

Answer

Temperature, atmospheric pressure field (isobars), and rainfall and snow accumulations

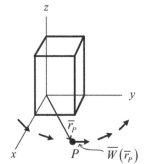

For a *vector field*, a vector function expresses the magnitude and direction of the field as a function of position. Within the vector field, a *field vector* is the vector quantity at a particular point that is located by a *position vector*.

Example 1.3.2

Consider the wind swirling around a building (see Figure 1.13). Let $\overline{W}$ be the wind velocity vector. If $\overline{W}(x, y, z) = 5xy\overline{a}_x - 2zz\overline{a}_y - 3\overline{a}_z$ at $P(1, 2, 3)$,

Figure 1.13   Wind velocity vector field.

Determine

a) the position vector, and
b) the field vector at P.
c) Is the result in part b) valid everywhere?

Solution

a) $P(1, 2, 3) \rightarrow \bar{r}_P = 1\bar{a}_x + 2\bar{a}_y + 3\bar{a}_z$

b) $\overline{W}(1, 2, 3) = 5(1)(2)\bar{a}_x - 2(3)\bar{a}_y - 3\bar{a}_z = 10\bar{a}_x - 6\bar{a}_y - 3\bar{a}_z$

c) No, this field vector result is valid *only* at $P(1,2,3)$.

The wind velocity field vector can be expressed in general as:

$$\overline{W}(x, y, z) = \overline{W}(\bar{r}_P) = W_x(x, y, z)\bar{a}_x + W_y(x, y, z)\bar{a}_y + W_z(x, y, z)\bar{a}_z \tag{1.34}$$

Mathematically, $W_x(x, y, z)$ is the magnitude of the wind velocity in the x direction as a function of position. Notice that W_x depends on *all three* position coordinates in general, but could be constant, as it was for the z-component of $\overline{W}$ in the previous example. The multiplication by $\bar{a}_x$ gives direction to W_x in order to make it a vector component. A field vector $\overline{W}$ *exists* at each location specified by each position vector $\bar{r}_P$.

Example 1.3.3

If $\overline{F} = 12\bar{a}_x + 16\bar{a}_y - 13\bar{a}_z$ at $P(3, 2, 4)$, what is $\overline{F}$ at $Q(-2, -1, 9)$?

Answer:

We don't know because we do not know how $\overline{F}$ depends on position.

The concept illustrated by the previous example is that a field vector that is evaluated at a specific position, without any additional information, cannot be used to determine the field vector at another position.

How do we visually represent a field vector $\overline{W}$ at the point located by $\bar{r}_P$?

- The field vector $\overline{W}$ represents the magnitude and direction of the field at the point located by $\bar{r}_P$.
- We draw a finite length vector to represent the magnitude and the direction of the field for visualization at each $\bar{r}_P$.

How do we draw "all" of the vectors to represent a vector field?

- Ideally, an infinite number of infinitely small arrows are needed (but impractical).
- Instead, "sample the field" with arrows, where length, boldness, and/or density of arrows indicates the field strength (magnitude) in that region.
- In addition to being practical for plotting vector field functions, electromagnetic field simulation software generates such plots to visualize the field behavior.

Examples of field sketches follow:

a) Draw a point charge with radial electric field lines – see Figure 1.14a. The *density* of the lines decreases as the distance from the charge increases; thus, the field strength also decreases versus distance from the charge.

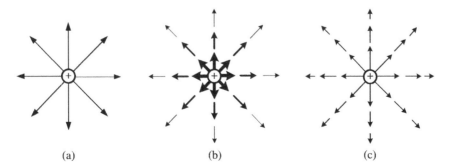

(a) (b) (c)

Figure 1.14 Three commonly used methods to indicate vector field strength (as well as direction). (a) Density of vectors. (b) Boldness of vectors. (c) Length of vectors.

b) Draw a point charge with radial arrows – see Figure 1.14b. The *boldness* of the arrows decreases as the distance from the charge increases; thus, the field strength also decreases versus distance from the charge.

c) Draw a point charge with radial arrows – see Figure 1.14c. The *length* of the arrows decreases as the distance from the charge increases; thus, the field strength also decreases versus distance from the charge.

1.4 Cylindrical Coordinate System, Vectors, and Conversions

A short review of the rectangular coordinate system will set the stage for the introduction to other coordinate systems.

1.4.1 Cartesian (Rectangular) Coordinate System: Review

Refer to Figure 1.15. In Cartesian coordinates, what is/are:

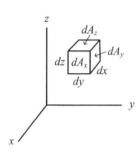

Figure 1.15 Cartesian differential length elements and areas.

- The general differential length elements?

$$dl_x = dx, \qquad dl_y = dy, \qquad dl_z = dz$$

- A general differential length? It is a *vector*!

$$\overline{dl} = dx\overline{a}_x + dy\overline{a}_y + dz\overline{a}_z$$

- The differential surface areas?

$$dA_z = dxdy, \quad dA_y = dxdz, \quad dA_x = dydz$$

The subscripts on dA refer to the "constant" variable; for example, $dxdy$ is in a constant z plane. (These subscripts will have more meaning when surface vectors are covered in a later chapter.)

- The differential volume?

$$dV = dx\,dy\,dz$$

Example 1.4.1

Derive the area of a rectangle of length a and height b.

Solution

$$A = \int dA = \int_0^b \int_0^a dxdy = \int_0^a dx \int_0^b dy = x|_0^a y|_0^b = (a-0)(b-0) = ab$$

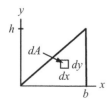

Figure 1.16 Triangle area set-up.

Example 1.4.2

Derive the area of a right triangle of base b and height h.

Solution

Set up a right triangle with the hypotenuse from the origin to (b,h) – see Figure 1.16. In this example, one limit depends on the other independent variable. What expression is needed for the variable limit?

The equation of the line that defines the variable limit: $y = \dfrac{h}{b}x$

$$A = \int dA = \int_0^b \int_0^y dy dx = \int_0^b \left[\int_0^{\frac{h}{b}x} dy \right] dx = \int_0^b y \Big|_0^{\frac{h}{b}x} dx = \int_0^b \frac{h}{b} x dx = \frac{h}{b} \frac{x^2}{2} \Big|_0^b = \frac{1}{2} bh$$

Position, distance, and field vectors have already been examined in the Cartesian coordinate system. The rest of this chapter will examine these vectors in two other frequently utilized coordinate systems, namely the cylindrical and the spherical coordinate systems.

1.4.2 Cylindrical Coordinate System

You are most likely already comfortable with Cartesian coordinates. Why complicate matters by using different coordinate systems?

1) Problems with cylindrical geometries are much easier to solve with cylindrical coordinates than with rectangular coordinates. Example: the double integral for the area of a circle in Cartesian coordinates cannot be separated, whereas the integrals can be separated in cylindrical coordinates.
2) Visualization of electromagnetic fields is often easier in the coordinate system that is closest to the geometry of the particular problem. Example: the magnetic field around a current in a straight wire.

What are the cylindrical coordinates? See Figure 1.17a: (ρ, ϕ, z).

- ρ is the radius parallel to the *x-y* plane. See Figure 1.17b.
- ϕ is the angle from the *x*-axis to the radius, in a plane parallel to the *x–y* plane. See Figure 1.17b.
- z is the same as in the Cartesian coordinate system.

> **Important note on notation:** Some textbooks and literature use *r* for radius in both the cylindrical and spherical coordinate systems. A large portion of the electromagnetics literature uses the Greek letter ρ for the radius in the cylindrical coordinate system and reserves *r* for the radius in the spherical coordinate system, and this notation will be used in this text.

What is the range of values for each cylindrical coordinate? $\rho \geq 0$; $0 \leq \phi < 2\pi$; $-\infty < z < +\infty$

How do cylindrical *coordinates* relate to rectangular coordinates? By the Pythagorean theorem and trigonometry. See Figure 1.17. Caution: ensure ϕ locates the position in the proper quadrant when using the arctangent function because the arctangent has a limited range of $-\pi/2$ to $\pi/2$.

What are the differential *length* elements in the cylindrical coordinate system? See Figure 1.18.

$$d\ell_\rho = d\rho, \quad d\ell_\phi = \rho d\phi, \quad d\ell_z = dz$$

Reminder: arc length = radius × change of angle $d\ell_\phi = \rho d\phi$

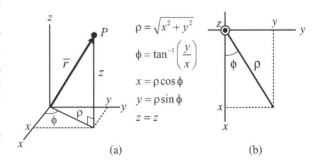

Figure 1.17 The cylindrical coordinate system and cylindrical – Cartesian coordinate relationships. (a) Three-dimensional (3D) view. (b) Top view.

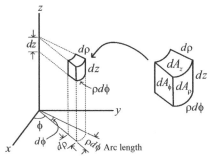

Figure 1.18 Cylindrical differential length elements and areas.

Example 1.4.3

Use calculus to derive the circumference of a circle of radius a.

Solution

$$\ell = \int d\ell = \int \rho d\phi = \int_0^{2\pi} a d\phi = a \int_0^{2\pi} d\phi = a\phi|_0^{2\pi} = a(2\pi - 0) = 2\pi a$$

Note that ρ is a *constant* that equals a along this path.

What are the differential *surface area* elements in the cylindrical coordinate system?

$$dA_\phi = d\rho dz, \qquad dA_z = \rho d\phi d\rho, \qquad dA_\rho = \rho d\phi dz$$

Example 1.4.4

Use calculus to derive the area of a circle of radius a.

Solution

$$A = \int dA = \int_0^{2\pi} \int_0^a d\rho \rho d\phi = \int_0^a \rho d\rho \int_0^{2\pi} d\phi = \frac{\rho^2}{2}\Big|_0^a \phi|_0^{2\pi} = \left(\frac{a^2}{2} - 0\right)(2\pi - 0) = \pi a^2$$

Note that ρ is a variable in this surface.

What is the differential *volume* element in the cylindrical coordinate system? $dV = \rho d\rho d\phi dz$

Example 1.4.5

Use calculus to derive the volume of a right cylinder of radius a and height h.

Solution

$$V = \int dV = \int_0^h \int_0^{2\pi} \int_0^a d\rho \rho d\phi dz = \int_0^a \rho d\rho \int_0^{2\pi} d\phi \int_0^h dz = \frac{\rho^2}{2}\Big|_0^a \phi|_0^{2\pi} z|_0^h = \left(\frac{a^2}{2}\right)(2\pi)(h) = \pi a^2 h$$

Note that ρ is a variable in this volume.

What are the *cylindrical unit vectors*? See Figure 1.19a. Note that the cylindrical coordinate system is right-handed: $\bar{a}_\rho \times \bar{a}_\phi = \bar{a}_z$. See Figure 1.19b.

What is the difference between cylindrical unit vectors and Cartesian unit vectors?

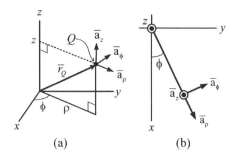

(a)   (b)

Figure 1.19 Cylindrical coordinates and unit vectors. (a) 3D view. (b) Top view ($\bar{a}_\phi$ is perpendicular to $\bar{a}_\rho$ and $\bar{a}_z$, and is parallel to the x–y plane).

1) All *Cartesian* unit vectors are always constant: they always point in the *same direction*.
2) The $\bar{a}_\rho$ and $\bar{a}_\phi$ *cylindrical* unit vectors are not constant: their *directions depend on position*.

Upon which coordinate(s) do each unit vector depend? The unit vectors in the ρ and ϕ directions depend on the coordinate ϕ. In functional notation, $\bar{a}_\rho(\phi)$ and $\bar{a}_\phi(\phi)$, that is, the direction of the unit vector in the ρ direction depends on ϕ and the direction of the unit vector in the ϕ direction depends on ϕ. One can see in Figure 1.20 that the directions of $\bar{a}_\rho(\phi)$ and $\bar{a}_\phi(\phi)$ clearly differ for different values of ϕ. Unit vector $\bar{a}_z$ does not depend on position.

What is the general cylindrical *position* vector with respect to the origin? Is it $\bar{r} = \rho\bar{a}_\rho + \phi\bar{a}_\phi + z\bar{a}_z$? No, because ϕ is *not* a length. Imagine moving from the origin to point Q. One moves ρ units in the $\bar{a}_\rho$ direction at an angle of ϕ and z units in the $\bar{a}_z$ direction, as illustrated in Figure 1.21. So the general **cylindrical position vector** is:

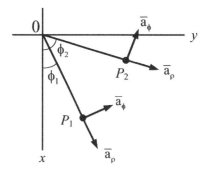

Figure 1.20 Cylindrical unit vectors for two different values of ϕ.

$$\boxed{\bar{r} = \rho\bar{a}_\rho + z\bar{a}_z} \qquad (1.35)$$

Any point can be located using these two vector components.

How is ϕ "present" in a cylindrical position vector? Unit vectors $\bar{a}_\rho$ and $\bar{a}_\phi$ are functions of ϕ, that is, $\bar{a}_\rho(\phi)$ and $\bar{a}_\phi(\phi)$. Angle ϕ must be specified to know the directions of $\bar{a}_\rho$ and $\bar{a}_\phi$. Thus, position vectors may be specified in Cartesian or cylindrical forms:

$$\bar{r} = \rho\bar{a}_\rho + z\bar{a}_z = (x\bar{a}_x + y\bar{a}_y) + z\bar{a}_z = x\bar{a}_x + y\bar{a}_y + z\bar{a}_z \qquad (1.36)$$

The key concept is that angle ϕ must be known for the cylindrical position vector to locate a specific position, otherwise the direction of $\bar{a}_\rho$ is unknown.

Example 1.4.6
Determine the position vector to $P(2, 5, 3)$ in cylindrical form.

Strategy
Convert the Cartesian coordinates to cylindrical coordinates, then write $\bar{r} = \rho\bar{a}_\rho + z\bar{a}_z$.

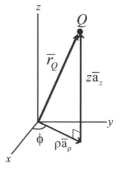

Figure 1.21 Position vector in the cylindrical coordinate system.

Solution

$$\rho = \sqrt{x^2 + y^2} = \sqrt{2^2 + 5^2} = 5.3852$$

$$\phi = \tan^{-1}\left(\frac{y}{x}\right) = \tan^{-1}\left(\frac{5}{2}\right) = 68.199°$$

$$\bar{r} = \rho\bar{a}_\rho + z\bar{a}_z = 5.39\bar{a}_\rho + 3\bar{a}_z \quad \text{at } \phi = 68.2°$$

Note that specifying the angle establishes the direction of unit vector $\bar{a}_\rho$, which is necessary for the cylindrical position vector to locate a specific position.

Example 1.4.7
Determine the position vector to $P(5.3852, 68.199°, 3)$ in rectangular form.

Strategy
Convert the cylindrical coordinates to Cartesian coordinates, then write $\bar{r} = x\bar{a}_x + y\bar{a}_y + z\bar{a}_z$.

Solution

$$x = \rho\cos\phi = 5.3852\cos 68.199° = 2.0000$$

$$y = \rho\sin\phi = 5.3852\sin 68.199° = 5.0000$$

$$\bar{r} = x\bar{a}_x + y\bar{a}_y + z\bar{a}_z = 2\bar{a}_x + 5\bar{a}_y + 3\bar{a}_z$$

This result matches the position of P in Example 1.4.6.

Just as there are Cartesian *distance* vectors, there are also cylindrical distance vectors. See Figure 1.22a–c for three-dimensional views and a two-dimensional view. Recall that lowercase $\bar{r}$ is a position vector (locates a point with respect to the origin), whereas uppercase $\bar{R}$ is a distance vector (locates a point with respect to another point,

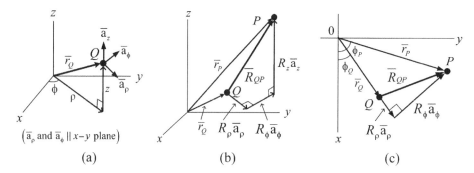

Figure 1.22 *Distance* vector in cylindrical coordinates. (a) Unit vector directions for point Q shown. (b) 3D view of the distance vector and its vector components from Q to P. (c) Top view of the distance vector and its vector components from Q to P.

neither necessarily the origin). The unit vector directions $\bar{a}_\rho$ and $\bar{a}_\phi$ depend on ϕ, so by convention the unit vector directions at the "start" point (Q in Figure 1.22) are utilized for the unit vectors in the distance vector. The important observation to make from the figure is that, in general, a cylindrical distance vector can have vector components in all three coordinate directions:

$$\bar{R} = R_\rho\bar{a}_\rho + R_\phi\bar{a}_\phi + R_z\bar{a}_z \tag{1.37}$$

where R_ϕ is *not* the angle ϕ, but instead is a length in the $\bar{a}_\phi$ direction. Cylindrical distance vectors are useful in electromagnetics. For example, they are often used in Coulomb's law and the Biot–Savart law, which are the main topics of Chapter 2.

Cylindrical *field* vectors are similar to cylindrical distance vectors in that they have, in general, all three cylindrical vector components:

$$\bar{A} = A_\rho\bar{a}_\rho + A_\phi\bar{a}_\phi + A_z\bar{a}_z \tag{1.38}$$

Each vector component of the field vector depends, in general, on all three coordinates, that is, A_ρ depends on $(\rho,\ \phi,\ z)$, and so forth:

$$\bar{A}(\rho,\phi,z) = A_\rho(\rho,\phi,z)\bar{a}_\rho + A_\phi(\rho,\phi,z)\bar{a}_\phi + A_z(\rho,\phi,z)\bar{a}_z \tag{1.39}$$

Although each vector component may depend on all three coordinates, the direction of the field depends on the relative magnitudes of the three vector components A_ρ, A_ϕ, and A_z at that location when they are vectorially summed, as indicated in Eq. (1.39).

An effective way to visualize field vectors in different coordinate systems is to *convert* a field vector from one form to another (Cartesian to cylindrical, or vice versa). The next example illustrates the conversion process and the logic behind the process.

Example 1.4.8

If field vector $\bar{A} = 3.29995\bar{a}_x + 0.98335\bar{a}_y + 3\bar{a}_z$ is located at $Q(2,\ 65°,\ -1)$, determine $\bar{A}$ in cylindrical form.

Strategy

- Convert each Cartesian vector component of $\bar{A}$ into cylindrical form.
- Sum the corresponding cylindrical vector components.

Solution

See Figure 1.23a for a sketch of $\bar{A}$ in Cartesian form and Figure 1.23b for a sketch of $\bar{A}$ in cylindrical form. The z-components are not shown.

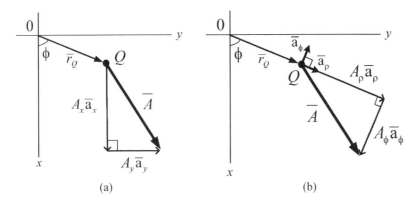

Figure 1.23 Approximate sketches of the vector components. (a) $\overline{A}$ in Cartesian form. (b) $\overline{A}$ in cylindrical form. (Note: the A_ϕ component is in the $-\overline{a}_\phi$ direction).

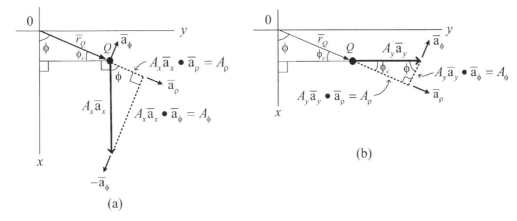

Figure 1.24 Decomposing Cartesian field vector components into cylindrical components. (a) Decompose the A_x component into A_ρ and A_ϕ components. (b) Decompose the A_y component into A_ρ and A_ϕ components (angles with a c subscript are complementary angles).

Convert the Cartesian vector components of $\overline{A}$ into cylindrical form using the cylindrical unit vector directions of point Q. The example is interrupted to discuss these conversions.

How are the Cartesian components of $\overline{A}$ converted from Cartesian form into cylindrical form? This is a situation where scalar and vector projections are very useful. Refer to Figure 1.24a and b for a general decomposition of each of the Cartesian vector components into cylindrical vector components. Note the unit vector directions of the location, point Q in this example, are utilized to determine the cylindrical vector components of *each* of the Cartesian vector components. Thus, the angle ϕ is that of point Q.

Then each cylindrical vector component can be determined by summing the projections from each of the Cartesian vector components. In general:

$$A_\rho = \overline{A}\cdot\overline{a}_\rho = \overbrace{A_x\overline{a}_x\cdot\overline{a}_\rho}^{\substack{\text{length of } A_x \text{ in}\\ \text{the }\rho\text{ direction}}} + \overbrace{A_y\overline{a}_y\cdot\overline{a}_\rho}^{\substack{\text{length of } A_y \text{ in}\\ \text{the }\rho\text{ direction}}} + \overbrace{A_z\overline{a}_z\cdot\overline{a}_\rho}^{\substack{\text{length of } A_z \text{ in}\\ \text{the }\rho\text{ direction}}} \tag{1.40}$$

The z-component is perpendicular to the ρ-component, so that term is zero, leaving

$$A_\rho = A_x\overline{a}_x\cdot\overline{a}_\rho + A_y\overline{a}_y\cdot\overline{a}_\rho \tag{1.41}$$

Table 1.1 Dot product results for unit vector conversions.

$\bar{a}_{(\ )} \cdot \bar{a}_{(\ )}$	$\bar{a}_\rho$	$\bar{a}_\phi$
$\bar{a}_x$	$\cos\phi$	$-\sin\phi$
$\bar{a}_y$	$\sin\phi$	$\cos\phi$
$\bar{a}_z$	0	0

Likewise for the other cylindrical vector components:

$$A_\phi = \bar{A} \cdot \bar{a}_\phi = A_x \bar{a}_x \cdot \bar{a}_\phi + A_y \bar{a}_y \cdot \bar{a}_\phi \tag{1.42}$$

$$A_z\Big|_{\text{Cylindrical}} = A_z\Big|_{\text{Cartesian}} \tag{1.43}$$

How are the dot products between unit vectors of different coordinate systems determined? Geometry and trigonometry! These determinations are addressed later in this section. The results of these determinations are reported in Table 1.1.

Example 1.4.8 now resumed: Use the results in Table 1.1 to calculate the cylindrical vector components:

$$\bar{A} = 3.29995\bar{a}_x + 0.98335\bar{a}_y + 3\bar{a}_z, \quad \phi = 65°,$$

$$A_\rho = A_x\bar{a}_x \cdot \bar{a}_\rho + A_y\bar{a}_y \cdot \bar{a}_\rho = A_x\cos\phi + A_y\sin\phi = 3.29995\cos 65° + 0.98335\sin 65°$$
$$= 2.28584 \approx 2.29$$

$$A_\phi = A_x\bar{a}_x \cdot \bar{a}_\phi + A_y\bar{a}_y \cdot \bar{a}_\phi = A_x(-\sin\phi) + A_y\cos\phi = 3.29995(-\sin 65°) + 0.98335\cos 65°$$
$$= -2.57519 \approx -2.58$$

Finally, sum the vector components to obtain the field vector in cylindrical form:

$$\bar{A} = A_\rho\bar{a}_\rho + A_\phi\bar{a}_\phi + A_z\bar{a}_z = 2.29\bar{a}_\rho - 2.58\bar{a}_\phi + 3.00\bar{a}_z$$

Note that coordinate conversion and field vector form conversion are *different* processes – do not confuse them! An example of converting a field vector from cylindrical form into Cartesian form follows.

Example 1.4.9

Convert field vector $\bar{W} = 11\bar{a}_\rho + 6\bar{a}_\phi - 7\bar{a}_z$, located at (8, 5, 2), into *rectangular* (*Cartesian*) form.

Strategy

Determine ϕ from the Cartesian coordinates.
- Use the dot product table to convert each cylindrical component of $\bar{W}$ into Cartesian components.
- Add corresponding Cartesian components to determine $\bar{W}$ in Cartesian form.

Solution

$$\phi = \tan^{-1}\left(\frac{y}{x}\right) = \tan^{-1}\left(\frac{5}{8}\right) = +32.005°;$$

W_z is the same in cylindrical and Cartesian form.

$$W_x = \bar{W} \cdot \bar{a}_x = W_\rho\bar{a}_\rho \cdot \bar{a}_x + W_\phi\bar{a}_\phi \cdot \bar{a}_x = W_\rho\cos\phi - W_\phi\sin\phi = 11\cos 32.005° - 6\sin 32.005° = 6.148$$
$$W_y = \bar{W} \cdot \bar{a}_y = W_\rho\bar{a}_\rho \cdot \bar{a}_y + W_\phi\bar{a}_\phi \cdot \bar{a}_y = W_\rho\sin\phi + W_\phi\cos\phi = 11\sin 32.005° + 6\cos 32.005° = 10.918$$

Sum the vector components to express $\bar{W}$ in rectangular form:

$$\bar{W} = W_x\bar{a}_x + W_y\bar{a}_y + W_z\bar{a}_z = 6.15\bar{a}_x + 10.92\bar{a}_y - 7\bar{a}_z$$

Now the determination of the dot products between unit vectors of different coordinate systems, per Table 1.1, is addressed. Geometry and trigonometry are applied, as illustrated in the next example. One might wonder why it is necessary to derive the dot product results if they are already provided in Table 1.1. This exercise builds critical visualization of the relationships between the vector components in different coordinate systems. Once these vector relationships are understood, the known dot product results can be used when converting field vectors[2] from one coordinate system form to another.

Example 1.4.10

Determine the dot product result for $\bar{a}_\phi \cdot \bar{a}_x$.

Strategy
- Sketch the geometry showing the unit vectors of concern at a point away from the origin.
- Use trigonometry to determine the dot product(s).

Solution

See Figure 1.25 for a sketch of the geometry.

- $\bar{a}_\phi$ was arbitrarily selected to start with and was decomposed into Cartesian components using scalar projections.
- ϕ_c is the complement (90°- angle) of ϕ.
- The dot product result is readily determined by trigonometry: $\bar{a}_\phi \cdot \bar{a}_x = -\sin\phi$.
- Note that in general $\bar{a}_\phi \cdot \bar{a}_x = -\sin\phi$ based on the $-\bar{a}_x$ direction of that vector.
- As a bonus, $\bar{a}_\phi \cdot \bar{a}_y$ is also readily determined by trigonometry: $\bar{a}_\phi \cdot \bar{a}_y = \cos\phi$.
- These results match the entries in Table 1.1.

Note that in the previous example, one could have started with $\bar{a}_x$ and decomposed it into cylindrical components – you should try this approach and verify the same result is obtained. You may be wondering how to keep track of signs in such developments. If the dot product relationships between the unit vectors are derived in general in the first octant, then the signs will "take care of themselves" in all octants.

You may be also wondering how a sine function appeared in a dot product expression. The angle between $\bar{a}_\phi$ and $\bar{a}_x$ is ϕ_c, so $\bar{a}_\phi \cdot \bar{a}_x = \cos\phi_c$. However, from trigonometry, $\cos\phi_c = \cos(90° - \phi) = \sin\phi$. Thus, $\bar{a}_\phi \cdot \bar{a}_x = -\sin\phi$ (again, the negative sign is due to the $-\bar{a}_x$ direction of the vector in Figure 1.25).

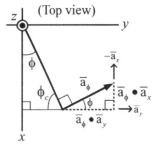

Figure 1.25 Decomposition of $\bar{a}_\phi$ into Cartesian components.

1.5 Spherical Coordinate System, Vectors, and Conversions

We now turn our attention to vectors in spherical coordinates. Just as with cylindrical coordinates for cylindrical geometries, spherical coordinates often simplify the analysis of and visualization of electric and magnetic fields in problems with spherical geometries. What are the spherical coordinates? Refer to Figure 1.26. The three spherical coordinates are radius, angle from the z axis, and angle in the x–y plane, that is, (r, θ, ϕ).

What is the range of values for each spherical coordinate? $r \geq 0;\quad 0 \leq \theta \leq \pi;\quad 0 \leq \phi < 2\pi$.

How do spherical *coordinates* relate to rectangular coordinates? Again, refer to Figure 1.26.

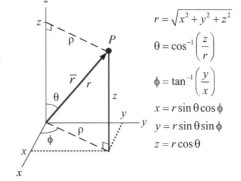

$$r = \sqrt{x^2 + y^2 + z^2}$$
$$\theta = \cos^{-1}\left(\frac{z}{r}\right)$$
$$\phi = \tan^{-1}\left(\frac{y}{x}\right)$$
$$x = r\sin\theta\cos\phi$$
$$y = r\sin\theta\sin\phi$$
$$z = r\cos\theta$$

Figure 1.26 Spherical coordinates.

- ρ is the projection of r into the x–y plane.
- ϕ is the same in both the cylindrical and spherical coordinate systems.
  - The $\rho - z - r$ triangle is in a "vertical" plane, that is, a $\phi = $ constant plane.

2 This vector conversion approach works for distance vectors also.

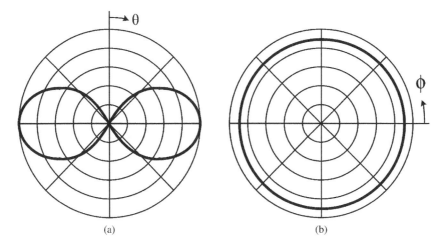

Figure 1.27 Example of spherical coordinate use in antenna radiation patterns. (a) Vertical radiation pattern. (b) Horizontal radiation pattern.

Some applications of spherical coordinates include antenna radiation patterns, as illustrated in Figure 1.27a, b, microphone sound sensitivity plots, and LED light intensity plots.

The differential length, area, and volume elements are shown in Figure 1.28a–d. What are the differential *length* elements in the spherical coordinate system?

$$d\ell_r = dr, \quad d\ell_\theta = r\,d\theta, \quad d\ell_\phi = r\sin\theta\,d\phi$$

Note: The arc length parallel to the *x*–*y* plane, $d\ell_\phi$, has a radius that depends on θ; thus:

$$d\ell_\phi = \rho d\phi = r\sin\theta\,d\phi$$

Example 1.5.1

Use calculus to derive the circumference of a circle parallel to the *x*–*y* plane on a sphere of radius *a* at (a) θ = 90° and (b) θ = 45° (you should be able to set up the problem and develop the results).

Answers:

a) $2\pi a$, b) $\sqrt{2}\pi a$

What are the differential *surface area* elements in the spherical coordinate system?

$$dA_\phi = r\,dr\,d\theta, \quad dA_\theta = r\sin\theta\,dr\,d\phi, \quad dA_r = r^2\sin\theta\,d\theta\,d\phi$$

Example 1.5.2

Use calculus to derive the surface area of a sphere of radius *a* (you should be able to set up the problem and develop the result).

Answer: $4\pi a^2$

What is the differential *volume* element in the spherical coordinate system? $dV = r^2\sin\theta\,dr\,d\theta\,d\phi$

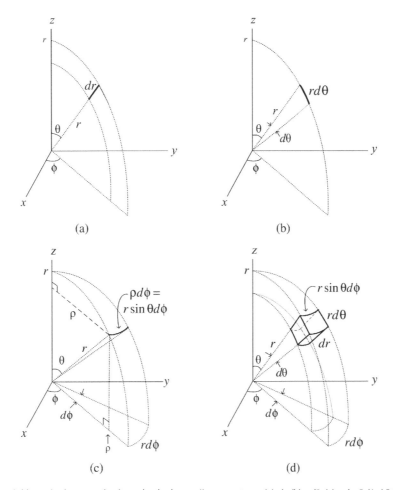

Figure 1.28 Differential length elements in the spherical coordinate system. (a) *dr*. (b) *r dθ*. (c) *r* sin *θdφ*. (d) Spherical differential length elements.

Example 1.5.3

Use calculus to derive the volume of a sphere of radius *a* (you should be able to set up the problem and develop the result).

Answer: $\frac{4}{3}\pi a^3$

What are the spherical unit vectors? See Figure 1.29. Note that $\bar{a}_r \times \bar{a}_\theta = \bar{a}_\phi$ (a right-handed coordinate system). What is the difference between spherical unit vectors and Cartesian unit vectors?

- The Cartesian unit vectors are always constant – they always point in the same direction.
- The spherical unit vectors are not constant – their directions depend on position.

Upon which coordinate(s) do each unit vector depend? $\bar{a}_r(\theta, \phi)$, $\bar{a}_\theta(\theta, \phi)$, $\bar{a}_\phi(\phi)$

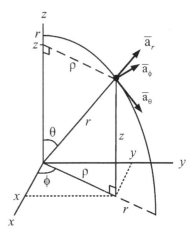

Figure 1.29 Spherical unit vectors ($\bar{a}_\phi$ is perpendicular to $\bar{a}_r$ and $\bar{a}_\theta$, and is parallel to the *x–y* plane).

Note that $\bar{a}_\phi$ is the same unit vector in both the cylindrical and spherical coordinate systems.

What is the general *spherical position vector* with respect to the origin?

$$\boxed{\bar{r} = r\bar{a}_r} \tag{1.44}$$

Note that θ and ϕ must be known to determine a position vector in spherical coordinates. Otherwise, the direction of $\bar{a}_r$ is unknown.

Example 1.5.4
Determine the position vector in spherical form to $P(2, 5, 3)$.

Strategy
Convert the Cartesian coordinates to spherical coordinates, then write $\bar{r} = r\bar{a}_r$.

Solution

$$r = \sqrt{x^2 + y^2 + z^2} = \sqrt{2^2 + 5^2 + 3^2} = 6.1644$$

$$\theta = \cos^{-1}\left(\frac{z}{r}\right) = \cos^{-1}\left(\frac{3}{6.1644}\right) = 60.878°$$

$$\phi = \tan^{-1}\left(\frac{y}{x}\right) = \tan^{-1}\left(\frac{5}{2}\right) = 68.199°$$

$$\bar{r} = r\bar{a}_r = 6.16\bar{a}_r \quad \text{at} \quad \theta = 60.9° \text{and } \phi = 68.2°$$

Note that specifying both angles establishes the direction of unit vector $\bar{a}_r$.

Just as there are Cartesian and cylindrical *distance* vectors, there are also spherical distance vectors. The spherical distance vector has, in general, all three spherical vector components:

$$\bar{R} = R_r\bar{a}_r + R_\theta\bar{a}_\theta + R_\phi\bar{a}_\phi \tag{1.45}$$

where each vector component depends on (r, θ, ϕ) in general. Spherical distance vectors are useful tools for electromagnetics analysis in areas such as spherical Coulomb's law and Biot–Savart law problems, antennas, and other advanced applications. For this text, spherical distance vectors are not used directly in analysis. They are briefly discussed here to emphasize that position, distance, and field vectors are applicable regardless of the particular coordinate system being utilized.

Spherical *field* vectors are similar to spherical distance vectors in that they have, in general, all three vector components:

$$\bar{A} = A_r\bar{a}_r + A_\theta\bar{a}_\theta + A_\phi\bar{a}_\phi \tag{1.46}$$

Each vector component of the field vector depends, in general, on all three coordinates. Note that θ and ϕ do *not* generally give the direction of the field. Why not? The direction of a field vector at a given position depends on (r, θ, ϕ) in general, but it is the relative magnitudes of the *vector* components that determine the direction of the field at that location.

How is a field vector in spherical form converted into Cartesian form, or vice versa? In the same manner as it was for cylindrical form, that is, by using dot products:

- The dot products between Cartesian unit vectors and cylindrical/spherical unit vectors are tabulated in Table 1.2.
- Example 1.5.5 illustrates the conversion technique with a spherical field vector.
- Example 1.5.6 illustrates how the dot product result between a spherical unit vector and a Cartesian unit vector is determined.

Table 1.2 Dot product results for unit vector conversions.

$\overline{a}_{(\ )} \cdot \overline{a}_{(\ )}$	$\overline{a}_\rho$	$\overline{a}_\phi$	$\overline{a}_r$	$\overline{a}_\theta$
$\overline{a}_x$	$\cos\phi$	$-\sin\phi$	$\sin\theta\cos\phi$	$\cos\theta\cos\phi$
$\overline{a}_y$	$\sin\phi$	$\cos\phi$	$\sin\theta\sin\phi$	$\cos\theta\sin\phi$
$\overline{a}_z$	0	0	$\cos\theta$	$-\sin\theta$

Example 1.5.5

If field vector $\overline{A} = 10\overline{a}_x + 13\overline{a}_y - 16\overline{a}_z$, convert $\overline{A}$ into spherical form.

Answer

This conversion cannot be performed without knowledge of the position of the vector because the spherical unit vector directions are unknown.

If $\overline{A}$ is located at $P(6, 4.7, 2.4)$, then one can proceed by first determining the spherical angles:

Solution

$$\theta = \cos^{-1}\left(\frac{z}{r}\right) = 72.521°, \quad \phi = \tan^{-1}\left(\frac{y}{x}\right) = 38.073°$$

Utilize the dot product table results to decompose the Cartesian vector components into spherical vector components:

$$\begin{aligned} A_r = \overline{A}\cdot\overline{a}_r &= A_x\overline{a}_x\cdot\overline{a}_r + A_y\overline{a}_y\cdot\overline{a}_r + A_z\overline{a}_z\cdot\overline{a}_r = 10\sin\theta\cos\phi + 13\sin\theta\sin\phi - 16\cos\theta \\ &= 10\sin72.521°\cos38.073° + 13\sin72.521°\sin38.073° - 16\cos72.521° \\ &= 7.5087 + 7.6465 - 4.8057 = 10.3 \end{aligned}$$

$$\begin{aligned} A_\theta = \overline{A}\cdot\overline{a}_\theta &= A_x\overline{a}_x\cdot\overline{a}_\theta + A_y\overline{a}_y\cdot\overline{a}_\theta + A_z\overline{a}_z\cdot\overline{a}_\theta = 10\cos\theta\cos\phi + 13\cos\theta\sin\phi - (-16)\sin\theta \\ &= 10\cos72.521°\cos38.073° + 13\cos72.521°\sin38.073° + 16\sin72.521° \\ &= 2.3649 + 2.4078 + 15.2612 = 20.0 \end{aligned}$$

$$\begin{aligned} A_\phi = \overline{A}\cdot\overline{a}_\phi &= A_x\overline{a}_x\cdot\overline{a}_\phi + A_y\overline{a}_y\cdot\overline{a}_\phi + A_z\overline{a}_z\cdot\overline{a}_\phi = 10(-\sin\phi) + 13\cos\phi + 0 \\ &= 10(-\sin38.073°) + 13\cos38.073° = -6.1666 + 10.234 = 4.07 \end{aligned}$$

Finally, sum the vector components to express field vector $\overline{A}$ in spherical form:

$$\overline{A} = A_r\overline{a}_r + A_\theta\overline{a}_\theta + A_\phi\overline{a}_\phi = 10.3\overline{a}_r + 20.0\overline{a}_\theta + 4.1\overline{a}_\phi$$

Just as in the cylindrical case, the table of known dot product results can be used when converting field vectors between spherical and Cartesian coordinates. However, to build critical visualization of the vector relationships, it is important to practice deriving these dot products graphically, using geometry and trigonometry, as shown in the next example.

Example 1.5.6

Determine the dot product result for $\overline{a}_\theta \cdot \overline{a}_y$.

Advice

Sketch and set up the vector conversions in the first octant with positive vector components for the given vector. The signs on the vector components that result from the vector conversion are more apparent (see the negative $\overline{a}_x \cdot \overline{a}_\phi$ component in the cylindrical conversions discussion, for example).

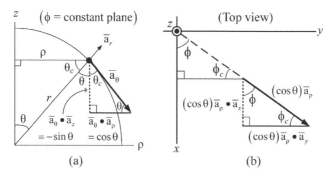

Figure 1.30 Sketches for the dot product $\bar{a}_y \cdot \bar{a}_\theta$. (a) Decomposition of the $\bar{a}_\theta$ unit vector into z and ρ components. (b) Decomposition of the $\bar{a}_\rho$ unit vector into x and y components.

Strategy
- Sketch the geometry showing the unit vectors of concern at a point away from the origin (Figure 1.30).
- Use trigonometry to determine the dot product(s).

Solution

Refer to Figure 1.30. This particular conversion is obtained in a two-step process:

- Decompose $\bar{a}_\theta$ into z and ρ components in the sketch (Figure 1.30a). Then determine the ρ-directed vector component:

$$\bar{a}_\theta \cdot \bar{a}_\rho = \cos\theta \qquad \text{(length of } \bar{a}_\theta \text{ in the } \rho \text{ direction)}$$

$$(\bar{a}_\theta \cdot \bar{a}_\rho)\bar{a}_\theta = (\cos\theta)\bar{a}_\rho \qquad \text{(vector component of } \bar{a}_\theta \text{ in the } \rho \text{ direction)}$$

- Decompose the $(\cos\theta)\bar{a}_\rho$ vector component into x and y components in the sketch (Figure 1.30b). Then determine the length of the y component:

$$(\cos\theta)\bar{a}_\rho \cdot \bar{a}_y = \cos\theta\sin\phi \quad \text{(length of the vector component of } \bar{a}_\theta \text{ in the } y \text{ direction)}$$

Note that this expression is equivalent to

$$\underbrace{\bar{a}_\theta \cdot \bar{a}_y}_{\substack{\text{fraction of} \\ \bar{a}_\theta \text{ in the} \\ y\text{-direction}}} = \underbrace{\left(\bar{a}_\theta \cdot \bar{a}_\rho\right)}_{\substack{\text{fraction of} \\ \bar{a}_\theta \text{ in the} \\ \rho\text{-direction}}} \underbrace{\left(\bar{a}_\rho \cdot \bar{a}_y\right)}_{\substack{\text{fraction of} \\ \bar{a}_\rho \text{ in the} \\ y\text{-direction}}} = \cos\theta\sin\phi$$

Thus, the result is $\bar{a}_\theta \cdot \bar{a}_y = \cos\theta\sin\phi$, which matches the entry in Table 1.2.

A few summary statements on vectors that are expressed in different coordinate systems are now given.

- In Cartesian coordinates, the unit vector directions are unique, that is, they point in the same direction at any location.
- Unit vectors in cylindrical and spherical coordinates are *not* unique unless certain coordinates are specified. The unit vector directions depend, in general, on the coordinates. For example, $\bar{r} = 6\bar{a}_\rho + 7\bar{a}_z$ does not specify a unique position vector because $\bar{a}_\rho$ depends on ϕ, which is unknown if unspecified.
- The angle ϕ must be specified along with the position vector in the cylindrical coordinate system to uniquely locate a point.

- Angles θ and φ must be specified along with the position vector in the spherical coordinate system to uniquely locate a point.
- Angle φ that locates a field vector must be known in order to convert the field vector from Cartesian into cylindrical form or vice versa.
- Angles θ and φ that locate a field vector must be known in order to convert the field vector from Cartesian into spherical form or vice versa.
- For the *dot* or *cross products* between vectors in cylindrical or spherical form:

1) If the vectors are defined (located) at the same point in space, treat them like Cartesian vectors in the dot and cross products because the corresponding unit vectors of both vectors align.
2) If two vectors are *not* defined at the same point but their coordinates are known, one can convert them to Cartesian form and then perform dot and cross products.
3) If two vectors are *not* defined at the same point and their coordinates are *not* known, then one can*not* convert to Cartesian form and, thus, cannot evaluate the dot or cross product.

Example 1.5.7

Can the dot product be performed between two field vectors in spherical form if their coordinates are unknown?

Answer:

Only if the field vectors are located at the same θ and φ so that the unit vectors of both field vectors align.

1.6 Summary of Coordinate Systems and Vectors (see Figures 1.31 and 1.32)

$$\bar{r} = x\bar{a}_x + y\bar{a}_y + z\bar{a}_z \qquad \bar{r} = \rho\bar{a}_\rho + z\bar{a}_z \qquad \bar{r} = r\bar{a}_r$$

$$\bar{R} = R_x\bar{a}_x + R_y\bar{a}_y + R_z\bar{a}_z \qquad \bar{R} = R_\rho\bar{a}_\rho + R_\phi\bar{a}_\phi + R_z\bar{a}_z \qquad \bar{R} = R_r\bar{a}_r + R_\theta\bar{a}_\theta + R_\phi\bar{a}_\phi \qquad \bar{R} = R\bar{a}_R$$

$$\bar{A} = A_x\bar{a}_x + A_y\bar{a}_y + A_z\bar{a}_z \qquad \bar{A} = A_\rho\bar{a}_\rho + A_\phi\bar{a}_\phi + A_z\bar{a}_z \qquad \bar{A} = A_r\bar{a}_r + A_\theta\bar{a}_\theta + A_\phi\bar{a}_\phi \qquad \bar{A} = A\bar{a}_A$$

$$A = |\bar{A}| = \sqrt{A_x^2 + A_y^2 + A_z^2} \quad A = |\bar{A}| = \sqrt{A_\rho^2 + A_\phi^2 + A_z^2} \quad A = |\bar{A}| = \sqrt{A_r^2 + A_\theta^2 + A_\phi^2}$$

$$\bar{R}_{QP} = \bar{r}_P - \bar{r}_Q \qquad\qquad \bar{a}_r = \frac{\bar{r}}{r} \qquad\qquad \bar{a}_R = \frac{\bar{R}}{R} \qquad\qquad \bar{a}_A = \frac{\bar{A}}{A}$$

$$\bar{A} \cdot \bar{B} = |\bar{A}||\bar{B}|\cos\theta = AB\cos\theta = A_xB_x + A_yB_y + A_zB_z$$

$$\bar{A} \times \bar{B} = AB\sin\theta\,\bar{a}_n = (+A_yB_z - A_zB_y)\bar{a}_x + (+A_zB_x - A_xB_z)\bar{a}_y + (+A_xB_y - A_yB_x)\bar{a}_z = |\bar{A} \times \bar{B}|\,\bar{a}_n$$

$$\bar{a}_x \times \bar{a}_y = \bar{a}_z \qquad \bar{a}_\rho \times \bar{a}_\phi = \bar{a}_z \qquad \bar{a}_r \times \bar{a}_\theta = \bar{a}_\phi$$

$\bar{a}_{(\ )} \cdot \bar{a}_{(\ )}$	$\bar{a}_\rho$	$\bar{a}_\phi$	$\bar{a}_r$	$\bar{a}_\theta$
$\bar{a}_x$	$\cos\phi$	$-\sin\phi$	$\sin\theta\cos\phi$	$\cos\theta\cos\phi$
$\bar{a}_y$	$\sin\phi$	$\cos\phi$	$\sin\theta\sin\phi$	$\cos\theta\sin\phi$
$\bar{a}_z$	0	0	$\cos\theta$	$-\sin\theta$

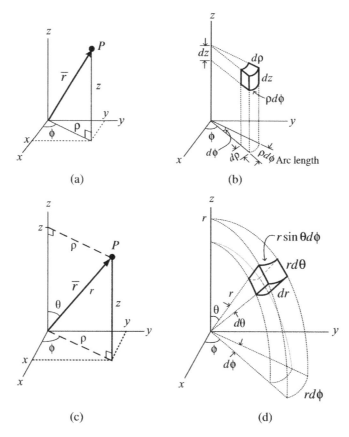

(a)

(b)

(c)

(d)

Figure 1.31 Coordinate system figures. (a) Cylindrical coordinates. (b) Cylindrical differential length elements. (c) Spherical coordinates. (d) Spherical differential length elements.

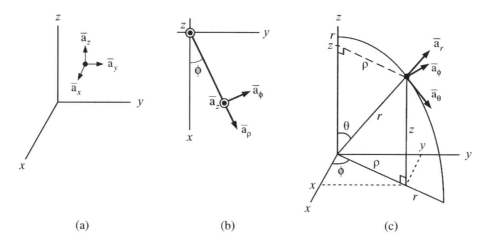

(a)

(b)

(c)

Figure 1.32 Unit vectors. (a) Cartesian unit vectors. (b) Cylindrical unit vectors. (c) Spherical unit vectors.

1.7 Homework

1. Given point A (−4, −5, 8) and point B (3, 0, 6), let $\bar{r}_A$ be the position vector from the origin to point A and $\bar{r}_B$ be the position vector from the origin to point B. Determine:
   a) the distance vector from point A to point B,
   b) sketch $\bar{R}_{AB}$ in the three-dimensional Cartesian coordinate system,
   c) the angle between $\bar{r}_A$ and $\bar{r}_B$,
   d) the vector projection of $\bar{r}_A$ onto $\bar{r}_B$, and
   e) a unit vector normal to the plane containing $\bar{r}_A$ and $\bar{r}_B$.

2. Repeat problem 1 given point A (6, 0, −7) and point B (9, −1, 2).

3. Given point A (8, 6, 7) and point B (−1, 4, 0), let $\bar{r}_A$ be the position vector from the origin to point A and $\bar{r}_B$ be the position vector from the origin to point B. Determine:
   a) the distance vector from point A to point B,
   b) the *vector* projection of $\bar{r}_A$ onto $\bar{r}_B$, and
   c) a unit vector normal to the plane containing $\bar{r}_A$ and $\bar{r}_B$.

4. Given point A (−3, 0, 7) and point B (6, 4, 5), let $\bar{r}_A$ be the position vector from the origin to point A and $\bar{r}_B$ be the position vector from the origin to point B. Determine:
   a) the distance vector from point A to point B,
   b) the angle between $\bar{r}_A$ and $\bar{r}_B$,
   c) the *scalar* projection of $\bar{r}_A$ onto $\bar{r}_B$, and
   d) a unit vector normal to the plane containing $\bar{r}_A$ and $\bar{r}_B$.

5. Derive the equation for the volume of a cylindrical tube (a hollow cylinder with a finite wall thickness) *using calculus*.
   - The inner radius is a, the outer radius is b, and the height is h.
   - Clearly show a sketch, the setup, integration steps, and the final result.

6. Derive the equation for the total surface area of half of a cylindrical tube (a hollow cylinder with a finite wall thickness that is cut in half parallel to its axis) *using calculus*.
   - The inner radius is a, the outer radius is b, and the height is h.
   - Clearly show a sketch, the setup, integration steps, and the final result.

7. Derive the equation for the volume of a spherical shell (a sphere with a hollow center and a finite wall thickness) *using calculus*
   - The inner radius is a and the outer radius is b.
   - Clearly show a sketch, the setup, integration steps, and the final result.

8. Derive the equation for the total surface area of a hemispherical shell (half of a sphere with a hollow center and a finite wall thickness) *using calculus*.
   - The inner radius is a and the outer radius is b.
   - Clearly show a sketch, the setup, integration steps, and the final result.

9. a) What is the general expression for a position vector in cylindrical form?
   b) How are each of the three coordinates incorporated into this position vector?

10. a) What is the general expression for a position vector in spherical form?
b) How are each of the three coordinates incorporated into this position vector?

11. Can the field vector $-7\bar{a}_\rho + 4\bar{a}_\phi + 2\bar{a}_z$ be converted into Cartesian form (with numerical values) if no other information is given? If yes, convert it. If not, explain why not.

12. Are coordinates needed to convert a field vector in Cartesian form to spherical coordinates? Explain why or why not.

13. A *field* vector $\overline{W} = 19\bar{a}_y + 18\bar{a}_z$ is located at (5, 2, 9).
a) Determine the position vector in both Cartesian and cylindrical form.
b) Convert the field vector into cylindrical form.
c) Why do the vector conversion processes in (a) and (b) differ?

14. A *field* vector $\overline{W} = 19\bar{a}_\phi - 7\bar{a}_z$ is located at (8, 5, 9).
a) Determine the position vector in both Cartesian and cylindrical form.
b) Convert the field vector into Cartesian form.
c) Why do the vector conversion processes in (a) and (b) differ?

15. Convert field vector $\overline{W} = 2\bar{a}_\rho - 8\bar{a}_\phi + 5\bar{a}_z$ located at (28, 21, 9) into rectangular form.

16. Convert field vector $\overline{W} = 2\bar{a}_\rho - 8\bar{a}_\phi + 5\bar{a}_z$ located at (−16, 8, −5) into rectangular form.

17. Convert field vector $\overline{F} = 894\bar{a}_x + 159\bar{a}_y - 368\bar{a}_z$ located at (27, 56°, 8) into cylindrical form.

18. Convert field vector $\overline{F} = 894\bar{a}_x + 159\bar{a}_y - 368\bar{a}_z$ located at (32, 221°, 3) into cylindrical form.

19. Convert field vector $\overline{H} = 20\bar{a}_r - 13\bar{a}_\theta - 19\bar{a}_\phi$ located at (3.6, 2.9, 4.5) into rectangular form.

20. Convert field vector $\overline{H} = 20\bar{a}_r - 13\bar{a}_\theta - 19\bar{a}_\phi$ located at (−5.3, 1.9, 8.7) into rectangular form.

21. Convert field vector $\overline{G} = 894\bar{a}_x + 159\bar{a}_y - 368\bar{a}_z$ located at (−9.4, 15.3, 6.5) into spherical form.

22. Convert field vector $\overline{G} = 894\bar{a}_x + 159\bar{a}_y - 368\bar{a}_z$ located at (7.5, 5.6, −4.3) into spherical form.

23. If $\overline{A} = 12xy\bar{a}_x + 16z^2\bar{a}_y - 5ye^z\bar{a}_z$ is a field vector located at (1, 1, 1), determine $\overline{A}$ in
a) rectangular form,
b) cylindrical form, and
c) spherical form.

24. If $\overline{A} = 12xy\bar{a}_x + 16z^2\bar{a}_y - 5ye^z\bar{a}_z$ is a field vector located at (−1, −3, −1), determine $\overline{A}$ in
a) rectangular form,
b) cylindrical form, and
c) spherical form.

25. Develop *every* entry in Table 1.1. Include detailed, labeled sketches that show and justify the angles.

26. Develop *every* spherical entry in Table 1.2. Include detailed, labeled sketches that show and justify the angles.

27. Determine the vector expression for the perpendicular component of one vector relative to another vector ($\overline{A_n}$ in the figure) in terms of appropriate vector algebra operations.

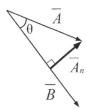

28. A triangle is defined by three arbitrary points M, P, and Q in three-dimensional space. Use vector algebra and geometry to develop the equation for the area of this triangle.

29. Given two vectors, does
    a) $\overline{A} \cdot \overline{B} = \overline{B} \cdot \overline{A}$?
    b) $\overline{A} \cdot \overline{a}_B = \overline{B} \cdot \overline{a}_A$?
       Justify each answer.

30. Given two vectors, does
    a) $\overline{A} \times \overline{B} = \overline{B} \times \overline{A}$?
    b) $\overline{A} \times \overline{a}_B = \overline{a}_A \times \overline{B}$?
       Justify each answer.

31. A parallelepiped has three intersecting vectors $\overline{A}$, $\overline{B}$, and $\overline{C}$ that are along the edges. Vectors $\overline{A}$ and $\overline{B}$ are along the edges in the base parallelogram of the parallelepiped. Vector $\overline{C}$ is along the third edge that extends at some angle from the plane of the base. Show that the volume can be written as: $Volume = \overline{C} \cdot (\overline{A} \times \overline{B})$.

32. Electromagnetic boundary conditions will be examined in subsequent chapters. They are the conditions on a vector at the boundary between two media. Given an arbitrarily oriented vector $\overline{A}$ at the boundary, it is often necessary to determine the magnitudes of the vector components of $\overline{A}$ that are parallel (A_{par}) and perpendicular ($A_\perp$) to the boundary. Define a unit vector $\overline{a}_\perp$ that is normal (perpendicular) to the boundary. Justify the following expressions. Hint: express $\overline{A}$ vectorially.
    a) $\overline{a}_\perp \cdot \overline{A} = A_\perp$
    b) $|\overline{a}_\perp \times \overline{A}| = A_{par}$

Part I

Static Electric and Magnetic Fields

2

The Superposition Laws of Electric and Magnetic Fields

Much of modern technology, including electronics, electric vehicles, communication systems, medical imaging, and so forth, is based upon the physics of electric and magnetic fields. As a result, it is essential for all electrical engineers to have a solid foundation of electromagnetics in order to understand and advance such technologies. This chapter focuses on developing insights into how static (constant with respect to time) electric and magnetic fields are produced by well-known configurations of electric charges and electric currents, respectively. Subsequent chapters will build on these concepts to develop time-varying electric and magnetic fields, which form the basis of electromagnetic waves in modern wireless communication systems, for example.

Electric charges create electric fields. Coulomb's law, which you may have studied in physics or another course, is a mathematical relationship that can be used to determine the static electric field from a single electric charge, a set of charges, or a continuous charge distribution. The mathematics in using Coulomb's law is important. It is based on unit vectors, including non-Cartesian unit vectors and unit vectors inside of an integral. Learning how to "integrate" a unit vector is a key mathematical aspect of this chapter. It will make your study of electromagnetics easier to understand, and it will "demystify" Coulomb's law, as well as many of the topics to come in subsequent chapters. More importantly, you will begin to visualize integrations and the resultant electric fields.

While electric charges create electric fields, electric currents (electric charges in motion) create magnetic fields. Current density and the static magnetic field are then introduced in this chapter. The Biot–Savart law is utilized to determine magnetic field intensity in a manner analogous to the utilization of Coulomb's law to determine electric field intensity.

Coulomb's law and the Biot–Savart law are often collectively called the **superposition laws**. Why? Think about this question as you cover this chapter.

The abbreviation "CL" will sometimes be used for Coulomb's law and similarly "BSL" for the Biot–Savart law. The fields in this chapter are assumed to be *static*, that is, the field has a constant value versus time at each position (the field value changes as a function of position but not time). A static electric field is often called an **electrostatic** field and a static magnetic field is often called a **magnetostatic** field. SI units are utilized in this text unless stated otherwise.

Electromagnetics and Transmission Lines: Essentials for Electrical Engineering, Second Edition.
Robert A. Strangeway, Steven S. Holland, and James E. Richie.
© 2023 John Wiley & Sons, Inc. Published 2023 by John Wiley & Sons, Inc.
Companion website: www.wiley.com/go/Strangeway/ElectromagneticsandTransmissionLines

2.1   Point Electric Charges, Coulomb's Law, and Electric Fields

The symbol for electric charge is Q. Examples of positive charges are protons and atoms with more protons than electrons. Conversely, examples of negative charges are electrons and atoms with more electrons than protons. The unit of electric charge is the coulomb (C). One coulomb of charge is equal to the charge of approximately 6×10^{18} charged particles (such as electrons), which is a large amount of charge. Practical charge quantities are typically much lower than one coulomb, often on the order of pC to mC (pico- to milli-coulombs). One electron has a charge of -1.602×10^{-19} C.

An electric charge exerts a force on other charges, and vice versa, and thus there are forces between electric charges: *like charges repel, unlike charges attract.* For example, if $+Q$ of charge flows onto one capacitor plate, this charge exerts a force that causes $+Q$ of charge to flow off of the other capacitor plate (leaving behind "$-Q$" of charge, as typically pictured in parallel plate capacitor figures).

Our study of the forces that electric charges exert on one another will begin with the electric charge in the form of a **point charge.** What is a point charge? Ideally, it is a model of electric charge with shrinking dimensions such that the volume of the charge approaches zero. Practically, this model is accurate if the finite dimensions of the charge are much smaller than the distance between charges and observation points. For example, a single proton or electron could be considered a point charge if the distance to the observation point is much larger than an atomic diameter. Similarly, a small disc of charge could be considered a point charge if the distance to the observation point is tens of times the disc diameter or more.

How much force exists between two point charges in free space? The force between charges is determined by **Coulomb's law**, shown in Eq. (2.1)(a). Interestingly, this law was developed from extensive experiments with charged objects in the eighteenth century. Coulomb's law happens to share the same mathematical form as Newton's law of gravitation, shown in Eq. (2.1)(b), and illuminating insights can be gained by comparing these two laws.

$$\text{(a)} \ \ F_{\text{electric}} = k \frac{Q_1 Q_2}{R^2} \quad \text{(b)} \ \ F_{\text{gravity}} = G \frac{m_1 m_2}{R^2} \tag{2.1}$$

In these equations, G is the gravitational constant (a proportionality constant), k is a proportionality constant, and R is the distance between the two masses or the two charges. The electric force is directly proportional to the amount of each charge, just as gravitational force is directly proportional to the amount of each mass. Note the inverse square law with respect to distance. This relationship can be visualized if one considers that the force due to one charge is spread out over a spherical surface centered on the charge. The surface area of a sphere is $4\pi R^2$. Hence, as the distance (radius) increases, the force per unit area decreases proportional to $1/R^2$. Finally, note that gravitational force is always attractive, whereas electric force can be either attractive (for unlike charges) or repulsive (for like charges).

The proportionality constant in Coulomb's law is $k = \dfrac{1}{4\pi\varepsilon_o}$; thus,

$$F = k \frac{Q_1 Q_2}{R^2} = \left(\frac{1}{4\pi\varepsilon_o} \right) \frac{Q_1 Q_2}{R^2} = \frac{Q_1 Q_2}{4\pi\varepsilon_o R^2} \tag{2.2}$$

What is ε_o? It is the free-space **permittivity** (permittivity of vacuum), the same quantity that is used in capacitance calculations. (Permittivity is examined in-depth in Chapter 4.) The value for ε_o in SI units is approximately 8.8542×10^{-12} farads/meter (F/m) (check the NIST website if more accuracy is required). Standard SI units of Newton (N), coulomb (C), and meter (m) apply to F, Q, and R, respectively.

How is the direction of the electric force specified? The magnitude of the force must be multiplied by a unit vector to express Coulomb's law in vector form:

$$\overline{F} = \frac{Q_1 Q_2}{4\pi\varepsilon_o R^2} \overline{a}_R \tag{2.3}$$

where the unit vector $\bar{a}_R$ is in the direction from Q_1 to Q_2 when determining the force on Q_2 due to Q_1 or is in the direction from Q_2 to Q_1 when determining the force on Q_1 due to Q_2. Equation (2.3) is Coulomb's law for two point charges. How is the unit vector determined?

$$\bar{a}_R = \frac{\bar{R}}{R} \tag{2.4}$$

Now visualize that Q_1 is held stationary and that Q_2 is moved to locations throughout the space surrounding Q_1. At each location, Eq. (2.3) gives the vector electric force $\bar{F}$ on Q_2 due to Q_1 and thus describes an *electric force field* in the space surrounding Q_1. It is important to emphasize that this electric force field depends on the values of *both* Q_1 and Q_2. It would be useful to be able to describe the electric force field that Q_1 exerts in the surrounding region more generally, without the dependence on Q_2.

How can the electric force field of one charge Q be analyzed without introducing a second charge? Let the second charge be a "sensing" charge to sample the field at various locations. It is called a *test charge*. A test charge is fictional in that it does not add its own electric field to the field of the charge under consideration. Practically, one could visualize a test charge that is much smaller (in terms of coulombs) than the charge under consideration so that the electric field of the test charge is insignificant relative to the field of the charge under consideration. (An electric circuit analogy is the ammeter, which should add an insignificant amount of series resistance relative to the resistances in the circuit in order to not significantly affect the current in the circuit.)

The force equation is divided by the test charge to give the definition of the **electric field intensity** $\bar{E}$, the force per unit test charge:

$$\bar{E} \equiv \frac{\bar{F}}{Q_{\text{test}}} = \frac{Q}{4\pi\varepsilon_o R^2}\bar{a}_R \tag{2.5}$$

where the unit vector is in the direction from Q to the observation point P, the test charge location. The units of electric field intensity are V/m ("volts per meter"; the reason for these units will become apparent in Chapter 4). This expression is often called Coulomb's law too, but technically Coulomb's law is the electric force equation. The force is easily obtained – just multiply $\bar{E}$ by the second charge Q_2. Thus, an electric field is a "force field," that is, a field that exerts forces on electric charges. It is quantified as force per unit charge.

The determination of the electric field intensity at some location due to a point charge at another location is straightforward: determine the distance vector $\bar{R}$ from the point charge to the observation point, and insert into Coulomb's law. How is the distance vector $\bar{R}$ determined? See Figure 2.1. It is the position vector of the observation point $\bar{r}$ minus the position vector of the point charge $\bar{r}'$:

$$\bar{r}' + \bar{R} = \bar{r} \rightarrow \bar{R} = \bar{r} - \bar{r}' \tag{2.6}$$

The position vector that locates the "source," a point charge here, is usually primed in electromagnetic fields notation (it is not a derivative). The distance vector $\bar{R}$ locates the observation point with respect to the source point (the charge). Often, Coulomb's law for a point charge is expressed with this notation substituted:

$$\boxed{\bar{E} = \frac{\bar{F}}{Q_{\text{test}}} = \frac{Q}{4\pi\varepsilon_o R^2}\bar{a}_R = \frac{Q}{4\pi\varepsilon_o R^2}\frac{\bar{R}}{R} = \frac{Q\bar{R}}{4\pi\varepsilon_o R^3} = \frac{Q(\bar{r} - \bar{r}')}{4\pi\varepsilon_o |\bar{r} - \bar{r}'|^3}} \tag{2.7}$$

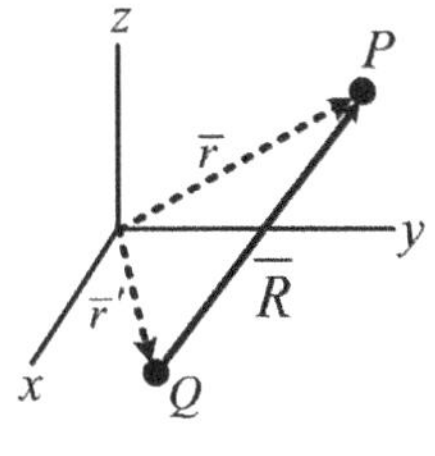

Figure 2.1 Position and distance vectors for setting up Coulomb's law.

The notations in the previous equation are all equally valid, and different authors use different but equivalent notations.

What is the electric field intensity at an observation point if there are two or more point charges? Just as superposition of voltages holds in a linear circuit, superposition of electric fields holds as long as the medium is linear,[1] since summing vector electric fields is actually summing vector forces per unit charge. Thus, add the electric field intensities (vectors) due to each point charge in order to determine the total electric field intensity at the observation point. The expression for the electric field intensity due to N point charges in summation notation is:

$$\overline{E} = \sum_{i=1}^{N} \overline{E}_i = \sum_{i=1}^{N} \frac{Q_i}{4\pi\varepsilon_o R_i^2}\overline{a}_{R_i} = \sum_{i=1}^{N} \frac{Q_i\overline{R}_i}{4\pi\varepsilon_o R_i^3} = \sum_{i=1}^{N} \frac{Q_i(\overline{r} - \overline{r}_i')}{4\pi\varepsilon_o|\overline{r} - \overline{r}_i'|^3} \tag{2.8}$$

Note that each charge and its position vector are indexed by subscript i (1, 2, ...) to distinguish them. The position vector to the observation point is not indexed – it is always the same in a given calculation.

Example 2.1.1

If $Q_A = 8$ C is located at the origin and $Q_B = 3$ C is located at (2, 1, −1), determine $\overline{E}$ at $P(1, 4, 2)$.

Strategy

Sketch and label P, each Q, and position and distance vectors (Figure 2.2)

- Use CL to determine $\overline{E}_A$ and $\overline{E}_B$
- Sum the vectors due to each charge for the total vector: $\overline{E}_T = \overline{E}_A + \overline{E}_B$

Solution

$$\overline{E}_A = \frac{Q_A}{4\pi\varepsilon_o R_A^2}\overline{a}_{R_A} = \frac{Q_A\overline{R}_A}{4\pi\varepsilon_o R_A^3} = \frac{Q_A(\overline{r} - \overline{r}_A')}{4\pi\varepsilon_o|\overline{r} - \overline{r}_A'|^3}$$

$$\overline{r}_A' = 0\overline{a}_x + 0\overline{a}_y + 0\overline{a}_z$$

$$\overline{r} = 1\overline{a}_x + 4\overline{a}_y + 2\overline{a}_z$$

$$\overline{R}_A = \overline{r} - \overline{r}_A' = (1-0)\overline{a}_x + (4-0)\overline{a}_y + (2-0)\overline{a}_z = 1\overline{a}_x + 4\overline{a}_y + 2\overline{a}_z$$

$$\overline{E}_A = \frac{Q_A(\overline{r} - \overline{r}_A')}{4\pi\varepsilon_o|\overline{r} - \overline{r}_A'|^3} = \frac{8(1\overline{a}_x + 4\overline{a}_y + 2\overline{a}_z)}{4\pi\varepsilon_o\left(\sqrt{1^2 + 4^2 + 2^2}\right)^3} = \frac{8(1\overline{a}_x + 4\overline{a}_y + 2\overline{a}_z)}{4\pi 8.8542 \times 10^{-12}(21)^{3/2}}$$

$$\overline{E}_A = 7.4716 \times 10^8(1\overline{a}_x + 4\overline{a}_y + 2\overline{a}_z) = (7.4716\overline{a}_x + 29.8863\overline{a}_y + 14.9431\overline{a}_z) \times 10^8 \text{ V/m}$$

$$\overline{r}_B' = 2\overline{a}_x + 1\overline{a}_y - 1\overline{a}_z \quad \text{and} \quad \overline{r} = 1\overline{a}_x + 4\overline{a}_y + 2\overline{a}_z$$

$$\overline{R}_B = \overline{r} - \overline{r}_B' = (1-2)\overline{a}_x + (4-1)\overline{a}_y + (2-(-1))\overline{a}_z = -1\overline{a}_x + 3\overline{a}_y + 3\overline{a}_z$$

$$\overline{E}_B = \frac{Q_B(\overline{r} - \overline{r}_B')}{4\pi\varepsilon_o|\overline{r} - \overline{r}_B'|^3} = \frac{3(-1\overline{a}_x + 3\overline{a}_y + 3\overline{a}_z)}{4\pi 8.8542 \times 10^{-12}(1^2 + 3^2 + 3^2)^{3/2}} = \frac{3(-1\overline{a}_x + 3\overline{a}_y + 3\overline{a}_z)}{4\pi 8.8542 \times 10^{-12}(19)^{3/2}}$$

$$\overline{E}_B = 3.2557 \times 10^8(-1\overline{a}_x + 3\overline{a}_y + 3\overline{a}_z) = (-3.2557\overline{a}_x + 9.7670\overline{a}_y + 9.7670\overline{a}_z) \times 10^8 \text{ V/m}$$

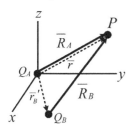

The total electric field intensity vector $\overline{E}_T$ at P is the sum (superposition) of the electric field intensity vectors due to Q_A and Q_B:

$$\overline{E}_T = \overline{E}_A + \overline{E}_B = (7.4716\overline{a}_x + 29.8863\overline{a}_y + 14.9431\overline{a}_z) \times 10^8$$

$$+ (-3.2557\overline{a}_x + 9.7670\overline{a}_y + 9.7670\overline{a}_z) \times 10^8$$

$$\overline{E}_T = (4.22\overline{a}_x + 39.65\overline{a}_y + 24.71\overline{a}_z) \times 10^8 \text{ V/m}$$

Figure 2.2 Two point charges.

Why is Q often on the order of pC to mC, not C? Note the huge order of the magnitude of the electric field intensity when Q is on the order of coulombs.

---

1 Vacuum is linear. The effect of materials on the electric field will be considered in the next two chapters.

2.2 Electric Charge Distributions and Charge Density

Electric charge can be distributed in a spatial region. Summing the electric fields due to each individual charge would be impossible. Hence, a method of considering distributed charge on a macroscopic basis is needed. The three types of charge distributions in free space (conductors are covered in a subsequent chapter), illustrated in Figure 2.3a–c, are:

- Line charge – one dimensional (example: a thin, charged string)
- Surface charge – two dimensional (example: a thin, charged sheet)
- Volume charge – three dimensional (example: a sphere filled with charge)

Volume charge density is most general because actual charge occupies a volume! Line charge density is an accurate approximation when the cross-sectional dimensions of a thin line of charge are much less than the distance to the observation point (imagine being 10 cm away from a long, charged cylinder of 1 mm diameter, for example). Surface charge density is similarly an accurate approximation when the distance to the observation point is much greater than the thickness of a charged surface. These approximations usually simplify the mathematics and lend greater insights relative to the more general volume charge density case.

How is the electric field intensity determined for charge distributions? First, the concept of **charge density** (charge per unit length, surface area, or volume) must be established. Let ΔQ be a small amount of charge in a charge distribution. In the limit as the dimensions of ΔQ approach zero, ΔQ becomes dQ:

$$Q \text{ per unit length}: \text{ line charge density} \left(\frac{\text{C}}{\text{m}} \right) = \rho_L = \lim_{\Delta \ell \to 0} \left(\frac{\Delta Q}{\Delta \ell} \right) = \frac{dQ}{d\ell} \to dQ = \rho_L d\ell \tag{2.9}$$

$$Q \text{ per unit area}: \text{ surface charge density} \left(\frac{\text{C}}{\text{m}^2} \right) = \rho_S = \lim_{\Delta A \to 0} \left(\frac{\Delta Q}{\Delta A} \right) = \frac{dQ}{dA} \to dQ = \rho_S dA \tag{2.10}$$

$$Q \text{ per unit volume}: \text{ volume charge density} \left(\frac{\text{C}}{\text{m}^3} \right) = \rho_V = \lim_{\Delta V \to 0} \left(\frac{\Delta Q}{\Delta V} \right) = \frac{dQ}{dV} \to dQ = \rho_V dV \tag{2.11}$$

Notation: The Greek letter ρ is used as a symbol in several different applications, such as:

- The radius in the cylindrical coordinate system
- Charge density (subscript L, S, or V indicates the type of charge density)
- Resistivity of a conductor
- Mass density

One must distinguish what ρ symbolizes from the context of the situation.

How is the total charge in a region determined from dQ? It is determined by integrating the differential charge elements (dQ) over the dimensions of the charge distribution:

$$Q = \int dQ: \quad Q = \int \rho_L d\ell \quad \text{or} \quad Q = \int \rho_S dA \quad \text{or} \quad Q = \int \rho_V dV \tag{2.12}$$

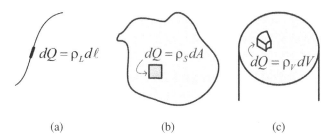

(a)   (b)   (c)

Figure 2.3   Charge Distributions. (a) Line charge. (b) Surface charge. (c) Volume charge.

Example 2.2.1

Determine the total charge inside a sphere of radius 2 m given a charge density of 24×10^{-6} C/m^3.

Solution

$$Q = \int dQ = \int \rho_V dV = \int_0^{2\pi} \int_0^\pi \int_0^2 24 \times 10^{-6} r^2 \sin\theta \, dr \, d\theta \, d\phi$$

$$Q = 24 \times 10^{-6} \int_0^2 r^2 dr \int_0^\pi \sin\theta d\theta \int_0^{2\pi} d\phi = 24 \times 10^{-6} \left(\frac{r^3}{3}\right)\Big|_0^2 (-\cos\theta)|_0^\pi \phi|_0^{2\pi}$$

$$Q = 24 \times 10^{-6} \left(\frac{2^3}{3}\right) \underbrace{(-\cos\pi + \cos 0)}_{2} 2\pi = 0.804 \text{ mC}$$

Note that the result is the same as multiplying the volume of the sphere and the volume charge density $[(24 \times 10^{-6})(\frac{4}{3}\pi 2^3) = 0.804 \text{ mC}]$ because the charge density is uniform.

Example 2.2.2

Determine the total charge inside a sphere of radius 2 m given a nonuniform (not constant) charge density of $\dfrac{24 \times 10^{-6}}{r(r+3)}$ C/m^3.

Solution

$$Q = \int dQ = \int \rho_V dV = \int_0^{2\pi} \int_0^\pi \int_0^2 \frac{24 \times 10^{-6}}{r(r+3)} r^2 \sin\theta \, dr \, d\theta \, d\phi$$

$$Q = 24 \times 10^{-6} \int_0^2 \frac{r \, dr}{(r+3)} \int_0^\pi \sin\theta d\theta \int_0^{2\pi} d\phi$$

One can use an integral table entry to evaluate the integral in r:

$$\int \frac{x \, dx}{ax+b} = \frac{x}{a} - \frac{b}{a^2} \ln(ax+b) \rightarrow x = r, a = 1, b = 3$$

$$Q = 24 \times 10^{-6} \left[\frac{r}{1} - \frac{3}{1^2} \ln(r+3)\right]\Big|_0^2 (-\cos\theta)|_0^\pi \phi|_0^{2\pi}$$

$$Q = 24 \times 10^{-6}[2 - 3\ln(2+3) - 0 + 3\ln(0+3)]4\pi = 0.141 \text{ mC}$$

Note that the result is *not* the same as multiplying the volume of a sphere and the volume charge density because the charge density is *not uniform* (charge density is not a constant value).

Practically, the charge density is usually unknown. However, it can often be assumed to be uniformly distributed in many applications. The usefulness of charge density as a variable will become more apparent as you learn more about electromagnetic fields. In the meantime, we will assume that the charge density is known as we learn how to apply charge density in Coulomb's law for charge distributions.

2.3 Coulomb's Law in Integral Form and Examples

How is Coulomb's law utilized for charge distributions? First, reconsider Coulomb's law for multiple point charges:

$$\overline{E} = \sum_{i=1}^N \overline{E}_i = \sum_{i=1}^N \frac{Q_i}{4\pi\varepsilon_o R_i^2} \overline{a}_{R_i} = \sum_{i=1}^N \frac{Q_i \overline{R}_i}{4\pi\varepsilon_o R_i^3} = \sum_{i=1}^N \frac{Q_i(\overline{r} - \overline{r}_i')}{4\pi\varepsilon_o |\overline{r} - \overline{r}_i'|^3} \tag{2.13}$$

The total electric field intensity is the sum of the electric field intensities from each point charge. What is the electric field intensity due to a differential amount of charge dQ? If a charge Q produces a finite electric field intensity $\overline{E}$ at the observation point, then a differential amount of charge dQ should produce a differential amount of electric field intensity $d\overline{E}$ at the observation point.

$$\overline{E}_i = \frac{Q_i}{4\pi\varepsilon_o R_i^2}\overline{a}_{R_i} \rightarrow d\overline{E} = \frac{dQ}{4\pi\varepsilon_o R^2}\overline{a}_R = \frac{dQ\overline{R}}{4\pi\varepsilon_o R^3} = \frac{dQ(\overline{r}-\overline{r}')}{4\pi\varepsilon_o |\overline{r}-\overline{r}'|^3} \tag{2.14}$$

If the differential electric field intensities $d\overline{E}$ (from the differential charge elements dQ) are being summed, the summation becomes integration:

$$\boxed{\overline{E} = \int d\overline{E} = \int \underbrace{\frac{dQ}{4\pi\varepsilon_o R^2}\overline{a}_R}_{d\overline{E}} = \int \frac{dQ\overline{R}}{4\pi\varepsilon_o R^3} = \int \frac{dQ(\overline{r}-\overline{r}')}{4\pi\varepsilon_o |\overline{r}-\overline{r}'|^3}} \tag{2.15}$$

Note why Coulomb's law is called one of the superposition laws: the differential electric field intensities $d\overline{E}$ are being summed (integrated) to determine the total $\overline{E}$. Can the vectors be removed from the integration in general? No, because *the position vector of the source point $\overline{r}'$ is a variable*. In other words, $\overline{r}'$, the position vector to dQ, varies as the location of dQ varies in the integration to sum the effects of all of the dQs in the charge distribution. Hence, in general, Coulomb's law involves the *integration of a vector*. How is a vector integrated?

1) If the unit vector is constant with respect to the variables being integrated, then it can be moved outside of the integral. Cartesian unit vectors are always constant, but cylindrical and spherical unit vectors may or may not be constant with respect to the variables being integrated.
2) If cylindrical and spherical unit vectors are functions of the variables being integrated, convert them into Cartesian form and then remove the Cartesian unit vector from each integral.
3) Vector components that integrate to zero due to the symmetry of the charge distribution can be eliminated. This approach requires visualization of the field in the appropriate coordinate system due to symmetry (another reason different coordinate systems are useful!).

Example 2.3.1
A finite length of uniform line charge density extends from $-h$ to $+h$ on the z-axis.

a) Determine the expression for the electric field intensity at $P(\rho, \phi, 0)$.
b) What does $\overline{E}$ approach if the length of the line charge is much greater than the distance to the observation point? Why?
c) What does $\overline{E}$ approach if the distance to the observation point is much greater than the length of the line charge? Why?
d) Plot the field behavior versus ρ. Plot and compare the infinitely-long line charge and point charge cases.

a) **Strategy**

- Draw and label a figure (see Figure 2.4)
- Assign ρ_L = uniform line change density (it is a constant in this example), units (C/m)
- Select an appropriate coordinate system
- Determine dQ and $\overline{R}$ in terms of the selected coordinate system
- Use the integral form of Coulomb's law

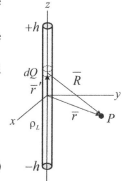

Figure 2.4 Line charge setup for Coulomb's law.

Solution

A cylindrical geometry and symmetry are present: use cylindrical coordinates

$$\overline{E} = \int d\overline{E} = \int \frac{dQ\overline{R}}{4\pi\varepsilon_o R^3} \quad \text{where } \overline{R} = \overline{r} - \overline{r}'$$

Given the line charge distribution, assign a line charge density and express $d\ell$ in terms of the cylindrical coordinate system:

$$dQ = \rho_L d\ell = \rho_L dz$$

Determine the position vectors that locate P and dQ. Note that $\overline{r}'$ must have a variable position coordinate z in order to be able to locate any dQ along the length of the line charge.

$$\left. \begin{array}{l} \overline{r} = \rho\overline{a}_\rho \\ \overline{r}' = z\overline{a}_z \end{array} \right\} \begin{array}{l} \text{Problem: How are the symbols for constants} \\ \text{distinguished from the symbols for variables?} \end{array}$$

One method is to mark the symbols that are constants with a subscript:

$$\overline{r} = \rho_1\overline{a}_\rho$$

$$\overline{r}' = z\overline{a}_z$$

Then determine the distance vector:

$$\overline{R} = \overline{r} - \overline{r}' = \rho_1\overline{a}_\rho - z\overline{a}_z$$

Note that the ϕ coordinate of the observation point is embedded in the $\overline{a}_\rho$ unit vector. Insert the expressions for R, $\overline{R}$, and dQ into CL:

$$R = |\overline{R}| = \sqrt{\rho_1^2 + z^2} \quad \text{and} \quad R^3 = \left(\sqrt{\rho_1^2 + z^2}\right)^3 = \left[\rho_1^2 + z^2\right]^{3/2}$$

$$\overline{E} = \int \frac{dQ\overline{R}}{4\pi\varepsilon_o R^3} = \int \frac{\rho_L dz\left[\rho_1\overline{a}_\rho - z\overline{a}_z\right]}{4\pi\varepsilon_o\left[\sqrt{\rho_1^2 + z^2}\right]^3} = \int \frac{\rho_L dz\left[\rho_1\overline{a}_\rho - z\overline{a}_z\right]}{4\pi\varepsilon_o\left[\rho_1^2 + z^2\right]^{3/2}}$$

Can ρ_L be removed from the integral? Yes, it is a constant with respect to z. How are the limits determined? Over the extent of the charge distribution dimensions, $-h$ to $+h$ for this charge distribution:

$$\overline{E} = \frac{\rho_L}{4\pi\varepsilon_o} \int\limits_{-h}^{+h} \frac{\left[\rho_1\overline{a}_\rho - z\overline{a}_z\right]}{\left[\rho_1^2 + z^2\right]^{3/2}} dz$$

How is the integral of a sum integrated? As the sum of the integrals:

$$\overline{E} = \frac{\rho_L}{4\pi\varepsilon_o} \left[\int\limits_{-h}^{+h} \frac{\rho_1\overline{a}_\rho dz}{\left[\rho_1^2 + z^2\right]^{3/2}} - \int\limits_{-h}^{+h} \frac{z\overline{a}_z dz}{\left[\rho_1^2 + z^2\right]^{3/2}} \right]$$

How are the unit vectors integrated? If they are constant with respect to the variables being integrated, then they can be removed from the integrals as a constant. In this case, the $\overline{a}_\rho$ unit vector is constant with respect to z. The $\overline{a}_z$ unit vector is always a constant.

$$\overline{E} = \frac{\rho_L}{4\pi\varepsilon_o} \left[\rho_1\overline{a}_\rho \int\limits_{-h}^{+h} \frac{dz}{\left[\rho_1^2 + z^2\right]^{3/2}} - \overline{a}_z \int\limits_{-h}^{+h} \frac{z dz}{\left[\rho_1^2 + z^2\right]^{3/2}} \right]$$

How are these integrals evaluated?

- Manually, using an integral table or calculator and sometimes lengthy manipulations.
- Mathematical software package, but little insight is gained if the expression is too unwieldy.

Manually, which integral table entries are appropriate? Recognize the mathematical forms of the integrands:

$$\int \frac{dx}{(x^2+a^2)^{3/2}} = \frac{x}{a^2\sqrt{x^2+a^2}} \xrightarrow[a=\rho_1]{x=z} \int_{-h}^{+h} \frac{dz}{[\rho_1^2+z^2]^{3/2}} = \frac{z}{\rho_1^2\sqrt{\rho_1^2+z^2}}\bigg|_{-h}^{+h}$$

$$\int \frac{x\,dx}{(x^2+a^2)^{3/2}} = \frac{-1}{\sqrt{x^2+a^2}} \xrightarrow[a=\rho_1]{x=z} \int_{-h}^{+h} \frac{z\,dz}{[\rho_1^2+z^2]^{3/2}} = \frac{-1}{\sqrt{\rho_1^2+z^2}}\bigg|_{-h}^{+h}$$

Insert these results into the electric field expression and evaluate the limits:

$$\overline{E} = \frac{\rho_L}{4\pi\varepsilon_0}\left[\rho_1\overline{a}_\rho \frac{z}{\rho_1^2\sqrt{\rho_1^2+z^2}}\bigg|_{-h}^{+h} - \overline{a}_z \frac{-1}{\sqrt{\rho_1^2+z^2}}\bigg|_{-h}^{+h}\right]$$

$$\overline{E} = \frac{\rho_L}{4\pi\varepsilon_0}\left[\overline{a}_\rho\left(\frac{+h}{\rho_1\sqrt{\rho_1^2+h^2}} - \frac{-h}{\rho_1\sqrt{\rho_1^2+h^2}}\right) + \overline{a}_z\left(\frac{1}{\sqrt{\rho_1^2+h^2}} - \frac{1}{\sqrt{\rho_1^2+h^2}}\right)\right]$$

Mathematically, the z-component integral equals zero. What is the physical reasoning that the z-component integral equals zero? Formulate a **symmetry argument**. Refer to Figure 2.5.

- For every dQ that creates a $d\overline{E}$ at the observation point, there is a symmetrically opposite dQ that creates another $d\overline{E}$.
- The ρ-components of the $d\overline{E}$ vectors add.
- The z-components of the $d\overline{E}$ vectors cancel each other.
- Thus, one could declare the z-integral to be zero without evaluating it (if the symmetry can be justified, as in Figure 2.5).

Only the ρ-component of $\overline{E}$ results due to the symmetry of the observation point with respect to the charge distribution. The resultant expression for the electric field intensity is

$$\overline{E} = \frac{\rho_L}{4\pi\varepsilon_0}\overline{a}_\rho\left(\frac{2h}{\rho_1\sqrt{\rho_1^2+h^2}}\right) = \left(\frac{\rho_L h}{2\pi\varepsilon_0\rho_1\sqrt{\rho_1^2+h^2}}\right)\overline{a}_\rho$$

The radius to the observation point was an unspecified constant when Coulomb's law was applied to determine $\overline{E}$. Once $\overline{E}$ is obtained, the constant radius for the observation point ρ_1 is replaced with ρ. Now the observation point radius can become a variable to study the behavior of the electric field intensity as a function of radial distance from the line charge density distribution. Thus, the electric field intensity at any point in the x–y plane for a line charge that extends from $+h$ to $-h$ on the z-axis is

$$\overline{E} = \left(\frac{\rho_L h}{2\pi\varepsilon_0\rho\sqrt{\rho^2+h^2}}\right)\overline{a}_\rho$$

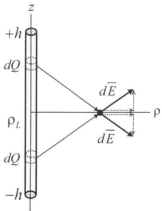

Figure 2.5 Symmetry argument for a straight line charge.

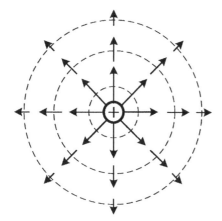

Figure 2.6 Visualization of $\overline{E}$ vs. radial distance (top view, 3 radii shown).

b) What does $\overline{E}$ approach if the length of the line charge is much greater than the distance to the observation point? Why?

When the line charge length is much greater than the distance to the observation point, $h \gg \rho$ and a limit can be taken:

$$\lim_{h \gg \rho} (\overline{E}) = \lim_{h \gg \rho} \left(\frac{\rho_L h}{2\pi\varepsilon_o \rho \sqrt{\rho^2 + h^2}} \right) \overline{a}_\rho$$

$$\rightarrow \left(\frac{\rho_L h}{2\pi\varepsilon_o \rho \sqrt{h^2}} \right) \overline{a}_\rho = \left(\frac{\rho_L}{2\pi\varepsilon_o \rho} \right) \overline{a}_\rho$$

This result is important. It describes the electric field when the observation point is very close to a long, but finite length line charge distribution, or equivalently, it describes the field observed due to an infinite length line charge. The electric field intensity falls off in inverse proportion to the radius ρ of the observation point. Why? The electric field intensity is spread out around the circumference of a circle. See Figure 2.6. The density of $\overline{E}$ decreases as the radius increases. The circumference of a circle is $2\pi\rho$. Thus, the field must "fall off" proportional to $1/\rho$ as the radius increases. Note the usefulness of cylindrical coordinates in interpreting the behavior of the electric field intensity as a function of radial distance.

Hence, the idealized result of the electric field intensity for an infinite line charge (infinitely long line charge density) in free space (vacuum) is

$$\boxed{\overline{E} = \frac{\rho_L}{2\pi\varepsilon_o \rho} \overline{a}_\rho} \tag{2.16}$$

c) What happens if the distance to the observation point is much greater than the length of the line charge? Why?

When the distance to the observation point is much greater than the line charge length, $\rho \gg h$, the limit becomes:

$$\lim_{\rho \gg h} (\overline{E}) = \lim_{\rho \gg h} \left(\frac{\rho_L h}{2\pi\varepsilon_o \rho \sqrt{\rho^2 + h^2}} \right) \overline{a}_\rho \rightarrow \left(\frac{\rho_L h}{2\pi\varepsilon_o \rho \sqrt{\rho^2}} \right) \overline{a}_\rho = \left(\frac{\rho_L h}{2\pi\varepsilon_o \rho^2} \right) \overline{a}_\rho$$

The electric field intensity falls off proportional to the radius-squared. Why? It reminds one of Coulomb's law for point charges. Can this electric field intensity expression be manipulated into Coulomb's law form?

$$\overline{E} = \left(\frac{\rho_L h}{2\pi\varepsilon_o \rho^2} \right) \overline{a}_\rho = \left(\frac{2\rho_L h}{4\pi\varepsilon_o \rho^2} \right) \overline{a}_\rho = \left(\frac{Q_T}{4\pi\varepsilon_o \rho^2} \right) \overline{a}_\rho$$

where $2\rho_L h$ is the total charge in the line charge of length $2h$. Thus, far away from the line charge distribution, the line charge "looks like" a point charge and the electric field intensity falls off proportional to $1/R^2$ because the field is spreading out over the surface area of a sphere ($4\pi R^2$).

d) Plot the field behavior versus ρ for cases a)–c) and compare the electric field behavior for the finite length line charge, the infinitely long line charge, and the point charge cases.

Figure 2.7 Comparison of the line charge electric field results.

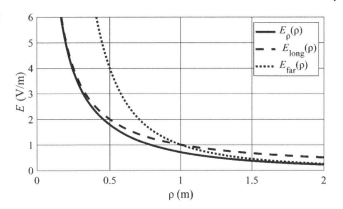

Plot $|\overline{E}|$ vs. ρ for P in the x–y plane for all three cases. Choose convenient mathematical values for ρ_L and h. The equations follow:

$$\rho_L = 2\pi\varepsilon_o \quad h = 1 \quad \varepsilon_o = 8.854 \times 10^{-12}$$

$$E_\rho(\rho) = \frac{\rho_L h}{2\pi\varepsilon_o\rho\sqrt{\rho^2 + h^2}} \quad E_{\text{long}}(\rho) = \frac{\rho_L}{2\pi\varepsilon_o\rho} \quad E_{\text{far}}(\rho) = \frac{\rho_L h}{2\pi\varepsilon_o\rho^2}$$

The "long" subscript refers to the result for $h \gg \rho$ in part b), and the "far" subscript refers to the result for $\rho \gg h$ in part c). The results are plotted in Figure 2.7. The general result for the ρ-component of $\overline{E}$, E_ρ, converges with the long line charge result as $h \gg \rho$. The general result converges with the "far away from the line charge" result as $\rho \gg h$.

This example illustrates how special limiting cases can add physical insight into a given problem. It also illustrates why developing general forms and being able to mathematically manipulate expressions are useful to gain insights into and predict the behavior of the field.

Example 2.3.2

a) Determine the electric field intensity along the z-axis due to a flat disc of uniform surface charge density of radius a lying in the x–y plane.
b) Evaluate the limit of the result as the distance to the observation point $h \ll a$. Interpret.
c) Evaluate the limit of the result as the distance to the observation point $h \gg a$. Interpret.
d) Plot the field behavior versus h. Plot and compare the infinite sheet of surface charge and point charge cases.

a) Strategy

- Draw a figure (see Figure 2.8).
- Assign $\rho_S = $ *uniform* surface charge density (a constant), units (C/m^2).
- Determine dQ and $\overline{R}$ in terms of the selected coordinate system.
- Use the integral form of Coulomb's law.

Solution

Start with Coulomb's law
Choose cylindrical coordinates for this cylindrical charge geometry

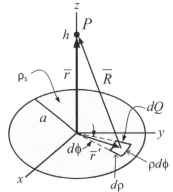

Figure 2.8 Disc of surface charge.

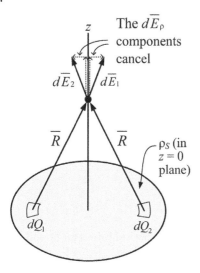

Figure 2.9 Symmetry argument for a disc of surface charge.

Determine dQ and $\overline{R}$:

$$\overline{E} = \int d\overline{E} = \int \frac{dQ\overline{R}}{4\pi\varepsilon_o R^3}$$

Given the surface charge distribution, use surface charge density and express dA in terms of the cylindrical coordinate system:

$$dQ = \rho_S dA = \rho_S d\rho\rho d\phi = \rho_S \rho d\rho d\phi$$

Determine the position vectors that locate P and dQ. Then determine the distance vector.

$$\overline{r} = h\overline{a}_z, \quad \overline{r}' = \rho\overline{a}_\rho$$

$$\overline{R} = \overline{r} - \overline{r}' = h\overline{a}_z - \rho\overline{a}_\rho$$

What is variable in $\overline{R}$ and what is constant? For this case, h is constant and ρ is variable because the observation point is fixed but the location of dQ will be varied throughout the charge distribution during the integration of Coulomb's law. Insert dQ and $\overline{R}$ into Coulomb's law, factor out constants, and then split the integral:

$$\overline{E} = \int_0^{2\pi}\int_0^a \frac{\rho_S \rho d\rho d\phi (h\overline{a}_z - \rho\overline{a}_\rho)}{4\pi\varepsilon_o (\rho^2 + h^2)^{3/2}} = \frac{\rho_S}{4\pi\varepsilon_o}\left[\int_0^{2\pi}\int_0^a \frac{\rho d\rho d\phi(h\overline{a}_z)}{(\rho^2 + h^2)^{3/2}} - \int_0^{2\pi}\int_0^a \frac{\rho d\rho d\phi(\rho\overline{a}_\rho)}{(\rho^2 + h^2)^{3/2}}\right]$$

Does symmetry exist? See Figure 2.9.

- For every dQ that creates a $d\overline{E}$ at the observation point, there is a symmetrically opposite dQ that creates another $d\overline{E}$.
- The z-components of $d\overline{E}$ add.
- The ρ-components of $d\overline{E}$ cancel each other, so the ρ-component integral is zero. Account for this symmetry and evaluate the surviving integral:

$$\overline{E} = \frac{\rho_S h\overline{a}_z}{4\pi\varepsilon_o}\int_0^{2\pi}\int_0^a \frac{\rho d\rho d\phi}{(\rho^2 + h^2)^{3/2}}$$

$$\overline{E} = \frac{\rho_S h\overline{a}_z}{4\pi\varepsilon_o}\int_0^{2\pi} d\phi \int_0^a \frac{\rho d\rho}{(\rho^2 + h^2)^{3/2}}$$

$$\overline{E} = \frac{\rho_S h\overline{a}_z}{4\pi\varepsilon_o} 2\pi \frac{-1}{\sqrt{\rho^2 + h^2}}\Bigg|_0^a$$

$$\overline{E} = \frac{\rho_S h\overline{a}_z}{2\varepsilon_o}\left(\frac{-1}{\sqrt{a^2 + h^2}} + \frac{1}{h}\right)$$

$$\overline{E} = \frac{\rho_S}{2\varepsilon_o}\left(1 - \frac{h}{\sqrt{a^2 + h^2}}\right)\overline{a}_z$$

Thus, the general solution for the electric field intensity along the z-axis is

$$\overline{E}_{\text{gen}} = \frac{\rho_S}{2\varepsilon_o}\left(1 - \frac{h}{\sqrt{a^2 + h^2}}\right)\overline{a}_z$$

b) Let the distance to the observation point be small relative to the radius of the disc, that is, $h \ll a$:

$$\lim_{h \ll a} (\overline{E}) = \lim_{h \ll a} \left[\frac{\rho_S}{2\varepsilon_o} \left(1 - \frac{h}{\sqrt{a^2 + h^2}} \right) \overline{a}_z \right] \rightarrow \frac{\rho_S}{2\varepsilon_o} \left(1 - \frac{h}{\sqrt{a^2}} \right) \overline{a}_z \rightarrow \frac{\rho_S}{2\varepsilon_o} (1 - 0) \overline{a}_z$$

$$\lim_{h \ll a} (\overline{E}) = \overline{E}_{\text{inf}} = \frac{\rho_S}{2\varepsilon_o} \overline{a}_z$$

Close to the charged disc, the electric field intensity becomes constant. As the observation point approaches the disc of charge, it "looks like" an infinite surface charge and the electric field intensity approaches the infinite sheet of surface charge density result.

c) Let the distance to the observation point be large relative to the radius of the disc, that is, $h \gg a$:

$$\lim_{h \gg a} (\overline{E}) = \lim_{h \gg a} \left[\frac{\rho_S}{2\varepsilon_o} \left(1 - \frac{h}{\sqrt{a^2 + h^2}} \right) \overline{a}_z \right] \rightarrow \frac{\rho_S}{2\varepsilon_o} \left(1 - \frac{h}{\sqrt{h^2}} \right) \overline{a}_z \rightarrow 0 \overline{a}_z$$

The field approaches zero as the observation point is moved far away from a finite-sized charge distribution. Is this interpretation very useful? Not really – it is fairly obvious that the field approaches zero if the observation point is moved far away from any finite charge distribution, but quantitatively this result is unusable. How can the result be made useful? Further mathematical manipulation is required in order to observe the trend in the field strength for $h \gg a$ but before h goes to infinity as it does in the limit taken above. First, make the total charge apparent by multiplying the numerator and the denominator by πa^2 to obtain the total charge in the disc $\rho_S \pi a^2$:

$$\overline{E} = \frac{\rho_S}{2\varepsilon_o} \left(1 - \frac{h}{\sqrt{a^2 + h^2}} \right) \overline{a}_z = \frac{\rho_S \pi a^2}{2\varepsilon_o \pi a^2} \left(1 - \frac{h}{\sqrt{a^2 + h^2}} \right) \overline{a}_z$$

$$\overline{E} = \frac{Q_T}{2\pi\varepsilon_o a^2} \left(1 - \frac{h}{\sqrt{a^2 + h^2}} \right) \overline{a}_z$$

Then work the expression into a form suitable for application of the *binomial series* to obtain an approximate limit of the electric field expression when $h \gg a$ but with finite h value:

$$(1 + x)^{-1/2} = \frac{1}{\sqrt{1 + x}} = 1 - \frac{1}{2}x + \frac{1 \cdot 3}{2 \cdot 4}x^2 - \dots \quad (-1 < x \leq +1) \qquad \text{(a result from the binomial series)}$$

$$\overline{E} = \frac{Q_T}{2\pi\varepsilon_o a^2} \left(1 - \frac{h}{\sqrt{a^2 + h^2}} \cdot \frac{1/h}{1/h} \right) \overline{a}_z = \frac{Q_T}{2\pi\varepsilon_o a^2} \left(1 - \frac{1}{\sqrt{1 + \frac{a^2}{h^2}}} \right) \overline{a}_z \quad \left[\text{so } x = \frac{a^2}{h^2} \right]$$

$$\overline{E} = \lim_{h \gg a} \left[\frac{Q_T}{2\pi\varepsilon_o a^2} \left(1 - \frac{1}{\sqrt{1 + \frac{a^2}{h^2}}} \right) \overline{a}_z \right] \rightarrow \frac{Q_T}{2\pi\varepsilon_o a^2} \left[1 - \left(1 - \frac{a^2}{2h^2} \right) \right] \overline{a}_z$$

Only two terms of the series were used because $x = a^2/h^2$ is very small as $h \gg a$. Simplifying:

$$\overline{E}_{\text{far}} = \frac{Q_T}{2\pi\varepsilon_o a^2} \left(\frac{a^2}{2h^2} \right) \overline{a}_z = \frac{Q_T}{4\pi\varepsilon_o h^2} \overline{a}_z = \frac{\rho_S \pi a^2}{4\pi\varepsilon_o h^2} \overline{a}_z$$

Once again, the finite-sized charge distribution looks like a point charge as the distance of the observation point from the charge distribution becomes much greater than the dimensions of the charge distribution.

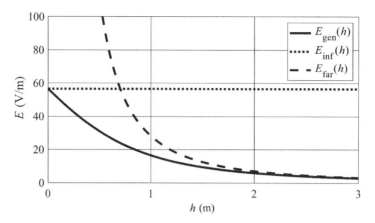

Figure 2.10 Behavior of *E* as the observation point height is increased.

d) Plot the field behavior for the field expressions obtained in parts a)–c) versus *h*. Plot and compare the infinite sheet of surface charge and point charge cases.

A plot of $|\overline{E}|$ vs. *h* for *P* on the *z*-axis is made in Figure 2.10. The following equations were used:

$$\rho_S = 10^{-9} \quad a = 1 \quad \varepsilon_o = 8.854 \times 10^{-12}$$

$$E_{\text{gen}}(h) = \frac{\rho_S}{2\varepsilon_o}\left(1 - \frac{h}{\sqrt{a^2 + h^2}}\right) \quad E_{\text{inf}}(h) = \frac{\rho_S}{2\varepsilon_o} \quad E_{\text{far}}(h) = \frac{\rho_S \pi a^2}{4\pi\varepsilon_o h^2}$$

In Figure 2.10, E_{gen} is the general result for a disc of surface charge density, E_{inf} is the result for an infinite sheet of surface charge density (or equivalently the observation point is very close to the disc), and E_{far} is the result for an observation point far away from the disc. The plot for the general result agrees with the analyses in parts b) and c) above, that is, when the distance to the observation point is very small or very large relative to the radius dimension of this charge distribution.

Two standard charge distributions were examined as limiting cases of the general expression to determine the electric field intensity. In the first case, we found that the *electric field intensity due to an infinitely long line charge* (see Example 2.3.1) is

$$\boxed{\overline{E} = \frac{\rho_L}{2\pi\varepsilon_o \rho}\overline{a}_\rho} \tag{2.17}$$

The previous equation is a good example of a well-known result. Although the result is precisely valid only for an infinitely long line charge, it is practically useful for reasoning the electric field intensity behavior close to a finite-sized, cylindrically symmetrical charge distributions, namely the $1/\rho$ dependence.

Another well-known result from the second case is the *electric field intensity due to an infinite sheet of surface charge density* (see Example 2.3.2):

$$\boxed{\overline{E} = \frac{\rho_S}{2\varepsilon_o}\overline{a}_n} \tag{2.18}$$

where $\overline{a}_n$ is a unit vector normal to the surface of charge density. Note that the electric field intensity is *constant* as the distance from the sheet of charge increases.

- What is the reason why *E* is constant?
  ○ The electric field vectors cannot "spread out" given the infinite size of the charged sheet.
- What are the practical limitations of Eq. (2.18)?
  ○ This result is valid when the distance to the observation point is much less than the size of the charged sheet and the observation point is not near an edge of the sheet.

We have developed key results for three fundamental source geometries and the trends in the electric field intensity magnitudes as distance from the charge distributions increases:

Ideal point charge: $\boxed{\overline{E} = \dfrac{Q}{4\pi\varepsilon_o R^2}\,\overline{a}_R}$ spherical source geometry: $E \propto \dfrac{1}{R^2} \to R^{-2}$ because the electric field intensity spreads out over a spherical surface area ($4\pi R^2$).

Infinitely long line charge: $\boxed{\overline{E} = \dfrac{\rho_L}{2\pi\varepsilon_o \rho}\,\overline{a}_\rho}$ cylindrical source geometry: $E \propto \dfrac{1}{\rho} \to R^{-1}$ because the electric field intensity spreads out over a cylindrical surface area ($2\pi\rho\ell$).

Infinite plane of surface charge: $\boxed{\overline{E} = \dfrac{\rho_S}{2\varepsilon_o}\,\overline{a}_n}$ planar source geometry: $E \propto$ constant $\to R^0$ because the electric field intensity does not spread out – it is uniform across the planar surface area.

These special, ideal cases are the "guidelines" that are used to visualize electric fields when the actual charge distribution geometry can be approximated as a point charge, a long line charge, or a large area of surface charge, again relative to the distance to the observation point.

Example 2.3.3

A rectangular bar with a square cross section of 0.1 mm by 0.1 mm, and 1 m length is charged. Identify and qualitatively explain the generally expected electric field behavior at the following three distances from the middle of one of the four side surfaces of the bar: (a) 1 μm, (b) 1 cm, and (c) 100 m.

Answer

a) At 1 μm, the surface of the bar appears to be an infinite sheet of surface charge density. The electric field intensity should be fairly uniform.
b) At 1 cm, the bar appears to be an infinite line charge. The electric field intensity should approach a $1/\rho$ dependence.
c) At 100 m, the bar appears to be a point charge. The electric field intensity should approach a $1/r^2$ dependence.

Motivation *Why study electromagnetic fields?* The previous examples demonstrate the insights that can be obtained from determining electric field intensity expressions. Sophisticated simulation software exists to visualize fields in all three dimensions simultaneously. Why go through the effort to determine field intensity expressions analytically if such software exists? The answer is crucial: *developing intuitive insight into field behavior so that the numerical results from electromagnetic simulation software can be interpreted.* As we analytically determine and interpret results, you will build an infrastructure of concepts and insights behind electromagnetic field behavior. These concepts and insights will give you the capability to further visualize and explain electromagnetic field behavior in electrical engineering applications. Enjoy your electromagnetic journey!

2.4 Introduction to Magnetostatics and Current Density

We have shown that an electric field is a field due to electric charges that exerts forces on other charges. What is a magnetic field? It is another type of field, created by *moving* electric charges. Can a magnetic field exert a force on another charge? If the charge is moving, *the force is exerted in a right-angled manner*, as expressed by the **Lorentz force law** in Eq. (2.19):

$$\boxed{\overline{F} = Q\left(\overline{E} + \overline{v} \times \overline{B}\right)} \tag{2.19}$$

where $\overline{v}$ is the velocity of the charge (m/s) and $\overline{B}$ is the **magnetic flux density**, a magnetic field quantity, with units of tesla (T) = webers/m^2 (Wb/m^2).

Example 2.4.1

Determine and justify the direction of the force on a moving positive charge if $\overline{B} = B_o\overline{a}_z$ and the charge moves in (a) the $\overline{a}_z$ direction, and (b) the $\overline{a}_y$ direction.

Answer

a) There is no force on the charge because the charge is moving in a direction parallel to the magnetic flux density. The cross product is zero.

b) The direction of $\overline{v}$ is $\overline{a}_y$, the $\overline{B}$ is $\overline{a}_z$, and the direction of $\overline{F}$ is $\overline{a}_y \times \overline{a}_z = +\overline{a}_x$.

What is the force on a current flowing in a wire that is immersed in a magnetic field? Consider a differential amount of moving charge and modify the Lorentz force law:

$$d\overline{F} = dQ\,\overline{v} \times \overline{B} = dQ\,\frac{d\ell}{dt}\overline{a}_I \times \overline{B} \tag{2.20}$$

where $\overline{a}_I$ is a unit vector and $d\ell$ is a differential length, both in the direction of the current. The differential length can be expressed as a vector $\overline{d\ell} = d\ell\,\overline{a}_I$, where the unit vector direction of the differential length is in the direction of the current in this development. The time derivative of displacement of the charge can equivalently be viewed as the rate of charge displacement past a point in the conductor. Both viewpoints express the rate of change of the same amount of differential charge:

$$d\overline{F} = \frac{dQ}{dt}d\ell\,\overline{a}_I \times \overline{B} = I\overline{d\ell} \times \overline{B} \tag{2.21}$$

$$\overline{F} = \int I\overline{d\ell} \times \overline{B} \tag{2.22}$$

The differential amount of moving charge $I\overline{d\ell}$ is called a **differential current element**. It can be thought of as a differential length of current that is the differential source of the magnetic field. The integral is required in general because the current and the magnetic field may not be uniform and/or at the same relative angle as a function of position. However, for a straight wire of length ℓ with a uniform current that is perpendicular to a uniform magnetic flux density, Eq. (2.22) reduces to a convenient rule-of-thumb (you should develop the next equation from Eq. (2.22)):

$$\boxed{F = I\ell B} \tag{2.23}$$

where the direction of the force is determined by the right-hand-rule per the Lorentz force law.

It is well known that currents generate magnetic fields. What is current? From circuit theory, it is the amount of charge crossing a fixed position in a circuit per unit time. Current can flow in other media besides conductors:

- *conduction current*: charges moving in a conductor or a semiconductor
- *electrolytic current*: charges moving in a solution
- *convection current*: charges moving in vacuum

What are the physical dimensions of current? This question leads us into the concept of **current density**. Flowing charge is most generally distributed over a spatial region (occupying a volume), which is where the volume current density concept applies. In many situations, current can be approximated to occupy a surface or a line. These three scenarios follow.

1) *Volume current density* $= \overline{J} = J\overline{a}_I$, see Figure 2.11a.

Note that the current is spread out over the cross-section of the media that the current crosses, as well as along the length of the current, that is, it occupies a volume. How does volume current density relate to the total current?

$$I = \int \overline{J} \cdot \overline{dA} \tag{2.24}$$

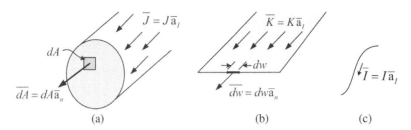

Figure 2.11 Current density types. (a) Volume current density. (b) Surface current density. (c) Line (filamentary) current.

where $\overline{dA}$ is a differential surface area, which is the differential area that the current density crosses with a vector direction $\bar{a}_n$ perpendicular to the differential surface, that is, $\overline{dA} = dA\bar{a}_n$.

What are the units of $\overline{J}$? Volume current density $\overline{J}$ is current per unit cross-sectional area, so the units are A/m², that is, ampere per square meter of area that the current crosses. Why is the dot product needed? To determine the component of $\overline{J}$ that is perpendicular to the cross-sectional area that $\overline{J}$ crosses. If $\overline{J}$ is perpendicular to the differential surface, the vector direction of $\overline{dA}$ is parallel to the direction of the current, so the integral in Eq. (2.24) reduces to

$$\boxed{I = \int J \, dA} \quad \text{when } \overline{J} \parallel \overline{dA} \tag{2.25}$$

Example 2.4.2

If the volume current density in a round wire is 10 A/m² (uniform), what is the total current if the wire has a 1 mm diameter?

Solution

For convenience, let $\overline{J}$ be in the $\bar{a}_z$ direction. Then $\overline{J}$ crosses a given constant z plane. The differential cross-sectional area in cylindrical coordinates in a given constant z plane is $\rho d\rho d\phi$ and the differential surface vector is $\rho d\rho d\phi \bar{a}_z$. Thus,

$$I = \int \overline{J} \cdot \overline{dA} = \int\int 10\bar{a}_z \cdot \rho d\rho d\phi \bar{a}_z = 10 \int\int \rho d\rho d\phi$$

The double integral is just the cross-sectional area of the circular wire:

$$I = 10\pi \left(\frac{0.001}{2}\right)^2 = 7.85 \times 10^{-6} = 7.85 \, \mu\text{A}$$

2) *Surface current density* $= \overline{K} = K\bar{a}_l = J_S\bar{a}_l$, see Figure 2.11b.

The key approximation (assumption) in $\overline{K}$ with respect to $\overline{J}$ is the negligible thickness of the current sheet relative to the distance to the observation point.

$$I = \int \overline{K} \cdot \overline{dw} \tag{2.26}$$

What is $\overline{dw}$? It is a differential width of the surface with a vector direction $\overline{a}_n$ assigned perpendicular to the width, that is, $\overline{dw} = dw\overline{a}_n$. Usually the surface current is perpendicular to the differential width, so the vector direction of $\overline{dw}$ is parallel to the direction of the current. Then the integral in Eq. (2.26) reduces to

$$\boxed{I = \int K\,dw} \quad \text{when } \overline{K} \parallel \overline{dw} \tag{2.27}$$

What are the units of $\overline{K}$? Surface current density $\overline{K}$ is current per unit cross-width, so the units are A/m, that is, amperes per meter of width that the current crosses.

3) *Line current* = filamentary current = $\overline{I} = I\,\overline{a}_l$, see Figure 2.11c.

What is the key approximation (assumption) in $\overline{I}$ with respect to $\overline{J}$? The cross-sectional dimensions of the current are negligible relative to the distance to the observation point. This assumption is normally made in circuit analysis. As a result, the filamentary current I has units of A, that is, amperes crossing a point in the circuit or in space.

> **Perspective**: When the distance to the observation point is significantly greater than the cross-sectional dimensions of the current distribution, then current densities are used to determine the line current value, which is then used to determine the magnetic field, the topic of the next section.

2.5 Biot–Savart Law and Examples for Line Currents

What is the vector field quantity for magnetic fields? You already know $\overline{B}$, the magnetic flux density. Another quantity is called the **magnetic field intensity** $\overline{H}$, with units of A/m. It is related to the magnetic flux density $\overline{B}$ in Eq. (2.19) by

$$\overline{B} = \mu\overline{H} \tag{2.28}$$

where $\mu = \mu_o\mu_r$ is the **permeability** and it is the same μ that is used to calculate the ideal inductance of a coil (N turns, cross-sectional area A, and length ℓ; coil inductance is examined in Chapter 4), μ_r is the relative permeability (unitless), and $\mu_o = 4\pi \times 10^{-7}$ H/m:

$$L = \frac{\mu N^2 A}{\ell} \tag{2.29}$$

What is the law for the magnetic field intensity due to a current distribution? Start with Coulomb's law for electric charge distributions:

$$\overline{E} = \int \frac{dQ\overline{a}_R}{4\pi\varepsilon_o R^2} \quad \text{where } \overline{R} = \overline{r} - \overline{r}' \tag{2.30}$$

Consider the general quantities in Coulomb's law:

$$(\text{total vector field}) = \int \frac{\begin{pmatrix} \text{differential} \\ \text{source of the field} \end{pmatrix}\begin{pmatrix} \text{direction of the} \\ \text{differential field} \end{pmatrix}}{4\pi\varepsilon_o R^2} \quad \text{where } \overline{R} = \overline{r} - \overline{r}' \tag{2.31}$$

Next, we need to interpret the general form of the law in Eq. (2.31) for the magnetic field. What is the differential source of the magnetic field? It is the **differential current element** (a nonstandard, convenient abbreviation: *dce*), the same quantity identified in Eq. (2.22):

$$dce = I\,d\ell \text{ (for line currents)} \tag{2.32}$$

The differential current element is the *line current multiplied by the differential length in the direction of the current.*[2] See Figure 2.12.

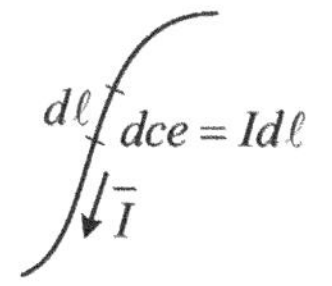

How is the right-angled nature of magnetic fields with respect to current expressed? What vector operation is involved? The cross product!

$$\overline{a}_I \times \overline{a}_R \tag{2.33}$$

where $\overline{a}_I$ is the unit vector in the current direction and $\overline{a}_R$ is the unit vector in the direction from the *dce* to the observation point. Notice that $\overline{a}_I \times \overline{a}_R$ is effectively the right-hand-rule (RHR) for the *dce*.

Figure 2.12 Differential current element for a line current.

What is the mathematical dependence in the denominator? As in Coulomb's law, it is an inverse distance-squared relationship from the differential source of the field to the observation point. Thus, the mathematical dependence is proportional to $1/R^2$.

Based on Eq. (2.31), the law for the magnetic field intensity due to a current distribution, the **Biot–Savart law** (BSL), is

$$\boxed{\overline{H} = \int d\overline{H} = \int \underbrace{\frac{(dce)\overline{a}_I \times \overline{a}_R}{4\pi R^2}}_{d\overline{H}} = \int \frac{(dce)\overline{a}_I \times \overline{R}}{4\pi R^3} = \int \frac{(dce)\overline{a}_I \times (\overline{r} - \overline{r}')}{4\pi |\overline{r} - \overline{r}'|^3}} \tag{2.34}$$

For line (filamentary) currents: $\boxed{\overline{H} = \int d\overline{H} = \int \frac{Id\ell \; \overline{a}_I \times \overline{R}}{4\pi R^3}}$ (2.35)

What is the significance of the BSL? First, it is a superposition law: the total magnetic field intensity $\overline{H}$ can be determined by summing (integrating) the differential magnetic field intensities $d\overline{H}$.[3] Second, the BSL can be used to determine $\overline{H}$ from a known current distribution, and current distributions are often known because conductors are used to establish current distributions. Thus, the BSL is practical and useful.

What is a difference in the analogy between Coulomb's law and the BSL (beyond the obvious field type and cross product)? There is no material property in the BSL for the magnetic field intensity $\overline{H}$. However, materials affect the magnetic flux density because $\overline{B} = \mu\overline{H}$.

What is the implication of the unit vector inside the integral? The unit vector $\overline{a}_I$ as well as $\overline{R}$ are functions of position and the cross-product result must be included in the integration in general.

The implementation strategy is similar to that in Coulomb's law, but the cross product inside the BSL integral sometimes eliminates vector components, often reducing the need to implement symmetry arguments. Use any symmetry as a solution check!

When the line current form of the BSL is used for volume or surface current distributions, the observation point must be sufficiently far away such that the current distribution "looks like a line current." If this is not the case, then the differential current element for that volume or surface charge distribution must be utilized, a topic for advanced electromagnetic studies.

Example 2.5.1

a) Determine the magnetic field intensity for a uniform line current I that flows on the z-axis in the $+z$ direction. See Figure 2.13a.
b) Interpret the result as the radius of the observation point increases.

---

2 There are also differential current elements for surface and volume current densities. In this text, we consider only line currents, which are prevalent, quite practical, and mathematically straightforward.

3 The linearity of the medium is assumed in superposition for $\overline{B} = \mu\overline{H}$ to be valid.

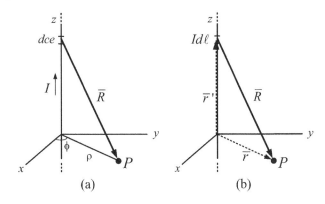

Figure 2.13 Infinitely long line current (a) problem sketch (b) distance vector setup.

(a) (b)

Solution

a) The observation point is at $P(\rho, \phi, 0)$.

Note: The line current is infinitely long, so locating P in the $z = 0$ plane is general.

Start with the Biot–Savart law: $\overline{H} = \int d\overline{H} = \int \dfrac{Id\ell\, \overline{a}_l \times \overline{R}}{4\pi R^3}$

The current is in the z direction: $\overline{a}_l = \overline{a}_z$

The differential length of current is along the z-axis, so $d\ell = dz$, and the differential current element (dce) is $Id\ell = Idz$.

Set up the distance vector from a typical dce to the observation point (P): see Figure 2.13b.

$$\overline{R} = \overline{r} - \overline{r}' = \rho\overline{a}_\rho - z\overline{a}_z$$

Note that ρ is a constant because the observation point is fixed, but z is a variable because the dce position is a function of z. Let $\rho = \rho_1$ to indicate it is a fixed quantity during the BSL evaluation:

$$\overline{R} = \overline{r} - \overline{r}' = \rho_1\overline{a}_\rho - z\overline{a}_z$$

Then the scalar distance is $R = \sqrt{\rho_1^2 + z^2}$. Insert these quantities into the BSL:

$$\overline{H} = \int \frac{Idz\overline{a}_z \times \left(\rho_1\overline{a}_\rho - z\overline{a}_z\right)}{4\pi(\rho_1^2 + z^2)^{3/2}}$$

Execute the cross product: $\overline{a}_z \times \left(\rho_1\overline{a}_\rho - z\overline{a}_z\right) = \rho_1\overline{a}_\phi - 0 = \rho_1\overline{a}_\phi$ and substitute it into the BSL:

$$\overline{H} = \int \frac{Idz\rho_1\overline{a}_\phi}{4\pi(\rho_1^2 + z^2)^{3/2}}$$

Then execute the integral. Remove constants from the integral, noting that $\overline{a}_\phi$ is a constant with respect to z:

$$\overline{H} = \frac{I\rho_1}{4\pi}\overline{a}_\phi \int\limits_{-\infty}^{+\infty} \frac{dz}{(\rho_1^2 + z^2)^{3/2}}$$

The integration result of the general form $\int \dfrac{dx}{(x^2 + a^2)^{3/2}}$, where x is the independent variable and a is a constant, is found in an integral table, software, or a calculator to be

$$\int \frac{dx}{(x^2 + a^2)^{3/2}} = \frac{x}{a^2\sqrt{x^2 + a^2}}$$

In this development, $x \to z$ and $a \to \rho_1$. Executing the integral: $\overline{H} = \dfrac{I\rho_1}{4\pi}\overline{a}_\phi \dfrac{z}{\rho_1^2\sqrt{\rho_1^2 + z^2}}\Bigg|_{-\infty}^{+\infty}$

$$\overline{H} = \frac{I}{4\pi\rho_1}\overline{a}_\phi\left[\lim_{z\to+\infty}\left(\frac{z}{\sqrt{\rho_1^2 + z^2}}\right) - \lim_{z\to-\infty}\left(\frac{z}{\sqrt{\rho_1^2 + z^2}}\right)\right]$$

$$\overline{H} = \frac{I}{4\pi\rho_1}\overline{a}_\phi[1 - (-1)] \to \frac{I}{2\pi\rho_1}\overline{a}_\phi \to \overline{H} = \frac{I}{2\pi\rho}\overline{a}_\phi \tag{2.36}$$

where ρ_1 is re-replaced with ρ. Although ρ_1 was a constant during the BSL evaluation, it is a variable in the general result because the magnetic field intensity depends on ρ. Equation (2.36) is a well-known result because straight line currents are prevalent in electrical engineering.

b) Interpretation: What happens to $\overline{H}$ as ρ increases? Why that proportionality?

Note that at any given radius ρ, $\overline{H}$ is spread around the circumference $2\pi\rho$. Thus, $|\overline{H}| \propto 1/\rho$, that is, the magnitude of the magnetic field intensity decreases inversely with respect to the radius from the line current.

Example 2.5.2

A uniform line current I flows in a circle of radius a in the $+\overline{a}_\phi$ direction and lies in the $z = 0$ plane. See Figure 2.14a.

a) Determine the general expression for $\overline{H}$ if the observation point lies on the z-axis.
b) Determine the behavior of $\overline{H}$ as the distance to the observation point on the z-axis increases.
c) Determine the behavior of $\overline{H}$ as the radius of the circle of current increases.

Solution

a) The observation point is at $P(0, 0, h)$. Set up the BSL. See Figure 2.14b.

$$\overline{a}_I = \overline{a}_\phi \qquad d\ell = ad\phi \qquad \overline{R} = \overline{r} - \overline{r}' = h\overline{a}_z - a\overline{a}_\rho \qquad R = \sqrt{h^2 + a^2}$$

Substitute these quantities into the BSL:

$$\overline{H} = \int d\overline{H} = \int \frac{Id\ell\overline{a}_I \times \overline{R}}{4\pi R^3} = \int \frac{Iad\phi\overline{a}_\phi \times (h\overline{a}_z - a\overline{a}_\rho)}{4\pi(h^2 + a^2)^{3/2}} \quad \text{where } \overline{a}_\phi \times (h\overline{a}_z - a\overline{a}_\rho) = +h\overline{a}_\rho + a\overline{a}_z$$

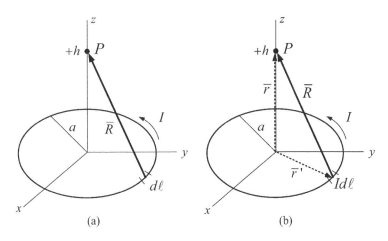

Figure 2.14 Circle of current. (a) Problem sketch. (b) Distance vector setup.

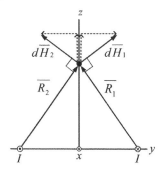

Figure 2.15 Symmetry argument for a circle of current. The ρ components of $d\overline{H}$ cancel. The z components of $d\overline{H}$ add.

$$\overline{H} = \int_0^{2\pi} \frac{Ia d\phi\left(+h\overline{a}_\rho + a\overline{a}_z\right)}{4\pi\left(h^2 + a^2\right)^{3/2}} = \frac{Ia}{4\pi\left(h^2 + a^2\right)^{3/2}} \int_0^{2\pi}\left(+h\overline{a}_\rho + a\overline{a}_z\right)d\phi$$

where the factor $(h^2 + a^2)$ has been removed from the integral because it is a constant. Also note that $\overline{a}_\rho$ is a function of ϕ, so it cannot be removed from the integral.

Symmetry argument

Sketch the $d\overline{H}$ components for each of two oppositely oriented *dce*s. A cross-section of the ring with two *dce*s is shown in Figure 2.15.

Use the cross product $\overline{a}_I \times \overline{R}$ to determine the direction of each $d\overline{H}$. Note that each $d\overline{H}$ is perpendicular to its corresponding $\overline{R}$ vector.

The ρ-component of $\overline{H}$ is zero by symmetry, leaving: $\overline{H} = \dfrac{Ia^2\overline{a}_z}{4\pi\left(h^2 + a^2\right)^{3/2}} \displaystyle\int_0^{2\pi} d\phi$.

Execute the integral and simplify:

$$\overline{H} = \frac{Ia^2}{2\left(h^2 + a^2\right)^{3/2}}\overline{a}_z \quad \text{A/m}$$

b) What happens to $\overline{H}$ as $h \gg a$? Why?

As the distance to the observation point becomes much greater than the radius of the ring of current, the denominator becomes much greater than the numerator. Hence, the magnetic field intensity mathematically approaches zero for very large distances h compared with the ring radius. More quantitatively, the limit can reveal the behavior as h becomes large compared with a:

$$\lim_{h \gg a}\left(\overline{H}\right) = \lim_{h \gg a}\left[\frac{Ia^2}{2\left(h^2 + a^2\right)^{3/2}}\overline{a}_z\right] \rightarrow \frac{Ia^2}{2\left(h^2\right)^{3/2}}\overline{a}_z = \frac{Ia^2}{2h^3}\overline{a}_z$$

Physically, as the distance from the ring of current increases, the magnetic field intensity decreases. However, unlike what we observed in the electric field examples, H does not decrease in an inverse square manner when the distance to the observation point is large because the current configuration is not a "point source."

c) What happens to $\overline{H}$ as $a \gg h$? Why?

As the radius of the ring of current becomes much greater than the height h to the observation point, the denominator approaches a^3. Thus, the magnetic field intensity mathematically approaches an a^{-1} dependence based on the resultant a in the denominator:

$$\lim_{a \gg h}\left(\overline{H}\right) = \lim_{a \gg h}\left[\frac{Ia^2}{2\left(h^2 + a^2\right)^{3/2}}\overline{a}_z\right] \rightarrow \frac{Ia^2}{2\left(a^2\right)^{3/2}}\overline{a}_z = \frac{I}{2a}\overline{a}_z$$

Physically, as the radius increases, the differential magnetic field due to any given *dce* has a $1/R^2$ dependence. However, as the radius increases, the circumference of the ring of current also increases. The length of current and the *dce*s in it increase proportional to a. Thus, the net effect is H is proportional to $1/a$.

Application: The Helmholtz Coil

The Helmholtz coil consists of two parallel circular conductors with a common axis, as shown in Figure 2.16a. Superposition of the BSL results for both rings of current gives the following result:

$$H(z_1) = \left(\frac{Ia^2}{2}\right)\left\{\frac{1}{\left[a^2 + (z_1 - h)^2\right]^{3/2}} + \frac{1}{\left[a^2 + (z_1 + h)^2\right]^{3/2}}\right\} \tag{2.37}$$

The magnetic field intensity along the z-axis is plotted in Figure 2.16b. What is the rate of change of the magnetic field around the midpoint between the two coils? Zero! Thus, the Helmholtz coil can present an almost constant magnetic field intensity over a small volume to a material sample. This apparatus is used in some scientific experiments.

Once again, note the insight that is achieved by analyzing this structure using the techniques developed in this chapter. Many practical problems can be solved either by superposition of known results (as shown here) or by interpretation of practical structures as the limiting cases when we are observing the field very close to or very far away from an object.

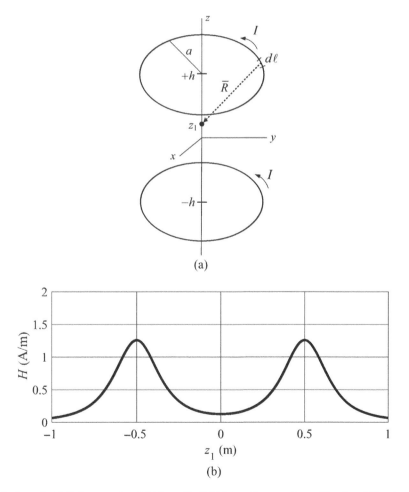

Figure 2.16 Helmholtz coil. (a) Coil geometry. (b) Magnetic field intensity vs. axis position (a = 0.2 m, h = 0.5 m, I = 0.5 A).

2.6 Summary of Important Equations

$$\bar{E} = \sum_{i=1}^{N} \bar{E_i} = \sum_{i=1}^{N} \frac{Q_i}{4\pi\varepsilon_o R_i^2}\bar{a}_{R_i} = \sum_{i=1}^{N} \frac{Q_i\bar{R_i}}{4\pi\varepsilon_o R_i^3} = \sum_{i=1}^{N} \frac{Q_i(\bar{r}-\bar{r_i}')}{4\pi\varepsilon_o|\bar{r}-\bar{r_i}'|^3}$$

$$\bar{E} = \int d\bar{E} = \int \frac{dQ}{4\pi\varepsilon_o R^2}\bar{a}_R = \int \frac{dQ\bar{R}}{4\pi\varepsilon_o R^3} = \int \frac{dQ(\bar{r}-\bar{r}')}{4\pi\varepsilon_o|\bar{r}-\bar{r}'|^3} \qquad \varepsilon_o = 8.8542 \times 10^{-12}\,\text{F/m}$$

$dQ = \rho_L d\ell$ where ρ_L = line charge density(C/m)

$dQ = \rho_S dA$ where ρ_S = surface charge density(C/m^2)

$dQ = \rho_V dV$ where ρ_V = volume charge density(C/m^3)

$\bar{E} = \dfrac{\rho_L}{2\pi\varepsilon_o\rho}\bar{a}_\rho$ (infinitely long line charge)

$\bar{E} = \dfrac{\rho_S}{2\varepsilon_o}\bar{a}_n$ (infinite plane of surface charge)

$$\bar{F} = Q\left(\bar{E} + \bar{v}\times\bar{B}\right) \qquad\qquad \bar{F} = \int I d\bar{\ell}\times\bar{B}$$

$$\bar{B} = \mu\bar{H} \qquad \mu = \mu_o\mu_r \qquad \mu_o = 4\pi\times10^{-7}\,\text{H/m}$$

$$I = \int \bar{J}\cdot\overline{dA}$$

$$I = \int \bar{K}\cdot\overline{dw}$$

$$\bar{H} = \int d\bar{H} = \int \frac{Id\ell\bar{a}_I\times\bar{R}}{4\pi R^3} \text{ (line currents)}$$

$$\bar{H} = \frac{I}{2\pi\rho}\bar{a}_\phi \text{ (infinitely long straight line current)}$$

2.7 Homework

1. a) What is an electric field?

    b) Why are the following forms of Coulomb's law equivalent?

$$\bar{E} = \frac{Q}{4\pi\varepsilon_o R^2}\bar{a}_R \quad \bar{E} = \frac{Q\bar{R}}{4\pi\varepsilon_o R^3} \quad \bar{E} = \frac{Q(\bar{r}-\bar{r}')}{4\pi\varepsilon_o|\bar{r}-\bar{r}'|^3}$$

2. a) What is the difference between $\bar{E} = \dfrac{Q}{4\pi\varepsilon_o R^2}\bar{a}_R$ and $\bar{E} = \displaystyle\int \dfrac{dQ}{4\pi\varepsilon_o R^2}\bar{a}_R$?

    b) Why does $d\bar{E}$ from dQ in $\bar{E} = \displaystyle\int \dfrac{dQ}{4\pi\varepsilon_o R^2}\bar{a}_R$ have a $1/R^2$ dependence?

3. a) Determine the electric field intensity at $P(2, 9, -3)$ due to a 4 μC point charge located at $(-4, 1, 0)$. Include a sketch showing the locations of Q and the observation point.

    b) If another point charge of -7 μC is located at $(2, -3, -1)$, determine the total $\bar{E}$ at P.

4. a) Determine the electric field intensity at $P(12, 7, 8)$ due to a $7\,\mu C$ point charge located at $(2, 1, 3)$. Include a sketch showing the locations of Q and the observation point.

b) If another point charge of $5\,\mu C$ is located at $(-2, 9, 5)$, determine the total $\bar{E}$ at P.

5. Determine the total charge in each of the following distributed charge configurations.

a) The *side* surface of a cylinder (ignore the ends), centered on the z-axis, that has a radius of 4 m, a height from 0 to $+3$ m, and $\rho_S = 2z\ C/m^2$.

b) Inside a cylinder, centered on the z-axis, that has a radius of 4 m, a height from 0 to $+3$ m, and $\rho_V = 2z\ C/m^3$.

6. Determine the total charge in each of the following distributed charge configurations.

a) Surface of a sphere, centered at the origin, that has a radius of 3 m and $\rho_S = 2r\ C/m^2$.

b) Inside a sphere, centered at the origin, that has a radius of 2 m and $\rho_V = 3r\ C/m^3$.

7. The objective of this problem is to develop the electric field intensity for an infinite sheet of uniform surface charge density: $\bar{E} = \dfrac{\rho_S}{2\varepsilon_o}\bar{a}_n$ (Eq. (2.18)).

a) Sketch and label the setup:
   - Infinite sheet of uniform surface charge density in the x–y ($z = 0$) plane
   - Observation point at $z = h$ on the z-axis
   - dQ at an arbitrary location in the first quadrant $(\rho, \phi, 0)$

b) The steps in the Coulomb's law development are listed below following the problem statement. Explain each step in this development.

c) Why doesn't the magnitude of the electric field intensity decrease as the distance from the plane of surface charge increases?

d) What is the practical interpretation of part c)?

Coulomb's law development steps for part b):

$$\bar{E} = \int \frac{dQ\bar{R}}{4\pi\varepsilon_o R^3}$$

$$dQ = \rho_S dA = \rho_S \rho d\rho d\phi$$

$$\bar{r} = h\bar{a}_z$$

$$\bar{r}' = \rho\bar{a}_\rho$$

$$\bar{R} = \bar{r} - \bar{r}' = h\bar{a}_z - \rho\bar{a}_\rho$$

What is variable in $\bar{R}$ and what is constant?

$$\bar{E} = \int_0^{2\pi}\int_0^\infty \frac{\rho_S \rho d\rho d\phi\left(h\bar{a}_z - \rho\bar{a}_\rho\right)}{4\pi\varepsilon_o\left(\rho^2 + h^2\right)^{3/2}}$$

The ρ-component of $\bar{E}$ is zero, why? (build a symmetry argument)

$$\bar{E} = \int_0^{2\pi}\int_0^\infty \frac{\rho_S \rho d\rho d\phi h\bar{a}_z}{4\pi\varepsilon_o\left(\rho^2 + h^2\right)^{3/2}}$$

$$\bar{E} = \frac{\rho_S h\bar{a}_z}{4\pi\varepsilon_o}\int_0^{2\pi}\int_0^\infty \frac{\rho d\rho d\phi}{\left(\rho^2 + h^2\right)^{3/2}}$$

$$\overline{E} = \frac{\rho_s h \overline{a}_z}{4\pi\varepsilon_o} \int\limits_0^{2\pi} d\phi \int\limits_0^\infty \frac{\rho d\rho}{\left(\rho^2 + h^2\right)^{3/2}}$$

$$\overline{E} = \frac{\rho_s h \overline{a}_z}{4\pi\varepsilon_o} 2\pi \frac{-1}{\sqrt{\rho^2 + h^2}}\Bigg|_0^\infty$$

$$\overline{E} = \frac{\rho_s h \overline{a}_z}{2\varepsilon_o}\left(0 + \frac{1}{h}\right)$$

$$\overline{E} = \frac{\rho_S}{2\varepsilon_o}\overline{a}_z$$

General : $\overline{E} = \dfrac{\rho_S}{2\varepsilon_o}\overline{a}_n$

8. The charge distribution is a circle of uniform *line* charge density ρ_L C/m that lies in the $z = h$ plane. The line charge density is centered on the z-axis and has a radius a.
   a) Determine the electric field intensity at $(0, 0, 0)$ for the charge distribution by performing or explaining each step in the following development.
   b) What does the result approach as $h \gg a$? Explain.
   c) What does the result approach as $h \to 0$? Explain.

a) $\overline{E} = \int \dfrac{dQ\overline{R}}{4\pi\varepsilon_o R^3}$

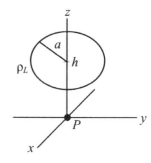

$$dQ = \rho_L d\ell = \rho_L \rho d\phi = \rho_L a d\phi$$

$$\overline{r} = 0$$

$$\overline{r}' = a\overline{a}_\rho + h\overline{a}_z$$

$$\overline{R} = \overline{r} - \overline{r}' = 0 - \left(a\overline{a}_\rho + h\overline{a}_z\right) = -a\overline{a}_\rho - h\overline{a}_z$$

What is variable in $\overline{R}$ and what is constant?

$$\overline{E} = \int\limits_0^{2\pi} \frac{\rho_L a d\phi\left(-a\overline{a}_\rho - h\overline{a}_z\right)}{4\pi\varepsilon_o\left(a^2 + h^2\right)^{3/2}}$$

The ρ-component of $\overline{E}$ is zero; why?

$$\overline{E} = \int\limits_0^{2\pi} \frac{\rho_L a d\phi\left(-h\overline{a}_z\right)}{4\pi\varepsilon_o\left(a^2 + h^2\right)^{3/2}}$$

$$\overline{E} = \frac{\rho_L a\left(-h\overline{a}_z\right)}{4\pi\varepsilon_o\left(a^2 + h^2\right)^{3/2}} \int\limits_0^{2\pi} d\phi$$

$$\overline{E} = \frac{\rho_L a\left(-h\overline{a}_z\right)2\pi}{4\pi\varepsilon_o\left(a^2 + h^2\right)^{3/2}}$$

$$\overline{E} = \frac{\rho_L a h}{2\varepsilon_o\left(a^2 + h^2\right)^{3/2}}\left(-\overline{a}_z\right)$$

9. a) Provide a labeled sketch, set up, and derive Eq. (2.17) using Coulomb's law.
    b) Why does the magnitude of the electric field intensity decrease inversely proportional to the radius as the radial distance from the line charge increases?
    c) What is the practical interpretation of part b)?

10. A finite-length uniform *line* charge density extends from $-h$ to $+h$ on the z-axis.
    The observation point is on the z-axis (assume the observation point location $z_1 > +h$).
    a) Determine the expression for the $\bar{E}$ field.
    b) How does the field intensity change as a function of z_1 for $z_1 \gg +h$? Explain.

11. Evaluate the following integrals and interpret each result (identify and properly implement symmetry *in the mathematics* for these problems).

    a) $\int_0^\pi \rho \bar{a}_\rho d\phi$

    b) $\int_0^\pi \rho \bar{a}_\phi d\phi$

12. Evaluate the following integrals and interpret each result (identify and properly implement symmetry *in the mathematics* for these problems).

    a) $\int_0^\pi r \bar{a}_r d\theta$

    b) $\int_0^\pi r \bar{a}_\theta d\theta$

    c) $\int_{\pi/2}^\pi r \bar{a}_r d\theta$ at $\phi = \dfrac{\pi}{2}$

13. A uniform *line* charge of density ρ_L C/m lies in the $z = 0$ plane on the semicircle of radius a from $90° < \phi < 270°$.
    a) Determine the electric field intensity at the origin.
    b) How does the electric field intensity depend on a? Why?

14. A cylinder of uniform *surface* charge density ρ_S C/m^2 extends over $0 < z < h$ (no ρ_S on the ends of the cylinder, only on the curved surface). The radius of the cylinder is a.
    a) Determine the electric field intensity at the origin.
    b) What happens to the electric field intensity as $h \ll a$? Why?

15. A semicylinder of uniform *surface* charge density ρ_S C/m^2 extends over $0 < z < h$ and $90° < \phi < 270°$ (no ρ_S on the ends of the semicylinder, only on the curved surface). The radius of the cylinder is a.
    a) Determine the electric field intensity at the origin.
    b) What happens to each component of the electric field intensity as $h \ll a$? Why (for each component)?

16. Show that the result of Problem 13 could be obtained from the result of Problem 15.

Note: Current density and magnetic field problems follow.

17. a) What is the difference between $\bar{B}$ and $\bar{H}$?
    b) What is the difference between I and $Id\ell$?

18. Under what two conditions does $I = \int \bar{J} \cdot \overline{dA} \approx JA$?

19. A wire with a *rectangular* cross section of dimensions 0.01 m × 0.02 m is centered on the z-axis. The volume current density is $J = 53$ A/m^2 in the $+z$ direction. Make and label a sketch. Determine I.

20. A wire with a *circular* cross section of radius 0.02 m is centered on the z-axis. The volume current density is $J = 53$ A/m^2 in the $+z$ direction. Make and label a sketch. Determine I.

21. Use the Biot–Savart law to determine the magnetic field intensity at $(0, a, 0)$ for the current distribution shown below. It is uniform current that extends from $z = -\infty$ to $z = 0$. Assume the current path is completed through paths that do not contribute to the magnetic field intensity at $(0, a, 0)$. Advice: Follow the setup of Example 2.5.1.

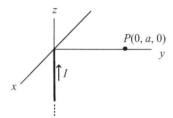

22. Use the Biot–Savart law to determine the magnetic field intensity at $(0, 0, h)$ for a uniform *line current* in the $+\bar{a}_\phi$ direction that lies in the $z = 0$ plane in two circular arcs of radius a: $0 \le \phi \le \pi/4$ and $\pi \le \phi \le 5\pi/4$. Assume the current path is completed through paths that do not contribute to the magnetic field intensity at $(0, 0, h)$. Include a labeled sketch. State and justify which vector components, if any, go to zero by symmetry. Advice: Follow the setup of Example 2.5.2.

23. Explain how a speaker converts an electronic signal into sound.

24. Explain how electricity that is applied to a DC motor produces mechanical rotation.

25. Why are Coulomb's law and the Biot–Savart law often collectively called the *superposition laws*?

3

The Flux Laws of Electric and Magnetic Fields

Our study of electric and magnetic fields in Chapter 2 focused on utilizing Coulomb's law and the Biot–Savart law to determine expressions for these fields. These analyses determined the fields at a desired observation point by vectorially summing the fields produced due to each differential source (charge or current). A large variety of general structures can be analyzed using these laws, with no requirements for symmetry or source uniformity. However, these methods can be mathematically involved, leading to lengthy analysis and more challenging visualization of the fields. We now introduce an intuitive approach to visualizing electric and magnetic fields using the concept of *flux*. An introduction to the concept of flux is an analogy to a familiar household object.

Imagine a light bulb in an otherwise dark, empty room. If the observer is close to the bulb, the light intensity is brighter relative to the light intensity if the observer is farther away. The light bulb is emitting the same quantity of light, called *flux*,[1] but the light density is higher when one is closer to the bulb because the light spreads out over a smaller area compared with when the observer is farther away. The light spreads out over a larger surface area as the distance from the light source increases. In other words, the *flux density* decreases as the distance from the source increases.

This flux model will be implemented with electric and magnetic fields in this chapter. The "quantity" of electric field from a charge distribution is called the **electric flux** (Ψ). The density of electric flux is called the **electric flux density** ($\overline{D}$). It is related to the electric field intensity ($\overline{E}$) that was used in Coulomb's law. In fact, the flux viewpoint is another way to determine the electric field from a charge distribution. These quantities will be developed from electric field intensity in Section 3.1. Moreover, the flux viewpoint is useful in visualizing electric fields (Section 3.2) and in determining charge (Section 3.3). The concepts developed from the flux viewpoint of electric fields in these sections are highly relevant to applications such as capacitors and transmission lines.

Similarly, the "quantity" of magnetic field, from a current distribution for example, is called **magnetic flux** (Φ). The density of magnetic flux is called **magnetic flux density** ($\overline{B}$). It is the magnetic field quantity in the Lorentz

---

1 In physics, the term "luminous flux" is used to describe the amount of visible light produced by a light source.

Electromagnetics and Transmission Lines: Essentials for Electrical Engineering, Second Edition.
Robert A. Strangeway, Steven S. Holland, and James E. Richie.
© 2023 John Wiley & Sons, Inc. Published 2023 by John Wiley & Sons, Inc.
Companion website: www.wiley.com/go/Strangeway/ElectromagneticsandTransmissionLines

Force law (that you probably examined in a physics course), and it is related to the magnetic field intensity $(\overline{H})$ that was used in the Biot–Savart law. The concepts developed from the flux viewpoint of magnetic fields in Section 3.4 are highly relevant to applications such as inductors, transformers, magnetic circuits, and transmission lines.

3.1 An Intuitive Development of Electric Flux and Gauss's Law

Motivational Questions   What is electric flux density $\overline{D}$? How does it relate to $\overline{E}$?

Gauss's law is introduced in this section using an inductive approach. A point charge is used as the first example case to develop expressions for the electric flux density and total electric flux. This result and a similar development of the line charge are then used to interpret the general form of Gauss's law and to identify its important properties. This background is then used to apply Gauss's law to other charge configurations in Section 3.2.

3.1.1   A First Look at *Electric Flux Density*

Surround a point charge Q at the origin with a closed "observation surface," which is formally named a **Gaussian surface**. A *closed* surface[2] is one that entirely encloses some volume. A Gaussian surface is a closed surface over which the electric field will be evaluated or determined. In this case, the Gaussian surface is a sphere of radius r as shown in Figure 3.1 (the reasoning for this surface selection will be examined in Section 3.1.2). How much "electric field," that is, electric flux, crosses this spherical Gaussian surface? An expression for electric flux density is needed with the reason soon to become apparent. Start with the electric field intensity of a point charge (from Chapter 2):

$$\overline{E} = \frac{Q}{4\pi\varepsilon_o R^2}\overline{a}_R \tag{3.1}$$

The point charge is at the origin ($\overline{r}' = 0$) so that the distance vector $\overline{R}$ becomes position vector $\overline{r}$:

$$\overline{E} = \frac{Q}{4\pi\varepsilon_o r^2}\overline{a}_r \tag{3.2}$$

The observation point is anywhere on a sphere of radius r, that is, on the Gaussian surface. Note that the surface area of this observation sphere, $4\pi r^2$, is contained in the denominator:

$$\overline{E} = \frac{Q}{\varepsilon_o(4\pi r^2)}\overline{a}_r \tag{3.3}$$

Rearrange the equation:

$$(\varepsilon_o\overline{E}) = \frac{Q}{4\pi r^2}\overline{a}_r \tag{3.4}$$

Interpretation: The new field quantity, $(\varepsilon_o\overline{E})$, equals the charge enclosed by the Gaussian surface divided by the entire surface area that the field crosses.

The quantity $(\varepsilon_o\overline{E})$ is defined to be $\overline{D}$, the **electric flux density** with SI units of coulomb per square meter (C/m^2). Thus, $\overline{D}$ equals the charge enclosed by the Gaussian surface divided by the entire surface area (A) that the field crosses:

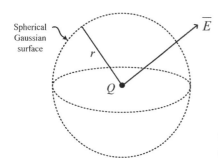

Spherical Gaussian surface

Figure 3.1   Gaussian surface setup for a point charge.

$$\overline{D} = \frac{Q}{4\pi r^2}\overline{a}_r = \frac{Q_{\text{encl}}}{A}\overline{a}_r \quad \text{(for a point charge at the origin)} \tag{3.5}$$

---

2 It is important to distinguish between *closed* and *open* surfaces. For a closed surface, there is no way to move from the inside of the surface to the outside of the surface without passing through the surface itself. Spheres, cubes, and cylinders are commonly used as closed surfaces in Gauss's law. In contrast, for an open surface such as a disc or a sheet, it is possible to move between the two sides of the surface without passing through the surface itself.

Does this result work for other charge distributions? Consider a cylindrical Gaussian surface of length ℓ centered on an infinitely long line charge on the z-axis with uniform line charge density ρ_L, as sketched in Figure 3.2. The electric field intensity (from Coulomb's law in Chapter 2) is

$$\overline{E} = \frac{\rho_L}{2\pi\varepsilon_o\rho}\overline{a}_\rho \tag{3.6}$$

This expression can be rearranged into a form similar to the point charge case of Eq. (3.4):

$$\varepsilon_o\overline{E} = \frac{\rho_L}{2\pi\rho}\overline{a}_\rho \tag{3.7}$$

The left-hand side of Eq. (3.7) is $\overline{D}$. Make this substitution and multiply the numerator and the denominator on the right-hand-side by ℓ, the length of the Gaussian surface:

$$\overline{D} = \frac{\rho_L\ell}{2\pi\rho\ell}\overline{a}_\rho \tag{3.8}$$

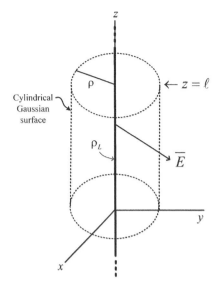

Figure 3.2 Gaussian surface setup for a line charge.

The numerator is the total charge inside (enclosed by) the Gaussian surface of length ℓ, while the denominator is the surface area of the cylindrical Gaussian surface (of radius ρ) that the field crosses. We ignore the surface areas of the ends of the cylinder because the field is completely in the $\overline{a}_\rho$ direction and does not cross through the ends. We observe once again:

$$\overline{D} = \frac{Q_{\text{encl}}}{A}\overline{a}_\rho \tag{3.9}$$

Thus, we observe that the electric flux density $\overline{D}$ equals, in general, the *charge enclosed divided by the surface area of the Gaussian surface* where $\overline{D}$ crosses. The guidelines for which this statement is applicable are examined in the next section (3.1.2).

The units of $\overline{D}$, C/m^2, are apparent from Eq. (3.9). Hence, the new field quantity, $\overline{D} = \varepsilon_o\overline{E}$, must be a *density*, that is, some "quantity" (flux!) per unit cross-sectional area that the field crosses, and this quantity is addressed in the next section.

3.1.2 Electric Flux and Gauss's Law

Motivational Questions   What is electric flux Ψ? How does it relate to $\overline{D}$? What is Gauss's law? How does Gauss's law relate to Ψ and $\overline{D}$?

What is the "quantity" per unit cross-sectional area in $\overline{D}$? (It is electric flux!) Starting with Eqs. (3.5) and (3.9), let the unit vector of $\overline{D}$ be $\overline{a}_D$. Move the surface area term to the left-hand-side of the relation:

$$\overline{D} = \frac{Q_{\text{encl}}}{A}\overline{a}_D \rightarrow (\overline{D})(A) = DA\overline{a}_D = Q\overline{a}_D \tag{3.10}$$

Let $\overline{a}_n$ be the unit vector normal (perpendicular) to the Gaussian surface. Note that $\overline{a}_n$ is also in the direction of $\overline{D}$ as it crosses the Gaussian surface in Figure 3.3. This arrangement is intentional: a Gaussian surface is selected such that $\overline{a}_n$ is parallel to $\overline{D}$ and both are perpendicular to the Gaussian surface wherever the flux density crosses it. For many practical situations, this selection enables determination of a closed-form analytic solution.

Thus, the Gaussian surface selection guidelines for obtaining analytic solutions to Gauss's law are as follows. Choose a Gaussian surface such that:

1) $\overline{D}$ is perpendicular to the surface wherever it crosses, that is, $\overline{a}_D$ is *normal* to the Gaussian surface ($\overline{a}_D \parallel \overline{a}_n$).
2) $\overline{D}$ is uniform (magnitude is constant) anywhere on the Gaussian surface that $\overline{D}$ crosses.

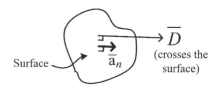

Figure 3.3 Electric flux density crossing a surface perpendicularly.

Regarding a detail pertaining to guideline 1), there are some Gaussian surface selections, such as a cylinder, for which there is a portion of the surface that $\overline{D}$ does not cross, and in those cases $\overline{a}_D \perp \overline{a}_n$. The parts of the Gaussian surface for which the flux does not cross do not contribute to the net flux calculation and can be ignored! When the direction of $\overline{D}$ as it crosses the Gaussian surface is parallel to $\overline{a}_n$, we only need to work with magnitudes:

$$(D)(A) = Q \text{ when } (\overline{a}_D \parallel \overline{a}_n) \tag{3.11}$$

Equation (3.11) reveals a profound concept: The electric flux density multiplied by the surface area of the Gaussian surface that D crosses equals the charge inside (enclosed) by the Gaussian surface. This "total" electric field quantity, DA is called **electric flux** and is assigned the symbol Ψ with the SI unit of coulomb (C):

$$\Psi = DA = Q \tag{3.12}$$

Note that *electric flux Ψ and electric flux density $\overline{D}$ are two different yet related quantities*! The area is the key relating factor. It is important to emphasize that the area A in Eq. (3.12) is sometimes less than the total surface area of the Gaussian surface, since A is only the area of the Gaussian surface that the flux crosses.

So what is electric flux? Electric flux is a model used to determine and visualize electric fields, usually by computer-generated plots or sketching of electric field lines. What does Eq. (3.12) tell us?

Q amount of charge creates Q amount of electric flux Ψ

This statement represents a concise expression of **Gauss's law** in the SI units system. Thus, if a charge distribution is surrounded by a closed (Gaussian) surface, then the total *net* electric flux, Ψ_{net}, leaving the closed surface equals the charge enclosed, Q_{encl}, by that Gaussian surface:

$$\boxed{\Psi_{\text{net}} = Q_{\text{encl}}} \quad \textbf{Gauss's law} \tag{3.13}$$

The unit of electric flux is the coulomb (C). By convention, the flux is positive if it is leaving the Gaussian surface and the flux is negative if it is entering. Although this result was developed only for the electric field of a point charge and of a line charge, Gauss's law is general and is true for any charge configuration.

Examples: A point charge Q creates Q amount of electric flux Ψ: $\Psi_{\text{net}} = Q$
A line charge of density ρ_L with a total charge of Q creates Q amount of flux: $\Psi_{\text{net}} = Q$
A surface charge of density ρ_S with a total charge of Q creates Q amount of flux: $\Psi_{\text{net}} = Q$
A volume charge of density ρ_V with a total charge of Q creates Q amount of flux: $\Psi_{\text{net}} = Q$

While Eq. (3.13) holds for any charge configuration, the development leading up to this general equation assumed the flux was either perpendicular to or parallel to the Gaussian surface, which allowed for the simplified analysis. What if the electric flux is not crossing the Gaussian surface perpendicularly? Then the vector component of $\overline{D}$ that crosses each differential surface area $\overline{dA}$ on the Gaussian surface must be determined. How? The dot product! The general form of Gauss's law that relates the electric flux to the electric flux density is

$$\boxed{\Psi_{\text{net}} = \oint \overline{D} \cdot \overline{dA} = Q_{\text{encl}}} \qquad (3.14)$$

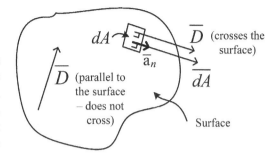

Figure 3.4 Visualization of vectors in Gauss's law.

where $\overline{dA} = dA\,\overline{a}_n$ is the **differential surface vector**, the differential surface area multiplied by $\overline{a}_n$. This unit vector is perpendicular (normal) to the surface. The circle around the integral symbol indicates that the integration is over a closed surface, the Gaussian surface in this context. The dot product between $\overline{D}$ and $\overline{dA}$ is used to determine the vector component of $\overline{D}$ that actually crosses $\overline{dA}$ by determining the *projection* of $\overline{D}$ in the direction of $\overline{dA}$. When $\overline{D} \parallel \overline{dA}$, the flux is crossing the surface perpendicularly and all of the flux is crossing that area; consequently, $\overline{D} \cdot \overline{dA} = D\,dA$. When $\overline{D} \perp \overline{dA}$, the flux is parallel to the surface and none of the flux is crossing that area; consequently, $\overline{D} \cdot \overline{dA} = 0$. See Figure 3.4 for both cases.

Perspective: The general form of Gauss's law simplifies to $\Psi_{\text{net}} = D(A) = Q_{\text{encl}}$ *if*

1) The direction of $\overline{D}$ is normal to the Gaussian surface wherever $\overline{D}$ crosses, *and*
2) $\overline{D}$ has a constant magnitude everywhere on the Gaussian surface where $\overline{D}$ crosses.

These two conditions must be met to enable use of the simplified form of Gauss's law. This approach allows D to be pulled outside of the integral and then determined. In practice, select a Gaussian surface that meets these two conditions. This analytic approach is only possible with highly symmetrical charge distributions, but solving these special symmetrical cases also gives significant insights and visualization into the electric fields from nonsymmetrical charge distributions. The use of Gauss's law to determine the electric field is illustrated in the next section.

Looking ahead: Electric flux is crucial to capacitance. Why? Recall the definition of capacitance and apply Gauss's law:

$$C = \frac{Q}{V} = \frac{\Psi}{V} \qquad (3.15)$$

The definition of capacitance is now expressed entirely in terms of electric field quantities, namely flux and voltage[3] (Ψ and V, respectively). Why is this fact important? Where is the energy stored in a capacitor? In the *electric field*!!! If the electric field can be determined, whether analytically, experimentally, or through simulation, then the capacitance of that structure can be determined.

3.2 Practical Determination of Electric Fields Using Gauss's Law

The previous section of this chapter developed the concepts of electric flux, electric flux density, and ultimately Gauss's law. Gauss's law expresses a fundamental principle of electromagnetics, namely that a charge configuration produces an amount of flux equal to the total of that enclosed charge. In fact, we will see in Chapter 5 that Gauss's law is one of the four Maxwell's equations, which along with the superposition laws and the Lorentz Force law, are used to describe classical (nonrelativistic) electromagnetism. With this in mind, one may still ask: How is Gauss's law useful? Gauss's law is used:

- To determine the electric field results of certain, highly symmetrical charge density distributions, including some cases that would be difficult to determine using Coulomb's law. This aspect is illustrated in this section.
- To determine electric flux and charge (Section 3.3) that will be used in the determination of capacitance (Chapter 4).

---

3 Voltage and capacitance will be related to the electric field intensity in Chapter 4.

- As a mathematical tool in vector calculus (divergence theorem, for example, in Chapter 5).
- As a tool in *visualizing and estimating vector fields.*

The statement $\Psi_{net} = Q_{encl}$ implies that if Q amount of charge creates Q amount of electric flux, then the electric flux must "go somewhere." How might the electric flux be visualized? If one can determine or estimate a Gaussian surface over which the electric field is either perpendicular or parallel to the surface and over which the magnitude of the perpendicular electric field is constant, then one can determine or estimate the electric field pattern. Several examples are presented in this section to demonstrate this approach.

Example 3.2.1
Determine $\overline{D}$ and $\overline{E}$ for a point charge.

Solution

Place a point charge at the origin. Select a spherical Gaussian surface. Why choose a spherical surface? By symmetry, the electric field is radial and would perpendicularly cross a spherical surface of radius r centered on the point charge (Figure 3.5). Although the magnitude of the electric field depends on r, it is uniformly distributed and therefore is constant in magnitude at a constant r as θ and ϕ are varied, everywhere on that spherical surface. Thus, a spherical Gaussian surface meets the surface selection criteria outlined in Section 3.1.2 for spherically symmetric charge configurations so that the simplified form of Gauss's law can be utilized. For this example, the charge enclosed by this spherical Gaussian surface is Q, that is, $Q_{encl} = Q$, the amount of charge of the point charge. Thus, Gauss's law can be utilized:

$$\Psi_{net} = \oint \overline{D} \cdot \overline{dA} = Q_{encl} \tag{3.16}$$

Given that the electric flux density is only in the radial direction ($\overline{D} = D_r \overline{a}_r$), that the surface vector is purely in the radial direction ($\overline{dA} = dA\,\overline{a}_r$), and that D_r has a constant magnitude at a constant radius, the integral in Gauss's law can be simplified:

$$\Psi_{net} = \oint \overline{D} \cdot \overline{dA} = \oint D_r \overline{a}_r \cdot dA\overline{a}_r = D_r \oint dA \tag{3.17}$$

where the $\oint dA$ is the closed surface area of a sphere, $4\pi r^2$. We then equate the net flux found in Eq. (3.17) and the enclosed charge $Q_{encl} = Q$, as shown in Eq. (3.18), and solve for the electric flux density. Hence, for this example, the *electric flux density* would be the total electric flux divided by the surface area of the spherical Gaussian surface:

$$\Psi = D_r 4\pi r^2 = Q \rightarrow D_r = \frac{\Psi}{4\pi r^2} = \frac{Q}{4\pi r^2} \tag{3.18}$$

Multiply D_r by the radial unit vector (which is the known direction of the electric flux density based on the charge symmetry) to determine the vector electric flux density:

$$\overline{D} = \frac{Q}{4\pi r^2}\overline{a}_r \tag{3.19}$$

Then determine the electric field intensity:

$$\overline{D} = \varepsilon_o \overline{E} \rightarrow \overline{E} = \frac{\overline{D}}{\varepsilon_o} \tag{3.20}$$

$$\overline{E} = \frac{Q}{4\pi\varepsilon_o r^2}\overline{a}_r \tag{3.21}$$

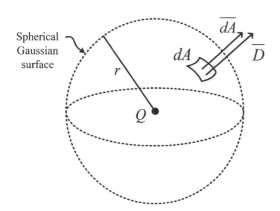

Figure 3.5 Point charge and Gaussian surface (only one flux line is shown).

which is Coulomb's law for a point charge (see Chapter 2), as expected. Note that in this example the flux is equal to the charge enclosed divided by the total surface area of the Gaussian surface because the flux is crossing the entire Gaussian surface.

Thus, Gauss's law is useful in visualizing the entire electric field of a point charge and how it is distributed in space around the charge. This well-known case was examined first so that one could see how Gauss's law works for a known scenario. The next example illustrates the determination of the electric field for a charge distribution that would be difficult to determine analytically using Coulomb's law.

Example 3.2.2

Determine $\overline{D}$ and $\overline{E}$ outside a sphere of uniform volume charge density ρ_V.

Solution

Assign radius a to the sphere of volume charge density. The field due to this charge distribution would be difficult to determine analytically using Coulomb's law. (Why? Try to write the expression for $\overline{R}$ in Coulomb's law for this spherical charge distribution.)

By symmetry, the entire electric field is radial and is uniformly distributed across a spherical Gaussian surface of radius r centered on the spherical volume of volume charge density. Both are centered at the origin (Figure 3.6).

Because the spherical surface conforms to the symmetry of the charge distribution, the integral in the general form of Gauss's law reduces, just as it did in the previous example:

$$\Psi_{\text{net}} = \oint \overline{D} \cdot \overline{dA} = \oint D_r \overline{a}_r \cdot dA \overline{a}_r = D_r \oint dA = D_r 4\pi r^2 \tag{3.22}$$

The charge enclosed in the spherical charge distribution must be determined from the volume charge density, which is uniform and can be pulled from the charge density integral:

$$Q_{\text{encl}} = \int \rho_V dV = \rho_V \int dV \bigg|_{\text{sphere}} = \rho_V \frac{4}{3}\pi a^3 \tag{3.23}$$

By Gauss's law, the net flux equals the charge enclosed; thus, we can equate the expression for the net flux in Eq. (3.22) and the enclosed charge in Eq. (3.23). Then solve for D_r and the resultant electric field:

$$D_r = \frac{\Psi_{\text{net}}}{4\pi r^2} = \frac{Q_{\text{encl}}}{4\pi r^2} = \frac{\rho_V \frac{4}{3}\pi a^3}{4\pi r^2} = \frac{\rho_V a^3}{3r^2} \tag{3.24}$$

$$\overline{D} = D_r \overline{a}_r = \frac{\rho_V a^3}{3r^2} \overline{a}_r \tag{3.25}$$

$$\overline{E} = \frac{\overline{D}}{\varepsilon_o} = \frac{\rho_V a^3}{3\varepsilon_o r^2} \overline{a}_r \tag{3.26}$$

Note how Gauss's law significantly simplified the determination of the electric field for this charge distribution (relative to using Coulomb's law). Furthermore, note the visualization of the electric field that is enabled through the application of Gauss's law.

The next example utilizes coaxial cable, commonly called **coax**, which consists of two concentric conductors of different radii, with an insulating material (called a dielectric) between them. Figure 3.7 shows an example of practical coax. No connector is shown in this image. For this particular sample of coax, the inner conductor is plated in copper,

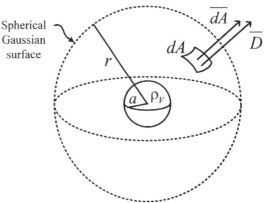

Figure 3.6 Volume charge distribution and Gaussian surface (only one flux line is shown).

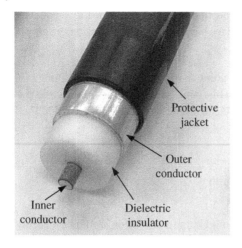

Figure 3.7 Construction of coax cable.

but the specific metals (and plating, if present) and dielectrics will vary depending on the specific cable model and manufacturer. Coax is commonly used in electrical systems, especially for laboratory measurements and signal transmission. You may have coax in your residence for connections to the internet or to connect cable, satellite, or over-the-air signals to your television. Vacuum is the insulator that is assumed for now. The main electrical property of insulating materials (dielectrics) is addressed in Chapter 4.

Example 3.2.3

Determine the $\overline{D}$ and $\overline{E}$ in a long length of coax of inner radius a and outer radius b. Assume a uniform surface charge density on the inner conductor. Also assume that the electric field is purely in the $\overline{a}_\rho$ direction (an appropriate assumption if we are not near an open end of the coax cable).

Solution

Given the cylindrical symmetry of coax, the electric field is radial in cylindrical coordinates. Assign a label to the uniform surface charge density on the inner conductor as ρ_S. There is also a charge density on the outer conductor, but it will not be enclosed by the Gaussian surface and can thus be ignored in our Gauss's law analysis. Select a cylindrical Gaussian surface of radius ρ and length ℓ, where $a < \rho < b$ and assume $\ell \ll$ length of the coax (see Figure 3.8) to avoid the effect of the ends of the coax. By symmetry, the electric flux density would perpendicularly cross the curved side of this cylindrical Gaussian surface. The electric field would be parallel to the surfaces on the ends of the cylinder and hence would not cross the ends. Although D_ρ depends on ρ, as expected for a cylindrical

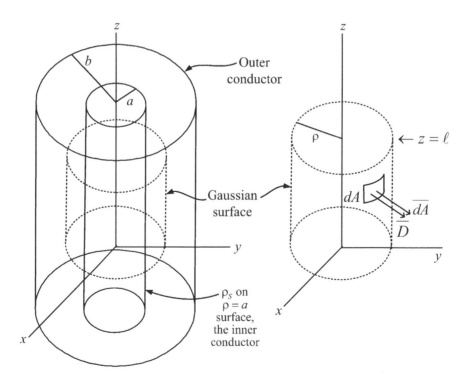

Figure 3.8 Gaussian surface for coax.

charge geometry, it is uniformly distributed, that is, constant in magnitude at a constant radius ρ as ϕ and z are varied.

$$\text{Start with Gauss's law}: \Psi_{\text{net}} = \oint \overline{D} \cdot \overline{dA} = Q_{\text{encl}} \tag{3.27}$$

Break up the closed cylindrical Gaussian surface into a top, a bottom, and a curved side:

$$\Psi_{\text{net}} = \oint \overline{D} \cdot \overline{dA} = \underbrace{\int \overline{D} \cdot \overline{dA}}_{\text{Top}} + \underbrace{\int \overline{D} \cdot \overline{dA}}_{\text{Bottom}} + \underbrace{\int \overline{D} \cdot \overline{dA}}_{\text{Side}} \tag{3.28}$$

where $\overline{dA}$ corresponds to the surface being integrated in each integral. The flux is not crossing the top and bottom flat surfaces of the cylindrical Gaussian surface, so those integrals are zero ($\overline{D} \cdot \overline{dA} = 0$ because $\overline{D} \perp \overline{dA}$). The flux is crossing only the curved side. Simplify the curved-side integral per the previous symmetry argument. The surface vector on the curved side of the cylindrical Gaussian surface is in the $\overline{a}_\rho$ direction:

$$\Psi_{\text{net}} = \underbrace{\int D_\rho \overline{a}_\rho \cdot dA \overline{a}_\rho}_{\text{Side}} = D_\rho \underbrace{\int dA}_{\text{Side}} = D_\rho 2\pi\rho\ell \tag{3.29}$$

The charge enclosed must be determined from the surface charge density on the inner conductor, which is uniform and can be pulled from the charge density integral:

$$Q_{\text{encl}} = \int \rho_S dA = \rho_S \int dA \bigg|_{\text{cylinder side}} = \rho_S 2\pi a\ell \tag{3.30}$$

Note that the surface charge density on the outer conductor ($\rho = b$) is not a factor when Gauss's law is applied between the conductors ($a < \rho < b$) because this Gaussian surface does *not enclose* that charge density. Equating the net flux found in Eq. (3.29) and the enclosed charge found in Eq. (3.30), we can then solve for electric flux density:

$$D_\rho = \frac{\Psi_{\text{net}}}{2\pi\rho\ell} = \frac{Q_{\text{encl}}}{2\pi\rho\ell} = \frac{\rho_S 2\pi a\ell}{2\pi\rho\ell} = \frac{\rho_S a}{\rho} \tag{3.31}$$

Finally, state the vector field quantities:

$$\overline{D} = \rho_S \frac{a}{\rho} \overline{a}_\rho \tag{3.32}$$

$$\overline{E} = \frac{\rho_S a}{\varepsilon_o \rho} \overline{a}_\rho \tag{3.33}$$

Looking ahead: The results for $\overline{D}$ and $\overline{E}$ in the previous example are used to determine the capacitance of coax in Chapter 4.

Gauss's law states that if the net charge inside a Gaussian surface is zero, then the net electric flux passing through the Gaussian surface is also zero. Does zero net electric flux from a Gaussian surface mean that there is zero electric field inside the Gaussian surface? Consider the following coax example.

Example 3.2.4
A Gaussian surface is set up inside a coax cable between the inner and outer conductors, but it does not surround the inner conductor, as sketched in Figure 3.9.

a) What is the net electric flux from this Gaussian surface? Why?
b) Is the electric field zero inside this Gaussian surface? Why or why not?
c) Explain how the answers to a) and b) are not contradictory.

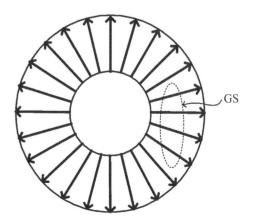

Figure 3.9 Gaussian surface for coax in Example 3.2.4 (end view).

Solution

a) The net electric flux Ψ_{net} from this Gaussian surface is zero because the charge enclosed is zero.
b) There is Ψ inside the Gaussian surface (this is *not* net flux), so the electric field is not zero inside the Gaussian surface.
c) Every flux line that enters the Gaussian surface also leaves the surface, resulting in a net flux of zero because the charge enclosed is zero. Thus, an electric field inside the Gaussian surface can exist and still produce zero net flux from a Gaussian surface.

The next example involves a perfect conductor.[4] Recall that charges can move freely in a conductor. A perfect conductor also has zero electric field inside it because charges redistribute on the conductor surface to cancel any electric field inside the conductor. Furthermore, electric flux lines are perpendicular to conductor surfaces. A qualitative justification is given here: the voltage across two points on a perfect conductor must be zero, by the definition of a perfect conductor. But voltage is electrical potential energy difference, and an electric field must be present to create forces on electric charges to do work (to have an energy difference). Hence, the electric field across (tangent to) a perfect conductor is zero and only the perpendicular component of the electric field is present at the surface of a perfect conductor. This explanation is formalized in the voltage discussion of Chapter 4.

Example 3.2.5
Determine the electric flux density and electric field intensity just outside a charged perfect conductor with a flat surface. Assume the surface is infinite in extent (or, practically, that the field is being observed close to the conducting surface and away from the edges of a finite-sized surface). The surface charge density lies uniformly on the surface of the conductor.

Solution

Let the surface be in the $z = 0$ (x–y) plane, with the conductor occupying $z \leq 0$ (Figure 3.10).

Gaussian surface selection: Choose a box with dimensions a and b and a small height h as the Gaussian surface. Let the conductor surface bisect the box height (half of the box is inside the conductor and the other half is outside the conductor). The electric flux is perpendicular to the conductor surface and normal to the a by b surface on the air-side of the conductor. The electric field is zero inside the conductor. The electric flux is parallel to the sides of the Gaussian surface and does not cross the sides. Electric flux crosses only the top of the Gaussian surface.

Set up Gauss's law and separate the closed surface integral into the top of the box Gaussian surface, the bottom, and the four sides:

$$\Psi_{\text{net}} = \oint \overline{D} \cdot \overline{dA} = \underbrace{\int \overline{D} \cdot \overline{dA}}_{\text{Top}} + \underbrace{\int \overline{D} \cdot \overline{dA}}_{\text{Bottom}} + \underbrace{\int \overline{D} \cdot \overline{dA}}_{\text{Sides}} \tag{3.34}$$

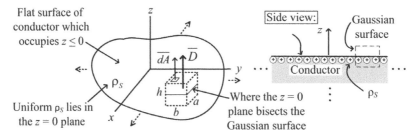

Figure 3.10 Gaussian surface for a flat conductor.

---

4 A perfect conductor has infinite conductivity and consequently has zero interior electric fields and zero voltage across any two points on its surface. These conditions usually simplify field analysis near the conductor. A perfect conductor is often a good approximation to practical, highly conductive metals such as copper, aluminum, silver, gold, and so forth. The conductivity of materials is discussed in Chapter 4.

Electric flux is parallel to the sides of the box and is zero inside the conductor where the bottom of the box is located, so flux crosses only the top of the Gaussian surface. Hence, only the integral for the top of the box survives. Both the flux and the surface vector are in the $+\bar{a}_z$ direction:

$$\Psi_{\text{net}} = \int D_z\bar{a}_z \cdot dA\bar{a}_z = D_z\int dA = D_z ab \tag{3.35}$$

On the conductor surface : $Q_{\text{encl}} = \int \rho_S dA = \rho_S \int dA\bigg|_{\text{rectangle}} = \rho_S ab \tag{3.36}$

Then, equating the net flux of Eq. (3.35) and the enclosed charge of Eq. (3.36), and solving for the flux density:

$$D_z = \frac{\Psi_{\text{net}}}{ab} = \frac{Q_{\text{encl}}}{ab} = \frac{\rho_S ab}{ab} = \rho_S \tag{3.37}$$

$$\overline{D} = D_z\bar{a}_z = \rho_S\bar{a}_z \text{ and } \overline{E} = \frac{\rho_S}{\varepsilon_o}\bar{a}_z \tag{3.38}$$

This result is as amazingly simple as it is important! The electric flux density just outside any flat charged conductor surface is equal to the surface charge density. This fact may not be so surprising if one visualizes $\Psi_{\text{net}} = Q_{\text{encl}}$ on a per unit area basis in this charged surface setup (Gauss's law is useful in visualizing electric fields!). Even more insight can be gained using the flux perspective. Electric flux emanates only on one side from the conductor surface. Consequently, the electric field next to the perfect conductor is *twice as large* as the electric field of an infinitely thin charged surface, with air on both sides, that was analyzed in Chapter 2 (Eq. 2.18). In that case, the flux emanates from both above and below the surface, resulting in half the field intensity on each side.

The result in the previous example is one of two general electric field **boundary conditions**, that is, the requirement on a field at the boundary of two different materials.[5] Using $\bar{a}_n$ as the unit vector normal to the conductor surface (Figure 3.11), the general boundary condition (BC) for the normal electric field component at the surface of a perfect conductor is:

$$\boxed{\overline{D}\cdot\bar{a}_n = D_n = \rho_S} \quad \text{(normal } E \text{ field BC)} \tag{3.39}$$

Insights into electric fields from these examples of Gauss's law:

There are only a few common, highly symmetric charge distributions for which the electric field can be exactly determined in closed form by Gauss's law, but the results of these ideal cases are important to many practical electronic applications. One primary set of results is the electric field behavior as a function of distance from three major classes of symmetric charge distributions (as developed in the previous examples):

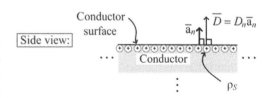

Figure 3.11 Normal electric field boundary condition at the surface of a perfect conductor.

1) For the point charge or sphere of uniform surface or volume charge density:

$$\text{(spherical geometry) } D_r = \frac{Q}{4\pi r^2} \rightarrow D_r \propto \frac{1}{r^2} \tag{3.40}$$

2) For the infinite line charge or the cylinder of uniform surface or volume charge density:

$$\text{(cylindrical geometry) } D_\rho = \frac{Q}{2\pi\rho\ell} \rightarrow D_\rho \propto \frac{1}{\rho} \tag{3.41}$$

3) For the infinite sheet of uniform surface charge density[6] (as opposed to the flat conductor in a previous example):

$$\text{(planar geometry) } D_z = \frac{\rho_S}{2} \rightarrow D_z \propto \frac{1}{z^0} = \text{constant} \tag{3.42}$$

---

5 The other electric field boundary condition, for *tangential* electric fields next to a conductor, will be examined in Chapter 4.
6 This case is analyzed using Gauss's law in Problem 6 at the end of this chapter.

The behavior of the electric fields surrounding many practical, charged structures can often be approximately modeled by the ideal behaviors of Eqs. (3.40)–(3.42). This approach is applicable when the structure symmetry and field observation distance together cause the structure to "look like" one of these scenarios from the perspective of the field observation point. The next example demonstrates this concept.

Example 3.2.6

A pipe (a long cylinder with non-zero wall thickness) in a factory is charged due to conveying a particulate at high velocity. Upon which coordinate variable does the electric field intensity depend and what is the mathematical dependence upon that coordinate outside of the pipe? Why?

Answer

A pipe with a circular cross section has a cylindrical geometry. The electric flux density will cross the side of a concentric cylindrical Gaussian surface ($\rho >$ pipe radius), which has surface area of $2\pi\rho\ell$, so the electric field depends on ρ ($\rho \ll \ell$ for a long pipe, so we assume the ends of the pipe are far enough away such that Ψ does not cross the ends of the cylindrical Gaussian surface). The magnitude of the electric field intensity should be inversely proportional to the area over which Ψ crosses, which is $2\pi\rho\ell$, so $D_\rho \propto 1/\rho$.

This discussion again illustrates the visualization of electric field patterns from Gauss's law. Other key concepts that are used to visualize static electric field patterns:

- *The flux must go somewhere*!
- Electric flux terminates (begins or ends) on electric charge.
- Electric flux is perpendicular to the surface of any perfect conductor.

The following example illustrates how Gauss's law and its consequences can be used to reason and visualize electric fields and charge in an electrical structure.

Example 3.2.7

Given coax (coax cable) with an inner conductor of radius a and a grounded outer conductor of radius b,

a) Sketch the electric flux in an end view of the coax.
b) Compare the charge on the outer conductor with respect to a total charge of Q on the inner conductor.
c) Repeat b) for the charge density on the outer conductor given ρ_{Sa} charge density on the inner conductor.
d) What is the electric field outside of the coax ($\rho > b$)? Why?

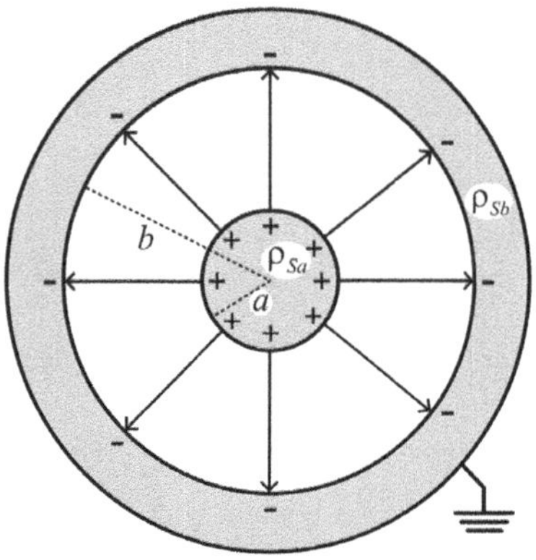

Solution

a) The end view of coax is sketched in Figure 3.12. The flux lines must be perpendicular to the inner and outer conductor surfaces.
b) Each Ψ line begins on a positive charge on the inner conductor and terminates on a negative charge on the outer conductor. Consequently, for every charge on the inner conductor, there must be a corresponding charge with opposite sign on the outer conductor. Where did the negative charge on the outer conductor come from? The outer conductor is grounded; hence, charge can flow onto or off of this conductor. Thus, the amount of charge on the inner conductor equals the amount of charge on the outer conductor: $|+Q|_{\rho = a} = |-Q|_{\rho = b}$.
c) Fewer flux lines *per unit area* terminate on the outer conductor relative to the inner conductor because the outer conductor has a greater surface area, so the flux density at the inner conductor must be greater than the flux density at the outer conductor: $D_n|_{\rho = a} > D_n|_{\rho = b}$. Given the boundary condition $D_n = \rho_S$, the surface charge density on the inner conductor

Figure 3.12 End view of coax.

must be greater than that on the outer conductor because the flux density is lower at the outer conductor: $\rho_{Sa} > \rho_{Sb}$.

d) If a Gaussian surface surrounds both the inner and outer conductors, the charge enclosed would be zero: $Q_{encl} = +Q - Q = 0$. Unlike the coax case in Figure 3.9, the Gaussian surface surrounds the *entire* coax structure. Also, due to cylindrical symmetry, there can only be radial flux entering or leaving this surface. Given there is zero enclosed electric charge, *the electric flux and electric field are both zero outside the coax*. This result is a primary reason that coax is used in electrical systems and equipment: *coax is self-shielding*.

Further thoughts on and uses of electric flux and Gauss's law:

1) How would the Gauss's law procedure change if the charge density were not uniform, for example, in a charge distribution with a cylindrical geometry,
   a) If the charge density had a radial dependence (was a function of ρ)?
   b) If the charge density had an angular dependence (was a function of ϕ)?

   Answers:

   a) The radial dependence of the charge density would not change the uniformity of $\overline{D}$ as it crosses the Gaussian surface. Hence, Gauss's law can be utilized and one would integrate the charge density to determine the total charge.
   b) The angular dependence of the charge density would *change* the uniformity of $\overline{D}$ as it crosses the Gaussian surface. Hence, *Gauss's law cannot be utilized* to solve for $\overline{D}$ analytically because a symmetrical Gaussian surface could not be identified for such a charge configuration. Gauss's law is still true but just not useful to solve for $\overline{D}$ in this situation.
2) Gauss's law can be useful with simulations because electromagnetic simulation programs can numerically determine the net flux crossing a surface defined by the user. By Gauss's law, the charge enclosed is then known, but its distribution, if nonuniform, would not be known. However, charge densities could be estimated from the electric flux density simulation results ($D_n = \rho_S$).
3) Gauss's law can be used to determine the electric flux that is then used to determine the capacitance of a structure $\left(C = \dfrac{Q}{V} = \dfrac{\Psi}{V} \right)$. This determination can be done analytically for highly symmetric capacitor geometries or, for complicated capacitor structures, a simulation program can be used to solve for the electric flux numerically.
4) If $\overline{D}$ is known (such as from $\overline{E}$, by solving Gauss's law, or from simulation), then the net electric flux crossing a non-closed surface S can be determined by integrating the flux density that crosses the defined surface:

$$\Psi = \int_S \overline{D} \cdot \overline{dA} \tag{3.43}$$

This flux could be useful in determining capacitive coupling between conductors in two adjacent circuits, for example, where some but not all of the flux from one conductor couples with the other conductor. Knowing $\overline{D}$ and $\overline{E}$ are useful in other ways, such as to avoid electric field breakdown in an insulator.

3.3 Determination of Charge from Electric Fields

Motivational Question   Can Gauss's law be practically used to determine enclosed charge?

So far we have been utilizing Gauss's law to determine the electric field due to a given charge distribution. However, as suggested in the previous section, a given electric field can be used to determine the total charge enclosed by a Gaussian surface. This aspect is especially useful in electromagnetic simulations, where $\overline{D}$ and Ψ are determined numerically. An analytic example follows, but keep in mind that this approach can be implemented within electromagnetic simulation software.

Example 3.3.1

Determine the charge enclosed by a cylinder of radius a and length ℓ (from the $z = 0$ plane) if the electric flux density is $\frac{K_1}{\rho}\bar{a}_\rho + K_2 z \bar{a}_z$ C/m^2 and K_1 and K_2 are constants.

Solution

Electric flux is crossing the side and the ends of the Gaussian surface as sketched in Figure 3.13. Start with the general form of Gauss's law and simplify the dot products:

$$\Psi_{\text{net}} = \oint \overline{D} \cdot \overline{dA} = Q_{\text{encl}} \tag{3.44}$$

$$\Psi_{\text{net}} = \oint \overline{D} \cdot \overline{dA} = \underbrace{\int \overline{D} \cdot \overline{dA}}_{\text{Side}} + \underbrace{\int \overline{D} \cdot \overline{dA}}_{\text{Top}} + \underbrace{\int \overline{D} \cdot \overline{dA}}_{\text{Bottom}} \tag{3.45}$$

$$\Psi_{\text{net}} = \int_0^\ell \int_0^{2\pi} \left[\left(\frac{K_1}{\rho}\bar{a}_\rho + K_2 z \bar{a}_z\right) \cdot \rho d\phi dz \bar{a}_\rho\right]\Bigg|_{\rho=a} + \int_0^a \int_0^{2\pi} \left[\left(\frac{K_1}{\rho}\bar{a}_\rho + K_2 z \bar{a}_z\right) \cdot \rho d\phi d\rho \bar{a}_z\right]\Bigg|_{z=\ell}$$

$$+ \int_0^a \int_0^{2\pi} \left[\left(\frac{K_1}{\rho}\bar{a}_\rho + K_2 z \bar{a}_z\right) \cdot \rho d\phi d\rho (-\bar{a}_z)\right]\Bigg|_{z=0} \tag{3.46}$$

Executing the dot products:

$$\Psi_{\text{net}} = \int_0^\ell \int_0^{2\pi} \left[\frac{K_1}{\rho}\rho d\phi dz\right]\Bigg|_{\rho=a} + \int_0^a \int_0^{2\pi} [K_2 z \rho d\phi d\rho]|_{z=\ell} + \int_0^a \int_0^{2\pi} [K_2 z \rho d\phi d\rho]|_{z=0} \tag{3.47}$$

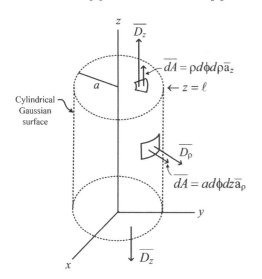

Figure 3.13 Electric flux crossing a cylindrical surface.

The last integral is zero because the z component of the electric field is zero at $z = 0$.

$$\Psi_{\text{net}} = K_1 \int_0^{2\pi} d\phi \int_0^\ell dz + K_2 \ell \int_0^{2\pi} d\phi \int_0^a \rho d\rho \tag{3.48}$$

$$\Psi_{\text{net}} = K_1 2\pi\ell + K_2 \ell 2\pi \frac{a^2}{2} \tag{3.49}$$

Hence, by Gauss's law, the charge enclosed by this Gaussian surface is

$$Q_{\text{encl}} = K_1 2\pi\ell + K_2 \pi a^2 \ell \text{ [C]} \tag{3.50}$$

Note that this result is the total enclosed charge, but it makes no statement about the symmetry, or lack thereof, of the charge distribution.

3.4 Magnetic Flux

Motivational Question How does magnetic flux differ from electric flux?

So far, this chapter has focused on electric flux, the insights gained from the flux perspective, and the analytic utility of Gauss's law. The flux concept is also applicable to magnetic fields and similarly yields insights into magnetic field behavior. **Magnetic flux** is assigned the symbol Φ and the SI unit is the weber (Wb). **Magnetic flux density**, $\overline{B}$, is related to the magnetic field intensity ($\overline{H}$) by the **permeability** μ:

$$\overline{B} = \mu\overline{H} = \mu_o\mu_r\overline{H} \tag{3.51}$$

The unit of $\overline{B}$ is tesla (T), and 1 T = 1 Wb/m². The unit of μ is henry/meter (H/m). The permeability is the same quantity that you may have utilized in previous circuits or physics courses for the calculation of the inductance of a long coil ($L = \mu N^2 A/\ell$). Magnetic material classifications are surveyed in Chapter 4. Note that these magnetic field quantities and units are similar in form to the electric field counterparts, revealing again the correspondence between the two field types.

In a manner analogous to Gauss's law and electric flux, the net magnetic flux Φ_{net} through a closed surface (Gaussian surface) can be simplified if two conditions are met:

$$\Phi_{net} = \oint \overline{B} \cdot \overline{dA} = 0 \rightarrow \left\{ \begin{array}{l} \text{If } \overline{B} \text{ is} \perp \text{GS} (\overline{B} \parallel \overline{dA}) \\ \text{If } |\overline{B}| \text{ is uniform across GS} \end{array} \right\} \rightarrow \Phi_{net} = BA = Q_{magnetic} = 0 \tag{3.52}$$

Again, the integral must be used if the two conditions are not met. This general "Gauss's law for magnetic fields" is formally stated in Eq. (3.53). Note that $\Phi_{net} = 0$. Why?

$$\Phi_{net} = \oint \overline{B} \cdot \overline{dA} = 0 \tag{3.53}$$

The net magnetic flux from a *closed* surface equals zero because there are *no* isolated magnetic charges (magnetic monopoles) upon which magnetic flux could terminate. Hence, the key ramification is that all magnetic flux lines must close in on themselves, that is, magnetic flux is **solenoidal** – magnetic flux lines must form closed (complete) paths. While Eq. (3.53) can be directly used in a few limited applications, the true importance of this equation is its consequence: the fundamental solenoidal behavior of magnetic fields. For example, Eq. (3.53) cannot be used to determine the magnetic fields around highly symmetric current distributions, the way that Gauss's law was useful in determining electric fields around highly symmetric charge distributions. Nevertheless, Eq. (3.53) is so important that it is included as one of Maxwell's four equations (to be studied in Chapter 5).

The concept of solenoidal magnetic flux is practically useful in many applications. An important application is that of *magnetic shielding* in the vicinity of stray magnetic fields, for example near an MRI machine or next to a high current motor. A perspective on magnetic shielding can be made by first considering electric shielding, where an enclosed conductive (metal) surface, called a Faraday shield (or Faraday "cage"), can shield an enclosed circuit from external electric fields. Electric shielding is easily achieved because an external interfering electric field will terminate on the charges on the conductor surface of the shield, thus preventing the field from entering the enclosure. In contrast for magnetic shielding, there are no magnetic charges and thus magnetic flux cannot be terminated. Instead, the external magnetic field must be guided around the circuit to shield it. This rerouting of magnetic flux is usually accomplished with a magnetic material such as iron.

Example 3.4.1

A sensitive electronic circuit must be shielded from a low-frequency magnetic field of known direction. Conceptually design an enclosure to accomplish this magnetic shielding.

Answer

See Figure 3.14. A magnetic material, such as cast iron, is used to route the magnetic flux around the electronic circuit. Note that the cast iron shield is also conductive, so electric field shielding is additionally present.

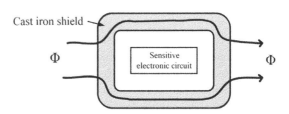

Figure 3.14 Conceptual sketch of a magnetic shield.

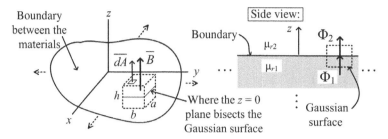

Figure 3.15 Normal magnetic field boundary condition setup.

The concept of solenoidal magnetic flux also leads to another boundary condition, this one for the normal component of the magnetic flux density. See Figure 3.15. Each flux line normal to the boundary between the two materials must "pass through" because there are no magnetic charges to terminate on. The area on the top of the Gaussian surface is the same as the area on the bottom of the Gaussian surface, and the flux is perpendicular to both surface areas. On a differential area basis:

$$d\Phi_1 = d\Phi_2 \rightarrow B_{n1}dA = B_{n2}dA \tag{3.54}$$

The differential areas on the top and the bottom of the Gaussian surface are the same; thus, the boundary condition for the normal component of the magnetic flux density is:

$$\boxed{B_{n1} = B_{n2}} \tag{3.55}$$

Note that per Eq. (3.55), a change of magnetic materials does not affect the magnetic flux density component that is normal to the surface. Is the magnetic field intensity H affected? Yes, because $B = \mu H$ (see homework problem 20 at the end of the chapter).

Looking ahead: We will utilize magnetic flux in Chapter 4 to develop the inductance for a given conductor configuration. Magnetic flux is crucial to inductance. Why? Recall the definition of inductance for a coil:

$$L = \frac{N\Phi}{I} \tag{3.56}$$

where N is the number of wire turns. The definition of inductance is expressed entirely in terms of magnetic field quantities (Φ and I). Why is this important? Where is the energy stored in an inductor? The *magnetic field*!!! If the magnetic field can be determined, whether analytically, experimentally, or through simulation, then the inductance of that coil structure can be determined. A more general definition of inductance in Chapter 4 will allow us to determine the inductance of general magnetic structures.

For inductor, magnetic circuit, and other applications, the magnetic flux is determined through a *non-closed* (non-Gaussian) surface by *integrating the magnetic flux density over the area that the flux is crossing*:

$$\Phi = \int \overline{B} \cdot \overline{dA} \tag{3.57}$$

where the circle has been removed from the integral to indicate the surface is not closed. As in Gauss's law, the dot product is needed to determine the component of the magnetic flux density that is crossing the surface of integration. The reason that integration is required in general is because the flux density may not be uniform as a function of position on the surface. If the flux density is uniform and perpendicular to the surface of integration, the integral reduces to $\Phi = BA$, a multiplication of the magnetic flux density and the cross-sectional area that the flux crosses. Often we can choose the surface to enable use of this simplified equation. The next example demonstrates this approach.

Example 3.4.2

Determine the magnetic flux crossing a $4\,\text{cm}^2$ area in the x-y plane if the magnetic flux density is uniform at $1.5\,\text{T}$ in the z-direction.

Strategy

Set up the dimensions and $\overline{B}$ in a coordinate system; evaluate $\Phi = \int \overline{B} \cdot \overline{dA}$.

Solution

$\overline{B} = 1.5\,\overline{a}_z \qquad \overline{dA} = dxdy\,\overline{a}_z$ (see Figure 3.16)

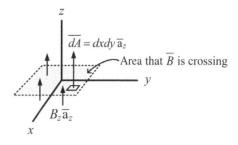

Figure 3.16 Setup for Example 3.4.2.

$$\Phi = \int \overline{B} \cdot \overline{dA} = \int_{-0.01}^{+0.01}\int_{-0.01}^{+0.01} 1.5\,\overline{a}_z \cdot dxdy\overline{a}_z = 1.5 \underbrace{\int_{-0.01}^{+0.01}\int_{-0.01}^{+0.01} dxdy}_{\text{area that } B \text{ crosses}} = 0.6\,\text{mWb}$$

Since the magnitude of the magnetic flux density is constant and crosses the surface area perpendicularly, the integral could have been reduced to a straightforward (flux density)(area) product, as it was with Gauss's law and electric flux density in similar circumstances:

$$\Phi = \int \overline{B} \cdot \overline{dA} = \int \underbrace{B_z\overline{a}_z \cdot dA\overline{a}_z}_{\overline{B}\,\|\,\overline{dA}} = \int \underbrace{B_z}_{\text{const}} dA = B\int dA = BA = (1.5\,\text{T})(4\,\text{cm}^2)\left(\frac{1\,\text{m}}{100\,\text{cm}}\right)^2 = 0.6\,\text{mWb}$$

Magnetic flux calculations are also useful in magnetic circuit calculations, such as iron-core transformers and toroid inductors. Magnetic circuits are covered in an appendix of Chapter 4. The following example illustrates the determination of how much magnetic flux from a current distribution "links" (crosses through a defined surface area) another conductor configuration.

Example 3.4.3

Determine the magnetic flux that crosses through ("links") the area of a thin rectangular coil area from a line current in the z-direction.

Strategy

This problem is not precisely defined. Thus, a reasonable geometry must be set up. Assume the current-carrying wire is on the z-axis and that the coil is in a $\phi = $ constant plane. See Figure 3.17. Assign coil dimensions and identify $\overline{B}$ in the cylindrical coordinate system. Evaluate $\Phi = \int \overline{B} \cdot \overline{dA}$

Solution

$\overline{B} = \mu\overline{H}\,\overline{a}_\phi = \dfrac{\mu I}{2\pi\rho}\overline{a}_\phi$ (from Chapter 2), $\overline{dA} = d\rho dz\overline{a}_\phi$ (see Figure 3.17)

$$\Phi = \int \overline{B} \cdot \overline{dA} = \int_c^d\int_a^b \frac{\mu I}{2\pi\rho}\overline{a}_\phi \cdot d\rho dz\,\overline{a}_\phi = \frac{\mu I}{2\pi}\int_a^b \frac{d\rho}{\rho}\int_c^d dz = \frac{\mu I}{2\pi}\ln\left(\frac{b}{a}\right)(d-c)\ (\text{Wb})$$

In this example, it was necessary to evaluate the magnetic flux integral because the magnetic flux density is not uniform over the area of the thin coil.

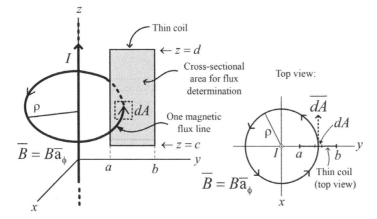

Figure 3.17 Geometry for Example 3.4.3.

Perspective: You probably noticed that we did not determine $\overline{B}$ as we did $\overline{D}$ in Gauss's law problems because there are no magnetic charge distributions that correspond to electric charge distributions. However, we already know how to determine $\overline{H}$ from the Biot–Savart law, and unlike charge distributions, we normally do know the current distribution because we designed the conductor configurations to carry currents. Thus, the Biot–Savart law and $\overline{B} = \mu\overline{H}$ are often used to determine the magnetic flux density for practical structures. As discussed in previous paragraphs, magnetic flux density is utilized in several electronic components and applications that will be examined in subsequent chapters.

3.5 Summary of Important Equations

$$\Psi_{\text{net}} = \oint \overline{D} \cdot \overline{dA} = Q_{\text{encl}} \qquad \overline{D} = \varepsilon_o \overline{E} \text{ (free space)} \qquad \overline{D} \cdot \overline{a}_n = D_n = \rho_S \text{ (perfect conductor)}$$

$$\Phi_{\text{net}} = \oint \overline{B} \cdot \overline{dA} = 0 \qquad \Phi = \int \overline{B} \cdot \overline{dA} \qquad \overline{B} = \mu\overline{H} = \mu_o\mu_r\overline{H} \qquad B_{n1} = B_{n2}$$

3.6 Homework

1. a) What is *electric flux*?
   b) What is the relationship between electric *field intensity* and electric *flux density*?
   c) A 25 μC, point charge is enclosed by a *spherical* Gaussian (observation) surface. What is the total electric flux crossing the Gaussian surface? Why?
   d) A 25 μC, point charge is enclosed by a *cubical* Gaussian (observation) surface. What is the total electric flux crossing the Gaussian surface? Why?

2. a) Why is electric flux density $D \propto 1/r^2$ outside a uniform charge distribution with a spherically symmetric geometry? Explain in terms of Gauss's law.
   b) What are the conditions on the electric field that must be present to use Gauss's law analytically to obtain a closed-form solution of the electric flux density?
   c) Is Gauss's law valid for any finite charge distribution? Explain.

3. An infinitely long line charge with uniform line charge density ρ_L (compare the result to the corresponding result of Coulomb's law).
   a) Use Gauss's law to determine $\overline{D}$ in the region outside of this charge distribution.
   b) Qualitatively explain the behavior of the field as a function of distance from this charge distribution.

4. A spherical surface of radius a and uniform surface charge density ρ_S.
   a) Use Gauss's law to determine $\overline{D}$ in the region outside of this charge distribution.
   b) Qualitatively explain the behavior of the field as a function of distance from this charge distribution.

5. An infinitely long cylindrical volume of radius a and uniform volume charge density ρ_V.
   a) Use Gauss's law to determine $\overline{D}$ in the region outside of this charge distribution.
   b) Qualitatively explain the behavior of the field as a function of distance from this charge distribution.

6. An infinite sheet of uniform surface charge density that lies in the $z = 0$ plane with air on both sides of the surface (see the example in the text for $\overline{D}$ just outside a flat perfect conductor.)
   a) Use Gauss's law to determine $\overline{D}$ in the region outside of this charge distribution.
   b) Qualitatively explain the behavior of the field as a function of distance from this charge distribution.

7. A rod of 1 mm diameter and 1 m length is charged. Identify and qualitatively explain the generally expected electric field behavior as a function of radial distance in the vicinity of (a) 1 μm, (b) 1 cm, and (c) 100 m from the surface of the rod. Assume that the field is being examined in a plane that intersects the midpoint of the rod lengthwise.

8. a) In Example 3.2.7, what is the electric field outside of the coax cable ($\rho > b$) if the outer conductor is *not* grounded. Why? Hint: The net charge on the outer conductor is zero.
   b) Sketch the charge distribution within the outer conductor (assume a finite thickness). Hint: The outer conductor is neutral. How do the charges within it redistribute?

9. An infinitely long cylindrical shell of uniform volume charge density ρ_V C/m^3 is centered on the z-axis and has an inner radius a and an outer radius b.
   a) Draw and label a sketch.
   b) Use Gauss's law to determine the *electric flux density* expression for $\rho < a$.
   c) Use Gauss's law to determine the *electric flux density* expression for $\rho > b$.
   d) Semi-challenge: Determine the expression for the *electric flux density* for $a < \rho < b$.

10. A spherical shell of uniform volume charge density ρ_V C/m^3 is centered at the origin and has an inner radius a and an outer radius b.
    a) Draw and label a sketch.
    b) Use Gauss's law to determine the *electric flux density* expression for $r < a$.
    c) Use Gauss's law to determine the *electric flux density* expression for $r > b$.
    d) Semi-challenge: Determine the expression for the *electric flux density* for $a < r < b$.

11. An infinitely long cylindrical shell of nonuniform volume charge density $\rho_V = K\rho$ C/m^3, where K is a constant, is centered on the z-axis, and has an inner radius a and an outer radius b.
    a) Draw and label a sketch.
    b) Use Gauss's law to determine the *electric flux density* expression for $\rho < a$.
    c) Use Gauss's law to determine the *electric flux density* expression for $\rho > b$.
    d) Semi-challenge: Determine the expression for the *electric flux density* for $a < \rho < b$.

12. A spherical shell of nonuniform volume charge density $\rho_V = Kr^2$ C/m^3, where K is a constant, is centered at the origin, and has an inner radius a and an outer radius b.
a) Draw and label a sketch.
b) Use Gauss's law to determine the *electric flux density* expression for $r < a$.
c) Use Gauss's law to determine the *electric flux density* expression for $r > b$.
d) Semi-challenge: Determine the expression for the *electric flux density* for $a < r < b$.

13. Determine the charge enclosed by a cylinder of radius a and length ℓ (from the $z = 0$ plane) if the electric flux density is $\dfrac{K_1}{\rho^2}\bar{a}_\rho + K_2 z^2 \bar{a}_z$ C/m^2 for $\rho > 0$.

14. Determine the charge enclosed by a sphere of radius a if $\overline{E}$ is $\dfrac{K}{r^3}\bar{a}_r$ V/m for $r > 0$.

15. a) What does $\oint \overline{B} \cdot \overline{dA}$ mean?
b) What does it equal? Why?

16. a) What does $\Phi = \int \overline{B} \cdot \overline{dA}$ mean? [Distinguish from the answer to Problem 15.]
b) When can the approximation $\Phi = \int \overline{B} \cdot \overline{dA} \approx BA$ be made?

17. The magnetic field intensity in a coaxial line of inner radius a, outer radius b, and length ℓ is given below. The current in the center conductor is I. Determine the total magnetic flux crossing a ϕ = constant plane for $a < \rho < b$ and $0 < z < \ell$.

$$\overline{H} = \frac{I}{2\pi\rho}\bar{a}_\phi \text{ for } a < \rho < b.$$

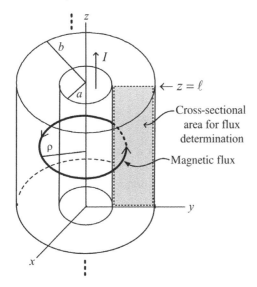

18. Two circles of thin wire of radii a and b, where $b > a$, lie in the $z = 0$ plane, and are centered on the origin. The magnetic field intensity is $\overline{H} = H_o \dfrac{e^{-k\rho}}{\rho}\bar{a}_z$ A/m.
a) Make and label a sketch.
b) Determine the magnetic flux that passes through the area between the circles of wire.
c) Would $\Phi = BA$ yield the same result as part b)? Why or why not?

19. A current in an ideal coil establishes a magnetic field intensity in the interior of the coil of NI/ℓ A/m, where N is the number of turns and ℓ is the length of the coil. Let A be the cross-sectional area of the coil perpendicular to the coil axis.

a) Make a sketch and assign a convenient unit vector direction to H.

b) Determine the total magnetic flux in the interior of the coil.

c) Identify four methods to increase the flux in a coil.

20. The magnetic flux density in a toroidal magnetic circuit (top view shown) is 1 T. The core is steel and the gap is air. Assume the gap length is small compared with other dimensions. The cross-sectional area A is circular. The relative permeability of the core is $\mu_r = 1000$ for the steel in use.

a) What is the magnetic field intensity H in the steel core and in the air gap?

b) Conclude on the effect of the magnetic material on H for a given flux density level.

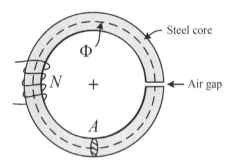

4

The Path Laws and Circuit Principles

Coulomb's law, the Biot–Savart law, Gauss's law, and magnetic flux have been introduced and examined to this point. This chapter introduces two more powerful laws of electromagnetism: the definition of voltage (electrical potential) and Ampere's circuital law (ACL). They are collectively known as the *path laws* because they involve line integrals along paths. Then these laws along with the other laws and concepts covered previously are utilized to relate electromagnetics to the important circuit quantities of resistance, capacitance, and inductance in electrical engineering. These electromagnetic-circuit relationships and other applications will enhance the context for and reinforce the understanding of the electromagnetic topics covered previously. as well as increase your insight into the fundamental circuit relationships and laws. *Enjoy*!

Electromagnetics and Transmission Lines: Essentials for Electrical Engineering, Second Edition.
Robert A. Strangeway, Steven S. Holland, and James E. Richie.
© 2023 John Wiley & Sons, Inc. Published 2023 by John Wiley & Sons, Inc.
Companion website: www.wiley.com/go/Strangeway/ElectromagneticsandTransmissionLines

4.1   Electric Potential (Voltage) and Kirchhoff's Voltage Law

4.1.1   Potential–Electric Field Relationship

Motivational Question   Where did the relationship between electric field intensity and electric potential (voltage), $V = -\int_A^B \overline{E} \cdot \overline{d\ell}$, come from?

Voltage[1] is an important circuit quantity with widespread usage in electrical engineering. We now will examine voltage at a fundamental level in the context of static electric fields. You may have seen the equation $V = -\int_A^B \overline{E} \cdot \overline{d\ell}$ in a previous physics course. It is the fundamental equation that relates voltage, also called potential, to the electric field. This equation will be developed next from the well-known definition of work. Visualize a charge Q being moved in an electrostatic field $\overline{E}$ from point A to point B as sketched in Figure 4.1a. How much work (energy) W is expended in moving Q from A to B? Start with the definition of work where the force is applied to an object in the direction of movement:

$$W = \int_A^B \overline{F} \cdot \overline{d\ell} \tag{4.1}$$

where $\overline{d\ell}$ is a differential length along the path taken from point A to B. The force in terms of electric field intensity is $\overline{F} = Q\overline{E}$ (recall Coulomb's law). However, this force is *the force of the electric field on the electric charge*. See Figure 4.1b. What force would be *applied to the charge* to move it in this electric field? Assume a positive charge and movement of the charge in a direction opposite to the vector direction of the electric field (doing positive work), as shown in Figure 4.1c. The force to be applied is then $\overline{F} = -Q\overline{E}$. Substitute the vector force $\overline{F} = -Q\overline{E}$ applied to move the charge in the electric field into the general work expression:

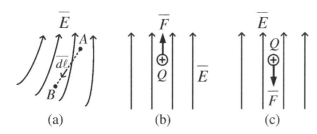

(a)   (b)   (c)

Figure 4.1   Charge movement and forces in an electric field. (a) Q movement in an $\overline{E}$ field. (b) $\overline{F}$ on Q due to $\overline{E}$. (c) $\overline{F}$ applied to move Q in the $\overline{E}$ field.

$$W = \int_A^B -Q\overline{E} \cdot \overline{d\ell} = -Q \int_A^B \overline{E} \cdot \overline{d\ell} \tag{4.2}$$

Why is a "−" sign required in front of the integral? It takes $+W$ to do work to move Q against the electric field. When the direction of motion and the electric field are in opposite directions, the dot product gives a negative result, thus the "−" sign makes this quantity positive. Why is the dot product required? It gives the vector component of motion parallel to the field.

Voltage (electric potential) is defined to be the work expended in moving a charge through an electric field per unit charge:

$$\boxed{V = \frac{W}{Q} = -\int_A^B \overline{E} \cdot \overline{d\ell}} \tag{4.3}$$

---

1 The quantity V is often referred to as *electric potential*, *potential difference*, and other similar phrases in the context of electrostatic fields. The term *voltage* is familiar from its use in circuits in electrical engineering, and it is used interchangeably with *potential* and similar phrases in this text.

What is *potential difference*? It is *electrical potential energy difference per unit charge*, which is voltage. Consider this viewpoint on potential energy: if a charge is released in the electric field, the potential energy in the electric field is converted into kinetic energy of the charge. Thus, the static electric field has potential energy stored in electrical form.

Significant points about Eq. (4.3):

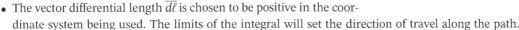

- Voltage is a scalar quantity (not a vector quantity!).
- Voltage can be determined from $\overline{E}$.
- It is a line (path) integral, so $\overline{E}$ must be known, usually as a function, along the path to use this relation.
- The vector differential length $\overline{dl}$ is chosen to be positive in the coordinate system being used. The limits of the integral will set the direction of travel along the path.
- Do *not* ignore the dot product between the field and the path.
- The voltage is independent of the integration path[2] (to be examined in Section 4.1.2). Thus, strategically choose an integration path to simplify the equation: choose the path between points A and B to be parallel to $\overline{E}$ (in which case $\overline{E} \cdot \overline{dl} = E\,dl$). Some parts of the path may be perpendicular to $\overline{E}$, in which case $\overline{E} \cdot \overline{dl} = 0$, that is, there is zero voltage across this part of the path. This strategy is analogous to choosing the correct Gaussian surface (GS) when using Gauss's law analytically!

Figure 4.2 Illustration of the voltage–electric field intensity relationship. (a) R of length ℓ. (b) R of length 2ℓ.

A formal notation for voltage is:
$$V_{BA} = -\int_{A}^{B} \overline{E} \cdot \overline{dl} = V_B - V_A \tag{4.4}$$

where V_{BA} is the potential at point B with respect to point A. Notice that $V_B - V_A$ explicitly shows that *voltage must exist between two distinct points*, just as it does in circuits. What is the basic difference between voltage and electric field intensity? Compare the electric field intensity inside a cylindrical resistor with the electric field intensity inside a cylindrical resistor of the same material but twice the length, as shown in Figure 4.2. The voltages across both resistors are the same. However, the voltage per unit length in (a) is twice that of (b). Hence, the electric field intensity is also twice as large in (a) relative to (b). One could interpret this comparison using the units of electric field intensity: the volts-per-meter in (a) is twice the V/m of (b).

Example 4.1.1

The electric field intensity in coax was previously determined from Gauss's law in Example 3.2.3 to be $\overline{E} = \dfrac{\rho_S a}{\varepsilon \rho} \overline{a}_\rho$, where ρ_S is the surface charge density on the inner conductor, in the positive $\overline{a}_\rho$ direction. Determine the potential between the inner conductor (radius a) and the outer conductor (radius b).

Solution

The setup is sketched in Figure 4.3. A radial path is chosen and a differential length along this path is shown. The negative direction along the path is accounted for by the integral limits, *not* by the differential length vector.

$$V = -\int_{A}^{B} \overline{E} \cdot \overline{dl} = -\int_{b}^{a} \frac{\rho_S a}{\varepsilon \rho} (\overline{a}_\rho) \cdot d\rho\,\overline{a}_\rho$$

$$V = -\frac{\rho_S a}{\varepsilon} \int_{b}^{a} \frac{d\rho}{\rho} = -\frac{\rho_S a}{\varepsilon} \ln \rho \big|_{b}^{a}$$

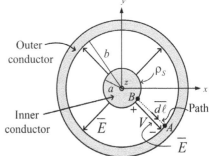

Figure 4.3 Path in coax for V (end view).

---

2 Technically, this result holds only for time-independent electric fields. We will examine the effects of time-varying fields in Chapter 5.

$$V = -\frac{\rho_S a}{\varepsilon}(\ln a - \ln b) = \frac{\rho_S a}{\varepsilon}(\ln b - \ln a)$$

$$V = \frac{\rho_S a}{\varepsilon} \ln\left(\frac{b}{a}\right)$$

These results will be utilized in the determination of the capacitance of coax (Section 4.2).

Further discussion on the path selection: The same voltage would be determined even if points A and B were not directly across from one another in the coax. How could this be? Examine the following scenario. Locate point A at $(b, 30°, 0)$ and point B at $(a, 80°, 0)$. A direct straight-line path between these points would be very difficult to evaluate analytically. However, noting that the voltage integral is independent of the integration path, one could break the path integral into two parts: path 1 along a constant radius $\rho = b$ between points $(b, 30°, 0)$ and $(b, 80°, 0)$, and path 2 along a radial path from $(b, 80°, 0)$ to $(a, 80°, 0)$. Path 1 contributes zero voltage because the path is everywhere perpendicular to the radially directed electric field, so the dot product equals zero (no change in energy along this part of the path!). The voltage over path 2 will give the same result found in this example. The important result to conclude from this discussion: *proper selection of the integration path(s) can greatly simplify the voltage analysis!*

4.1.2 Kirchhoff's Voltage Law (KVL)

Motivational Question Upon what principle is KVL based? Does it relate to $\overline{E}$ fields?

What happens if the path in $\int \overline{E} \cdot \overline{d\ell}$ is closed (same start and end points)? This scenario leads to the concept of a **conservative field**: the net energy expenditure in moving a charge around any closed path in an electrostatic electric field is zero. In other words, energy is conserved around a closed path in a static electric field. See Figure 4.4a. Equation (4.4) becomes:

$$\oint \overline{E} \cdot \overline{d\ell} = 0 \tag{4.5}$$

for a conservative electric field, where the circle around the path integral indicates a closed path.[3]

What is the consequence of conservative electric fields in circuits? **Kirchhoff's voltage law**! Equation (4.5) is the electric field basis for the sum of the voltages around any closed path equaling zero. Apply this relationship by going clockwise around the path in the circuit shown in Figure 4.4b:

$$\oint \overline{E} \cdot \overline{d\ell} = -\int \underbrace{\overline{E}_S \cdot \overline{d\ell}}_{\text{antiparallel}} - \int \underbrace{\overline{E}_L \cdot \overline{d\ell}}_{\text{parallel}} = +v_S - v_L = 0 \tag{4.6}$$

The first integral represents a gain in potential energy (energy in the source is converted into electrical energy that results in the electrical potential difference), and the second integral represents the conversion of electrical energy into another form of energy in the load. The net energy around the complete circuit path is zero, that is, the sum of the voltages around the closed circuit path is zero, and KVL is satisfied.

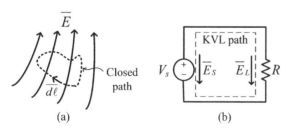

(a) (b)

Figure 4.4 Closed path in an electric field and in a circuit. (a) Closed path in an $\overline{E}$. field in general. (b) Closed path for application of KVL in a circuit.

4.1.3 Dielectric–Conductor Electric Field Boundary Conditions

Motivational Question What happens to $\overline{E}$ at the boundary of two different materials?

---

3 We will examine the effects of time-varying fields in Chapter 5.

Boundary conditions were introduced in Chapter 3. Again, what are boundary conditions? They are mandatory conditions on a field at the boundary of two different materials. Why are boundary conditions needed? They are essential in solving differential equations and in electromagnetic field simulations. Boundary conditions are requirements on the variables, $\overline{D}$ and $\overline{E}$, usually along a surface that defines a boundary between materials, and are used to determine the particular solution from the general solution of a differential equation (a topic in advanced electromagnetic studies). Another

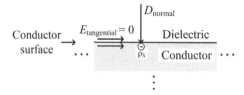

Figure 4.5 Electric fields at a dielectric–conductor boundary.

important use of boundary conditions is to visualize field patterns without formally solving for the field, and this aspect is our emphasis here.

For electric fields, the general types of material boundaries are dielectric–conductor[4] and dielectric–dielectric interfaces. The former is considered here. The latter is developed in Appendix 4.A. Recall the normal electric field boundary condition at a dielectric–conductor interface that was developed from Gauss's law in Chapter 3: $D_n = \rho_S$. See Figure 4.5. Electric flux lines begin or end (terminate) on electric charges.

In general, the $\overline{E}$ field is decomposed into vector components that are perpendicular (normal) and parallel (tangential) to a material interface (see Figure 4.5). Recall that the electric field inside a perfect conductor is zero because the charges on the surface redistribute to cancel out any internal electric fields. In the same manner, the tangential electric field intensity E_t must be zero at the surface of a perfect conductor because charges are free to redistribute and cancel any electric field. Consequently, the voltage between any two points on a perfect conductor is zero, as you know from the voltage across a short circuit in circuit theory:

$$V_{BA} = -\int_A^B \overline{E_t} \cdot \overline{d\ell} = V_B - V_A = 0 \rightarrow E_t = 0 \tag{4.7}$$

Thus, the *tangential electric field boundary condition at the surface of a perfect conductor* is:

$$\boxed{E_t = 0} \tag{4.8}$$

If the tangential electric field component is zero at the perfect conductor interface, then the $\overline{E}$ field *must be purely perpendicular* to the conductor at the boundary (visualization!). The electric field perpendicular to a perfect conductor was developed using Gauss's law in Chapter 3. The result was the normal electric flux density equaled the surface charge density.

$$D_{\text{normal}}|_{\text{dielectric side}} = \varepsilon E_{\text{normal}}|_{\text{dielectric side}} = \rho_S|_{\text{conductor side}} \rightarrow \boxed{D_n = \rho_S} \tag{4.9}$$

As another example of visualizing electric fields from boundary conditions, reinspect the figures for coax (Chapter 3) and note the electric field is perpendicular to both the inner and outer conductors.

4.2 Capacitance

Capacitors are the first of three important passive circuit components that we will now examine from a fundamental electric field perspective. This discussion provides revealing insights into the behavior of this common component, particularly with respect to the effect of the dielectric material and the storage of energy in the electric field.

---

4 This boundary condition can also be developed from a more general dielectric–dielectric tangential electric field boundary condition that is developed in Appendix 4.A.

4.2.1 Determination of Capacitance

Motivational Questions   Where did the expression for the capacitance of a parallel plate capacitor, $C = \varepsilon A/d$, come from? How is the capacitance of other conductor configurations (such as coax) determined?

A good place to start the development of the capacitance of a given structure is to express capacitance in terms of electric field quantities. Capacitance was defined in circuits courses to be the charge stored on the capacitor plates per volt of electric potential across the plates:

$$C = \frac{Q}{V} \tag{4.10}$$

This equation is a general definition because it does not specify the type of capacitor (parallel-plate, coaxial, etc.). Note that Q is defined as the magnitude of the charge on each of the conductors, where there is $+Q$ charge on one conductor and $-Q$ charge on the other conductor. The voltage in Eq. (4.10) is directly related to the electric field quantities by

$$V = -\int_A^B \overline{E} \cdot \overline{d\ell} \tag{4.11}$$

Hence, the general definition of capacitance in terms of the electric field is:

$$C = \frac{Q}{-\displaystyle\int_{A(-)}^{B(+)} \overline{E} \cdot \overline{d\ell}} \tag{4.12}$$

where the path integral for the voltage is between the conductors of the capacitor. Normally, "A" is the negative side of the potential and "B" is the positive side. The electric field intensity can be determined from the charge distribution.

One general approach to develop the capacitance for a given capacitor:

- Determine the $\overline{D}$ or $\overline{E}$, often with Gauss's law for symmetrical conductor configurations or Coulomb's law for asymmetrical conductor configurations.
  - Assume a surface charge density ρ_S on the positive conductor. If using Gauss's law, this charge is enclosed by a GS.
- Determine the voltage between the conductors using the previous $\overline{E}$ result.
  - The line integral should begin on the negative conductor and end on the positive conductor for a positive definition of voltage polarity.
    - Insert the results into Eq. (4.12).
      - The charge can be determined by integrating surface charge density (ρ_S) on the positive conductor.
      - The ρ_S in the numerator should cancel the ρ_S in the denominator, leaving an expression for capacitance in terms of only the capacitor dimensions and the dielectric constant.

Example 4.2.1

Develop the expression for the capacitance of a *parallel-plate capacitor*.

Solution

Setup: See Figure 4.6. The plate separation is d and each plate has area A.
    Let the bottom plate have negative charge $-Q$ and lie in the $z = 0$ plane, and let the upper plate have positive charge $+Q$ and lie in the $z = +d$ plane.

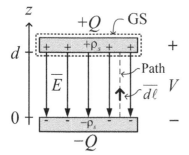

Figure 4.6   Parallel-plate capacitor setup.

The electric field goes from the upper positive plate to the lower negative plate. Assumptions:

- The charge density and the electric field intensity are uniform.
- The electric field intensity is parallel to the z-direction only, due to planar symmetry, assuming the plates are large compared with the separation ($d \ll \sqrt{A}$).
- The flux is completely contained between the plates, with negligible field outside the plates, for which there is no electric field *fringing*; hence, the electric field cross-sectional area is also A.

The GS selection encloses the entire upper plate. The net flux from the GS equals the charge enclosed:

$$\Psi_{net} = Q_{encl} \tag{4.13}$$

The charge density is uniform, so the charge density-total charge relationship simplifies to:

$$Q = \int \rho_S dA \rightarrow Q = \rho_S A \tag{4.14}$$

and algebraic simplification of the electric flux relationship is also valid based on the assumptions listed above (uniform field, flux passing only through the bottom of the GS):

$$\Psi = Q_{encl} = \oint \overline{D} \cdot \overline{dA} \rightarrow D_z = \frac{\Psi}{A} = \frac{Q}{A} = \frac{\rho_S A}{A} = \rho_S \tag{4.15}$$

which you hopefully recognized as the boundary condition at the surface of a perfect conductor! The surface charges on the bottom of the top plate "see" a perfect conductor above and a dielectric below. Thus, the flux can only flow downward into the dielectric. Form the electric field intensity vector[5] and substitute into the general voltage – $\overline{E}$ field relationship:

$$E_z = \frac{\rho_S}{\varepsilon} \rightarrow \overline{E} = \frac{\rho_S}{\varepsilon}(-\overline{a}_z) \tag{4.16}$$

$$V = -\int_{A(-)}^{B(+)} \overline{E} \cdot \overline{d\ell} = -\int_0^d \frac{\rho_S}{\varepsilon}(-\overline{a}_z) \cdot dz\overline{a}_z \tag{4.17}$$

Note the two negative signs cancel each other in the previous equation. Evaluate the integral and simplify:

$$V = \frac{\rho_S}{\varepsilon} \int_0^d dz = \frac{\rho_S}{\varepsilon}d \tag{4.18}$$

Insert the charge and voltage results into the general capacitance definition:

$$C = \frac{Q}{V} = \frac{\rho_S A}{\frac{\rho_S}{\varepsilon}d} = \frac{\varepsilon A}{d} \tag{4.19}$$

The well-known result for the capacitance of an ideal parallel-plate capacitor results:

$$\boxed{C = \frac{\varepsilon A}{d}\,[F]} \quad \text{(ideal parallel – plate capacitor)} \tag{4.20}$$

Example 4.2.2

Develop the expression for the capacitance of a *coaxial capacitor*.

Solution

Setup: See Figure 4.7. The inner radius is a, the outer radius is b (actually the radius to the inner surface of the outer conductor), and let the coax length be L (to distinguish it from the ℓ in $d\ell$). The inner conductor has a positive charge and the outer conductor has a negative charge.

---

5 ε is used here instead of ε_o because dielectrics (Section 4.2.2) are utilized in most capacitors.

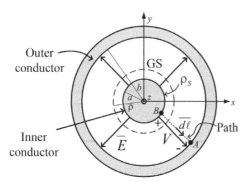

Figure 4.7 Coax capacitor setup (end view).

Assumptions:

- Positive surface charge density is present on the inner conductor, at $\rho = a$, and negative surface charge density is present on the inner surface of the outer conductor at $\rho = b$.
- The charge densities and, at any given radius, the electric field intensity are uniform.
- $\overline{E}$ is in the ρ-direction only (due to symmetry).
- $L \gg (b - a)$, so electric field fringing at the ends of the coax capacitor is negligible.

Choose a GS to be a cylinder of radius ρ, where $a < \rho < b$, and length L that encloses the inner conductor. The first step of determining the electric field inside the coax was previously done in Example 3.2.3, but the steps are summarized here.

Using the same approach as the previous example, the uniform ρ_S results in:

$$Q = \int \rho_S dA \rightarrow Q = \rho_S 2\pi a L \tag{4.21}$$

The flux crosses only the side of the cylindrical GS:

$$\Psi = Q_{encl} = \oint \overline{D} \cdot \overline{dA} \rightarrow D_\rho = \frac{\Psi}{2\pi\rho L} = \frac{Q}{2\pi\rho L} = \frac{\rho_S 2\pi a L}{2\pi\rho L} = \rho_S \frac{a}{\rho} \tag{4.22}$$

$$E = \frac{\rho_S a}{\varepsilon \rho} \rightarrow \overline{E} = \frac{\rho_S a}{\varepsilon \rho} \overline{a}_\rho \tag{4.23}$$

Choose a radial path so that the dot product simplifies to a scalar multiplication in the $V - \overline{E}$ relationship (this voltage was found previously in Example 4.1.1):

$$V = -\int_{A(-)}^{B(+)} \overline{E} \cdot \overline{d\ell} = -\int_b^a \frac{\rho_S a}{\varepsilon \rho} \left(\overline{a}_\rho\right) \cdot d\rho \overline{a}_\rho \tag{4.24}$$

$$V = -\frac{\rho_S a}{\varepsilon} \int_b^a \frac{d\rho}{\rho} = -\frac{\rho_S a}{\varepsilon} \ln \rho \big|_b^a \tag{4.25}$$

$$V = -\frac{\rho_S a}{\varepsilon} \left(\ln a - \ln b\right) = \frac{\rho_S a}{\varepsilon} \left(\ln b - \ln a\right) = \frac{\rho_S a}{\varepsilon} \ln \left(\frac{b}{a}\right) \tag{4.26}$$

$$C = \frac{Q}{V} = \frac{\rho_S 2\pi a L}{\frac{\rho_S a}{\varepsilon} \ln \left(\frac{b}{a}\right)} \tag{4.27}$$

The coax length is set to $L = 1$ m for "per unit length" (indicated by the prime), giving the well-known result of capacitance per unit length of coax:

$$\boxed{C'_{coax} = \frac{2\pi\varepsilon}{\ln \left(\dfrac{b}{a}\right)} \left[\frac{F}{m}\right]} \tag{4.28}$$

Looking ahead: The capacitance per unit length of a transmission line, such as coax, will be utilized in the modeling of transmission lines in Chapter 7.

4.2.2 Dielectrics and Permittivity

Motivational Question   Why does a dielectric between the conductors increase capacitance?

Capacitors typically have an insulating material, called a **dielectric**, between the conductors. This dielectric is used to mechanically keep the conductors separated and to increase the capacitance. In this section, we gain insight into permittivity (the dielectric constant[6]) $\varepsilon = \varepsilon_o \varepsilon_r$ and how it affects capacitance. The following assumptions are made about the dielectric material since they are accurate for typical dielectrics and simplify our discussion:

- Linear: The proportionality of D and E does not change versus the D or E level.
- Isotropic: The material properties are uniform with respect to direction.
- Homogeneous: The material properties are uniform with respect to position.
- Time-invariant: The material properties are constant with respect to time.

The following discussion is presented to develop insight into how the dielectric material affects the electric field in a capacitor. Start with a charged parallel-plate capacitor with no dielectric between the plates. Let Q be the free charge on the positive plate. Without a dielectric, $\overline{D} = \varepsilon_o \overline{E}$ between the plates. Then disconnect the source. See Figure 4.8a.

Next, let a dielectric be inserted between the plates. What is the effect of the electric field on the molecules in the dielectric? When the electric field is not present, charge is distributed within molecules per its normal molecular structure. When an electric field is applied to the molecule, the electron orbit within each molecule shifts to effectively separate the charges within the molecule,[7] as simplistically sketched in Figure 4.8b. This separated charge is called an electric **dipole**. Thus, the polarized molecules in the dielectric form dipoles, as sketched in Figure 4.8c. The charge in a dipole is **bound charge**, that is, it is not free to flow – it is bound to the molecule.

What is the effect of bound charge on the total electric field between the plates? One way to visualize this effect follows. Adjacent bound positive and negative charges in neighboring dipoles *within* the *interior* of the dielectric cancel, so their net macroscopic effect is zero. However, the bound charges at the *edges* of the dielectric (adjacent to each plate) are not canceled – see Figure 4.8d. The bound charges on the edges create an internal E field within the dielectric that *opposes* the applied E field due to free charge on the parallel plates, as shown in Figure 4.8e. Thus, the total E field within the dielectric *decreases*. However, the flux Ψ and flux density D must be the same as before the dielectric insertion because the free charge (Q) and the plate area are the same and there are no free charges within the dielectric. How is this discrepancy resolved?

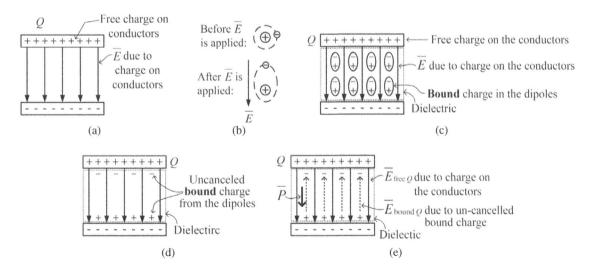

Figure 4.8 Dipoles and $\overline{E}$ in the dielectric of a parallel-plate capacitor. (a) Air dielectric capacitor, source disconnected. (b) Dipole formation. (c) Dielectric inserted, source still disconnected. (d) Uncanceled bound charge due to dipole formation. (e) Internal electric field due to dipole formation.

---

6 The term "dielectric constant" is commonly used to refer to $\varepsilon_o \varepsilon_r$ or ε_r. The order of magnitude of the number indicates which quantity applies in a given situation.
7 A simple model for nonpolar molecules.

A new electric flux quantity, called the **polarization** $\overline{P}\,(\text{C/m}^2)$, is introduced to model the opposing effect of the dipoles on the electric field. *Polarization is defined to be in the direction that corresponds to the negative-to-positive direction within a dipole, as shown in Figure 4.8e.* It is proportional to and parallel to the applied electric field intensity (the dipoles polarize more as E increases):

$$\overline{P} = \chi_e \varepsilon_o \overline{E} \tag{4.29}$$

where $\overline{E}$ is the total electric field intensity within the dielectric, and a material-dependent parameter, the **electric susceptibility** χ_e (dimensionless), is defined as a factor in the proportionality constant. Polarization is a flux density-type quantity (Eq. (4.29) has the same form as $\overline{D} = \varepsilon \overline{E}$) that is parallel to the applied electric field intensity by the definition of its direction, but its effect is to *oppose* the applied electric field intensity. Thus, $\overline{P}$ is included in the electric flux density $\overline{D}$:

$$\overline{D} = \varepsilon_o \overline{E} + \overline{P} = \varepsilon_o \overline{E} + \chi_e \varepsilon_o \overline{E} = \varepsilon_o (1 + \chi_e) \overline{E} = \varepsilon_o \varepsilon_r \overline{E} \tag{4.30}$$

where $(1 + \chi_e)$ is defined to be the **relative dielectric constant** ε_r (also called relative permittivity). Note that the decrease in the electric field intensity E is compensated by the polarization P to maintain the same electric flux density D. Hence, the aforementioned discrepancy is resolved. The main engineering result is:

$$\boxed{\overline{D} = \varepsilon \overline{E} = \varepsilon_o \varepsilon_r \overline{E}} \tag{4.31}$$

This result is the background of the well-known $\varepsilon = \varepsilon_o \varepsilon_r$ used in capacitance calculations and other applications involving dielectrics. Relative dielectric constant (relative permittivity) values are tabulated for numerous materials in standard references and handbooks. A more detailed development of relative permittivity is given in Appendix 4.B.

What if *the source was still connected* to the capacitor? The electric field intensity inside the dielectric $\overline{E}$ must be the same (on a macroscopic, not microscopic, scale) as without the dielectric because the applied voltage $V = -\int \overline{E} \cdot d\overline{\ell}$ and the distance (plate separation) are the same with or without the dielectric. What compensates for the opposing electric field $\overline{E}_{\text{bound } Q}$ from the dipoles? *More charge Q must flow onto the plates* from the source to increase $\overline{E}_{\text{free } Q}$. Relative to the air dielectric case, the capacitance has been *increased* by the presence of the dielectric because the stored charge on the plates increased for the same applied voltage.

Why does a dielectric increase the energy stored?

- It takes energy to "separate" the charge within the dipoles (polarize the molecules) and align them with the applied electric field. As a mechanical analogy, energy is required to stretch a spring, and potential energy is stored in the stretched spring.
- The separated charge in the dipoles return to their original positions (molecules become unpolarized) when the capacitor discharges. Hence, the *energy stored* in the dielectric material is returned to the circuit.

If the electric field intensity is too large in a dielectric, **dielectric breakdown** occurs. Dielectric breakdown occurs when the electric field intensity is strong enough to "pull off" (strip) bound electrons from the molecules in the dielectric, making them free electrons that can then flow as current (example: sparks in air from static electricity). Note that the **dielectric strength** of a dielectric material is specified in V/m, the units of E, not just volts, because it takes so many *volts per meter* to create dielectric breakdown.

The concepts behind dielectrics can be used to explain a change in capacitance due to a change of the dielectric, as illustrated in the next example.

Example 4.2.3
Initially, a parallel-plate capacitor *with air* between the plates is connected to a voltage source. Then a *dielectric is inserted* between the plates (the source is still connected). Does Q, D, E, V, Ψ, and C increase, decrease, or remain the same relative to the initial situation? Justify *each* answer.

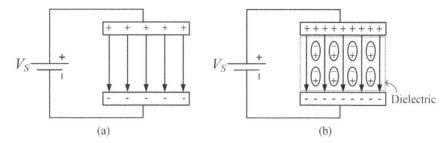

Figure 4.9 Parallel-plate capacitor for Example 4.2.3. (a) Before dielectric insertion. (b) After dielectric insertion.

Answer

The problem is sketched in Figure 4.9a before insertion of the dielectric and in Figure 4.9b after insertion of the dielectric.

V: remains the same, because the voltage source is still connected
E: remains the same, because the voltage and plate separation are the same
Q: increases, to overcome the opposing effect of the dipoles in the dielectric
Ψ: increases, because Q increased, by Gauss's law
D: increases, because Ψ increased and the plate area is the same
C: increases, because stored charge increased for the same voltage

4.2.3 Energy Storage in Electric Fields

Motivational Question   Capacitors store energy in electric fields. How is the amount of energy stored in an E field determined?

A capacitor stores energy in an electric field created by separated charge. We will use the capacitor to illustrate the energy stored in an electric field in this section. Let $w(t)$ be the energy stored in a capacitor as a function of time t. Start with standard circuit relations:

$$w(t) = \int_0^t p(\tau)d\tau = \int_0^t v(\tau)i(\tau)d\tau = \int_0^t v(\tau)C\frac{dv(\tau)}{d\tau}d\tau = C\int_{v(0)}^{v(t)} v(\tau)dv(\tau) = \frac{1}{2}Cv^2\Big|_{v(0)}^{v(t)} \tag{4.32}$$

where τ is being used as a dummy variable for time because the limits are in terms of time. Complete the integration:

$$w(t) = \frac{1}{2}Cv^2(t) - \frac{1}{2}Cv^2(0) = \frac{1}{2}Cv^2(t) - w(0) = \frac{1}{2}Cv^2(t) \tag{4.33}$$

where $w(0) = 0$, the initial energy stored in the capacitor is zero. The equation $\frac{1}{2}Cv^2(t)$ expresses the energy stored (accumulated) from a zero energy state. Once the voltage is nonzero, energy is accumulating. For DC steady state, $w(t)$ is constant, W, and

$$w(t)|_{\text{DC}} = W = \frac{1}{2}CV_{DC}^2 \tag{4.34}$$

The energy stored in the electric field of a parallel-plate capacitor can be generalized (this development is not a proof, just an illustration of a concept). Substitute the expression for the capacitance of an ideal parallel-plate capacitor:

$$W = \frac{1}{2}CV^2 = \frac{1}{2}\frac{\varepsilon A}{d}V^2 = \frac{1}{2}\varepsilon\left(\frac{V}{d}\right)^2 Ad \tag{4.35}$$

Recognize that (V/d) is the uniform electric field intensity (magnitude) in the ideal parallel-plate capacitor and that Ad is the volume between the plates:

$$W = \frac{1}{2}\varepsilon E^2 (\text{volume}) \tag{4.36}$$

Thus, $\frac{1}{2}\varepsilon E^2$ is an electric field **energy density** (J/m^3). Under the assumption that the development for the ideal parallel-plate capacitor is valid in general, the total energy stored in an electric field can be determined by integrating the energy density over the volume that contains E:

$$\boxed{W = \frac{1}{2}\varepsilon \int E^2 dV = \frac{1}{2}\varepsilon_o \varepsilon_r \int E^2 dV} \tag{4.37}$$

where dV is a differential volume in the region of the electric field.[8] The integration is necessary in general because the electric field may be nonuniform. Thus, an electric field stores energy and is the means through which a capacitor stores energy. A dielectric material ($\varepsilon_r > 1$) increases the energy stored.

4.3 Resistance

Motivational Question   Where did the expression for the resistance of a wire or a resistor, $R = \rho\ell/A$, come from?

Just as we did with capacitance, we can approach resistance using a field viewpoint. Visualize charges flowing in a conductor with a finite resistance as sketched in Figure 4.10. Per Ohm's law in circuit theory, $I = V/R = GV$, where G is the conductance in Siemens (S). For reasons soon to become clear, we refer to this equation as the **macroscopic form of Ohm's law**. We can make an analogy to this fundamental circuit relationship to qualitatively formulate a corresponding electromagnetic relationship:[9]

$$\begin{array}{ccc} I & = & G & V \\ \downarrow & & \downarrow & \downarrow \\ \overline{J} & = & \sigma & \overline{E} \end{array} \tag{4.38}$$

where $\overline{J}$ is the volume current density, and σ is the well-known **conductivity** (units: S/m). Note that an electric field can exist within a conductor that has *finite* conductivity. Thus, electric field intensity within a resistive material produces charge movement, that is, current density:

$$\boxed{\overline{J} = \sigma\overline{E}} \tag{4.39}$$

This relation relates charge flow to the force that causes the charge flow as a function of position. For this reason, it is called the **point form of Ohm's law**, or equivalently, the **microscopic form of Ohm's law**. This relationship will be used to develop the expression for the resistance of a conducting material in Example 4.3.1.

Example 4.3.1
Develop the expression for the resistance of a wire or resistor. Assume the conductivity and the current density are uniform throughout the volume of the material. Interpret the meaning of the dependencies in the resulting equation.

Solution

See Figure 4.10. Strategy: relate voltage and current in $R = V/I$ to E.

The current density is uniform across and is perpendicular to the cross-sectional area, which simplifies the current expression:

$$I = \int \overline{J} \cdot \overline{dA} = JA = \sigma EA \tag{4.40}$$

---

8 It has been assumed that the permittivity is constant in this simplified, intuitive development of the energy stored in an electric field.
9 This relationship is developed analytically in Appendix 4.C.

The electric field intensity is uniform along the length of the material because the conductivity is uniform:

$$V = -\int \overline{E} \cdot \overline{d\ell} = E\ell \tag{4.41}$$

Substitute Eq. (4.40) and Eq. (4.41) into the voltage-to-current ratio:

$$R = \frac{V}{I} = \frac{E\ell}{\sigma EA} = \frac{\ell}{\sigma A} = \frac{\rho\ell}{A} \rightarrow \boxed{R = \frac{\rho\ell}{A}} \tag{4.42}$$

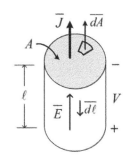

Figure 4.10 Figure for developing resistance.

where ρ = **resistivity** = $1/\sigma$ (ugh! Another quantity that uses the symbol ρ). Thus, the well-known expression for the resistance of a conducting component, such as a resistor or wire, results.

Equation (4.42) describes important resistance concepts:

- Resistance increases proportional to material length. There is increased voltage drop for the same amount current.
- Resistance increases as resistivity increases. A higher resistivity means it is more difficult for charges (current) to flow in the bulk material.
- Resistance decreases as cross-sectional area increases. Increased current results for the same amount of applied voltage.

Notice the assumptions that were made in the previous example:

1) The resistive material is uniform: linear, homogeneous, and isotropic.
2) J is uniform across A.
3) The voltage per unit length along the conductor is uniform based on 1).

Practically, these assumptions normally hold at DC and low frequencies. However, if these assumptions are not valid, then the fields approach must be utilized to determine resistance:

$$R = \frac{V}{I} = \frac{-\int_b^a \overline{E} \cdot \overline{d\ell}}{\int \sigma \overline{E} \cdot \overline{dA}} \tag{4.43}$$

Can an electromagnetic relationship for the dissipated power in a resistance be determined? From circuit theory, $P = V^2/R = GV^2$. An analogy is again utilized to qualitatively formulate a corresponding electromagnetic relationship:[10]

$$
\begin{array}{ccc}
P = & GV^2 \\
\downarrow & \downarrow\downarrow \\
P = & \int \sigma E^2 dV
\end{array}
\tag{4.44}
$$

where a volume integral is required because σE is a point-form relationship (σ and E are defined at each point and are functions of position when nonuniform). Thus, the electric field intensity within the resistive material causes current density, per $\overline{J} = \sigma\overline{E}$, which causes power dissipation due to the finite conductivity:

$$\boxed{P = \int \sigma E^2 dV} \tag{4.45}$$

So what is the power dissipated in the ideal resistor sketched in Figure 4.10? The conductivity and E are uniform, so they can be pulled from the integral:

$$P = \sigma E^2 \int dV \tag{4.46}$$

---

10 This relationship is also developed analytically in Appendix 4.C.

The volume of the cylinder of conductive material is $V = A\ell$. The electric field intensity within the resistor is $E = V/\ell$ because the resistive material is uniform. Substituting,

$$P = \sigma\left(\frac{V}{\ell}\right)^2 A\ell = V^2\left(\frac{\sigma A}{\ell}\right) \tag{4.47}$$

The quantity $\dfrac{\ell}{\sigma A}$ is the resistance, so the well-known result $P = \dfrac{V^2}{R}$ is indeed obtained!

4.4 Ampere's Circuital Law (ACL)

Motivational Questions Does a closed path integral such as $\oint \overline{E} \cdot \overline{d\ell} = 0$ for electric fields also apply to magnetic fields, that is, does $\oint \overline{H} \cdot \overline{d\ell} = 0$? What does $\oint \overline{H} \cdot \overline{d\ell}$ physically mean?

Ampere's circuital law (ACL) is a path integral involving the magnetic field intensity $\overline{H}$ similar to the integral in the voltage–electric field intensity relationship. However, it has a special property when the path is *closed*, and can be used, in a manner similar to Gauss's law, to solve for $\overline{H}$ from symmetrical current distributions. ACL will be intuitively developed and applied to some key current distributions in this section.

4.4.1 An Intuitive Development of ACL

Recall that the application of the Biot–Savart law to a uniform current on the z-axis in the $+\overline{a}_z$ direction resulted in the following "well-known" result:

$$\overline{H} = \frac{I}{2\pi\rho}\overline{a}_\phi \tag{4.48}$$

What if this magnetic field intensity were used in a closed path integral like that used for a closed path in an electric field? Would the result be zero?

$$\oint \overline{H} \cdot \overline{d\ell} = 0? \tag{4.49}$$

Let the path be a circle of radius ρ in the $\overline{a}_\phi$ direction: See Figure 4.11a

$$\oint \overline{H} \cdot \overline{d\ell} = \oint \frac{I}{2\pi\rho}\overline{a}_\phi \cdot \rho \, d\phi \overline{a}_\phi \tag{4.50}$$

$$\oint \overline{H} \cdot \overline{d\ell} = \frac{I}{2\pi}\oint d\phi = \frac{I}{2\pi}\int_0^{2\pi} d\phi = \frac{I}{2\pi}2\pi = I \tag{4.51}$$

The result is ACL: $\boxed{\oint \overline{H} \cdot \overline{d\ell} = I_{encl}}$ $\tag{4.52}$

What does the enclosed current I_{encl} mean? It is the current that "pierces" a surface that "stretches" over the closed path that is utilized in the ACL integral. See Figure 4.11b. This result was developed for a specific case, but it is a general result. The key aspect of this result is that the closed-path integral is *not zero* in general. Hence, *the magnetic field is not conservative.*

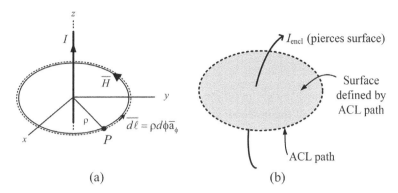

Figure 4.11 Path (dashed) for ACL and enclosed current. (a) ACL path (dashed). (b) Enclosed current concept.

In an analogy to Gauss's law, one can think of the closed path as an "observation path" because the magnetic field is being "observed" along the selected closed path. Be sure to make a clear distinction: the closed path is *not* the current path. It is the path along which $\overline{H}$ is evaluated, which will be described shortly. Note that ACL in Eq. (4.52) is general and holds for *any* path shape around *any* static current distribution. However, in further analogy to Gauss's law, a strategically chosen path will simplify the expression dramatically when applying ACL to symmetric current distributions.

An interesting viewpoint on the nonconservative nature of a magnetic field can be visualized as follows. Imagine a magnetic monopole (a fictitious particle with one magnetic pole type, either North or South) that is moved completely around a current carrying conductor (see Figure 4.12). The magnetic field of the wire would either push the pole all the way around, resulting in net negative energy, or it would be required to apply a force to push the pole all the way around against the magnetic field of the wire, resulting in a net positive energy. Either way, the net energy expended in moving the pole around the closed path is nonzero, and the magnetic field is not conservative.

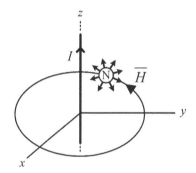

Figure 4.12 Movement of a magnetic monopole around a closed path.

4.4.2 Using ACL to Determine $\overline{H}$

Motivational Question   How can ACL be practically used in magnetic situations?

Ampere's circuital law can be analytically solved in closed form for the magnetic field intensity of highly symmetric current distributions. A path is chosen such that $\overline{H}$ is parallel to $\overline{d\ell}$ (it may be perpendicular along some parts of the path – those parts will be zero in the integral because of the dot product). Then $\overline{H} \cdot \overline{d\ell}$ reduces to $Hd\ell$. If the path for ACL is chosen such that the magnitude of $\overline{H}$, that is, H, is constant along the path, then H can be removed from the integral, and the integral over the path is just the path length ℓ. Thus,

$$\underbrace{\oint \overline{H} \cdot \overline{d\ell}}_{\substack{\text{general integral} \\ \text{in ACL}}} \;\rightarrow\; \underbrace{\oint H \, d\ell}_{\substack{\overline{H} \text{ is parallel} \\ \text{to the path}}} \;\rightarrow\; \underbrace{H \oint d\ell}_{\substack{H \text{ is constant} \\ \text{along the path}}} \;\rightarrow\; \underbrace{H\ell = I_{\text{encl}}}_{\substack{\text{ACL for highly} \\ \text{symmetric current} \\ \text{distributions}}} \qquad (4.53)$$

General strategy for the application of ACL:

- Choose a path that is parallel to $\overline{H}$ (some perpendicular portions are acceptable) and along which the magnitude of $\overline{H}$ is constant.
- The current that is enclosed by the path must be determined from the given type of current density or line current.

- Solve for H. Convert it into vector quantity $\overline{H}$ through multiplication by the appropriate unit vector.

Note the implementation of this general strategy in Examples 4.4.1, 4.4.2, and 4.4.3.

Example 4.4.1

a) Determine the magnetic field intensity for an infinitely long uniform line current on the z-axis using ACL.
b) Interpret the result as the radius of the observation point increases.

Solution

a) Create a sketch showing the current distribution, $\overline{H}$, a path, and an observation point P, as shown in Figure 4.11a.

Reasoning for choice of path: $\begin{cases} \overline{H} \| \ \overline{d\ell} \ (\overline{H} \text{ is parallel to the path}) \\ |\overline{H}| \text{ is constant along the path} \end{cases}$

ACL: $\oint \overline{H} \cdot \overline{d\ell} = I_{\text{encl}}$

Express $\overline{H}$ and $\overline{d\ell}$ in cylindrical coordinates

$I = \oint H_\phi \overline{a}_\phi \cdot \rho d\phi \overline{a}_\phi$

Note that the current enclosed by the path is I.

Simplify: $I = \oint H_\phi \rho d\phi$

H_ϕ and ρ are constant with respect to ϕ (due to the strategic path selection).

$I = H_\phi \rho \oint d\phi = H_\phi \rho \int_0^{2\pi} d\phi$

$I = H_\phi \rho 2\pi \rightarrow H_\phi = \dfrac{I}{2\pi\rho}$

$\boxed{\overline{H} = \dfrac{I}{2\pi\rho} \overline{a}_\phi}$ Key result: the magnetic field intensity of an infinitely long uniform line current

b) The magnitude of the magnetic field intensity is inversely proportional to the radius ρ to the observation point P. For any given observation point, the magnetic field is spread along the circumference at that radius. The circumference is directly proportional to ρ. Hence, the strength of the magnetic field intensity should be inversely proportional to ρ.

Example 4.4.2

a) Determine the magnetic field intensity at the midpoint of a long, tightly wound coil using ACL.
b) Interpret the result.

Solution

a) Sketch in Figure 4.13; note the strategic path selection: either $\overline{H} \| \ \overline{d\ell}$ or $\overline{H} \perp \overline{d\ell}$

When $\overline{H} \| \ \overline{d\ell}$, $|\overline{H}|$ is constant along the path.
When $\overline{H} \perp \overline{d\ell}$, $\overline{H} \cdot \overline{d\ell} = 0$.
ACL: $\oint \overline{H} \cdot \overline{d\ell} = I_{\text{encl}}$

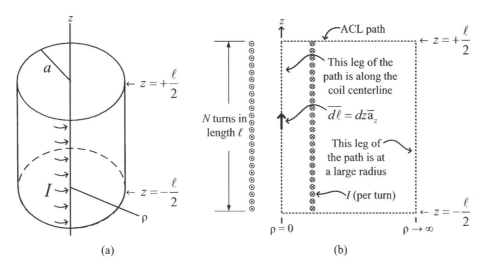

Figure 4.13 Long-coil ($\ell \gg a$) model for application of ACL. (a) Basic coil setup (turns not shown). (b) Cross section of the coil and the ACL path.

The total current enclosed is $I_{\text{encl}} = NI$, which is the current that "pierces" the surface defined by the closed path N times. Set up ACL around the selected closed path:

$$NI = \int_{-\ell/2}^{+\ell/2} H_z|_{\rho=0} \bar{a}_z \cdot dz\bar{a}_z + \int_{0}^{\infty} \underbrace{H_z\bar{a}_z \cdot d\rho\bar{a}_\rho}_{0} + \int_{+\ell/2}^{-\ell/2} H_z|_{\rho\to\infty} \bar{a}_z \cdot dz\bar{a}_z + \int_{\infty}^{0} \underbrace{H_z\bar{a}_z \cdot d\rho\bar{a}_\rho}_{0}$$

The second and fourth integrals are zero because $\overline{H} \perp \overline{d\ell}$. The third integral is zero because the magnetic field approaches zero as the distance from the coil approaches infinity. H_z is assumed to be constant along the axis inside the coil:

$$NI = H_z \int_{-\ell/2}^{+\ell/2} dz = H_z\ell \rightarrow H_z = \frac{NI}{\ell} \rightarrow \overline{H} = \frac{NI}{\ell}\bar{a}_z$$

b) The magnetic field intensity is uniform along the z-axis due to the uniformity of the current excitation. Practically, this assumption will not hold true at the ends of the coil but is valid for the middle region of the coil length (hence, the long-coil assumption). The strength of H is directly proportional to N and I because they both increase I_{encl} in a given length of the coil, whereas an increase in ℓ without increasing the number of turns effectively decreases the current enclosed per unit length.

Example 4.4.3

a) Determine the magnetic field intensity for a toroid (donut-shaped, highly magnetic[11] solid) using ACL. Assume the magnetic flux is confined to the core and uniformly distributed across the circular cross section.
b) Interpret the result.

---

11 A highly magnetic material is one with a large permeability, μ. These magnetic materials are a "path of least resistance" for magnetic flux, and thus magnetic flux is contained because it prefers to flow inside this material rather than air. This topic is addressed in Section 4.5.

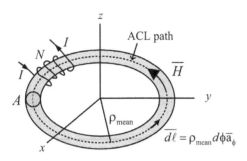

Figure 4.14 Toroid setup for ACL.

Solution

a) See the sketch in Figure 4.14

Select a path for which $\overline{H} \| \overline{d\ell}$ and $|\overline{H}|$ is constant.

$$\oint \overline{H} \cdot \overline{d\ell} = I_{\text{encl}}$$

The current enclosed by the selected path is NI. The integration path is along a circular path of radius ρ_{mean}, the average of the inner and outer radius of the toroid, since the magnetic field is assumed constant over the toroid's cross section. Further, this circular path is parallel everywhere to the expected magnetic field path, based on confinement of the magnetic flux in the magnetic toroid.

$$NI = \oint H_\phi \overline{a}_\phi \cdot \rho_{\text{mean}} d\phi \overline{a}_\phi$$

$$NI = H_\phi \rho_{\text{mean}} \oint d\phi = H_\phi \rho_{\text{mean}} 2\pi$$

$$H_\phi = \frac{NI}{2\pi\rho_{\text{mean}}} \rightarrow \overline{H} = \frac{NI}{2\pi\rho_{\text{mean}}} \overline{a}_\phi$$

b) With more turns, there is a larger H because the enclosed current is larger; a larger mean radius produces a smaller H because the H is spread out over a longer circumference (path) for the same I_{encl}.

4.5 Inductance

Motivational Questions Where did the expression for the inductance of a long coil, $L_{\text{coil}} = \mu N^2 A/\ell$, originate? How is the inductance of other current-carrying configurations (such as coax) determined?

Capacitance and resistance have been examined from a field's perspective. We now address the third fundamental passive circuit element, the inductor. While capacitor operation is based upon electric field principles, inductor operation is based upon magnetic field principles. We will identify many helpful analogies between the two devices and how we can approach them with a field's perspective. In this section, we will discuss the determination of inductance for basic structures, develop insight into the effects of permeability on inductance, and examine the storage of energy in the magnetic field.

4.5.1 Determination of Inductance

Recall that capacitance was fundamentally defined in terms of electric flux to be

$$C = \frac{\Psi}{V} \tag{4.54}$$

Inductance is similarly defined in terms of magnetic flux to be

$$\boxed{L = \frac{\Lambda}{I}} \tag{4.55}$$

where Λ (Greek uppercase lambda) is magnetic **flux linkage**, also called **linked magnetic flux**. What is flux linkage? It is the flux that "pierces" a surface that "stretches" over the closed path that is defined by the current, for example a typical wire turn in a coil. In Figure 4.15, the gray shading is the surface defined by one of the turns.

Three turns are shown for this N-turn coil. Six flux lines that represent the magnetic flux Φ are shown piercing this surface. These six flux lines are piercing (linking) all three coil turns. The total linked magnetic flux is the flux crossing the surface defined by a single turn multiplied by the total number of turns. It is equal to $N\Phi$ for an N-turn coil if all of the magnetic flux Φ links all N turns of the coil. Thus, the definition of inductance can be applied to N-turn coils:[12]

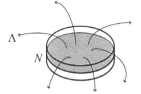

$$L = \frac{N\Phi}{I} \qquad (4.56)$$

Figure 4.15 Magnetic flux linking an N-turn coil.

We began this section by comparing the definitions of capacitance and inductance. The analogous role of flux in these definitions is now apparent:

$$C = \frac{\Psi}{V} = \frac{\text{electric flux}}{\text{voltage}} \leftrightarrow L = \frac{\Lambda}{I} = \frac{\text{linked magnetic flux}}{\text{current}} \qquad (4.57)$$

How is the determination of inductance approached for a given current-carrying conductor configuration? The *linked magnetic flux must be determined in terms of the current.*

Strategy

- ACL or the BSL is used to determine $\overline{H}$ and $\overline{B}$ in terms of the current.
- The linked magnetic flux Λ must be determined from $\overline{B}$ using $\int \overline{B} \cdot \overline{dA}$ over the surface defined by the current path (area A).

Example 4.5.1

Determine the expression for the inductance of an ideal N-turn coil of length ℓ with cross-sectional area A and core permeability μ.

Solution

From Example 4.4.2: $H_z \approx \dfrac{NI}{\ell}$; assume H_z is constant across and perpendicular to the cross section

$$\Lambda = N\Phi = N\int \overline{B} \cdot \overline{dA} \approx NBA = N\mu HA = N\mu\left(\frac{NI}{\ell}\right)A = \frac{\mu N^2 AI}{\ell}$$

$$L = \frac{\Lambda}{I} = \frac{N\Phi}{I} = \frac{\mu N^2 AI}{\ell I} \rightarrow L_{\text{coil}} = \frac{\mu N^2 A}{\ell} \qquad (4.58)$$

This result is well known: the inductance of an ideal long solenoid (coil).

Physical reasoning interpretations: Why is $L \propto N^2$? Why is $L \propto A$? Why is $L \propto 1/\ell$? (See Homework problem 25).

Example 4.5.2

Determine the expression for the inductance per unit length (H/m) of coax.

Solution

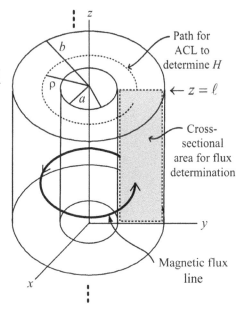

Figure 4.16 Coax setup for inductance determination.

See Figure 4.16 for setup; using a well-known ACL result:

$$\overline{H} = \frac{I}{2\pi\rho}\overline{a}_\phi \text{ for } a < \rho < b$$

---

12 This expression of inductance is applicable to linear cores only, where the magnetic flux linkages are proportional to the current. See Hayt, W.H. and Buck, J.A. (2006). *Engineering Electromagnetics*, 7e, 293. New York: McGraw-Hill.

$$\overline{B} = \mu\overline{H}$$

$$\overline{B} = \frac{\mu I}{2\pi\rho}\overline{a}_\phi$$

Linked flux: $\Lambda = \Phi = \int \overline{B} \cdot \overline{dA}$

The surface is selected such that the flux density is perpendicular to the surface (see Figure 4.16). *B* is *not* constant vs. radius, so the integral must be evaluated.

$$\Lambda = \int\int \frac{\mu I}{2\pi\rho}\overline{a}_\phi \cdot d\rho dz \overline{a}_\phi$$

$$\Lambda = \frac{\mu I}{2\pi}\int_a^b \frac{d\rho}{\rho}\int_0^\ell dz$$

$$\Lambda = \frac{\mu I}{2\pi}\ln\rho\Big|_a^b \ell = \frac{\mu I}{2\pi}\ln\left(\frac{b}{a}\right)\ell$$

$$L = \frac{\Lambda}{I} = \frac{\mu I}{2\pi I}\ln\left(\frac{b}{a}\right)\ell = \frac{\mu}{2\pi}\ln\left(\frac{b}{a}\right)\ell$$

The coax length is set to $\ell = 1$ m for "per unit length" (indicated by the prime), giving the well-known result of inductance per unit length of coax:

$$\boxed{L'_{\text{coax}} = \frac{\mu}{2\pi}\ln\left(\frac{b}{a}\right)\left[\frac{\text{H}}{\text{m}}\right]} \tag{4.59}$$

Looking ahead: The inductance per unit length of a transmission line, such as coax, will be utilized in the modeling of transmission lines in Chapter 7.

4.5.2 Magnetic Materials and Permeability

Motivational Question   Why does a magnetic material with larger permeability increase magnetic flux?

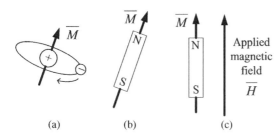

Figure 4.17 Magnetic material visualization ($\overline{M}$ is actually a macroscopic quantity, not microscopic). (a) Magnetic molecule. (b) Bar magnet viewpoint. (c) Bar magnet in alignment in an applied magnetic field.

Many if not most inductors utilize a magnetic material as the core in order to mechanically support the conductor in the windings, as well as to increase the inductance by increasing the magnetic flux. The permeability $\mu = \mu_o\mu_r$ of magnetic materials was utilized in the discussion of magnetic flux in Chapter 3. Here, we gain insight into permeability and how it affects magnetic fields. In a magnetic material, atomic and/or molecular effects contribute **magnetization** $\overline{M}$ (A/m) to the total magnetic field. Magnetization represents the effect of the molecules on the magnetic field. For example, visualize an electron orbiting an atomic nucleus. By the right-hand rule, a magnetic field is produced,[14] as sketched in Figure 4.17a. The molecule can be visualized as a small bar magnet (a "magnetic dipole"), as simplistically sketched in

---

14 Dominant atomic magnetic effects are due to electron spin and orbiting electrons.

Figure 4.17b. If this molecule is immersed in an applied magnetic field, the molecule will experience a torque toward aligning the two magnetic fields, as sketched in Figure 4.17c, due to repulsion of the north pole (outgoing magnetic flux) of the bar magnet by the applied magnetic field and attraction of the south pole. The magnetism of the molecule adds to the applied magnetic field. Thus, the magnetization of magnetic materials increases the magnetic flux relative to the no magnetic materials case.

The magnetization $\overline{M}$ in magnetic materials relates to the magnetic field intensity by

$$\overline{M} = \chi_m \overline{H} \tag{4.60}$$

where χ_m is a material property named the **magnetic susceptibility** (dimensionless). As the magnetic field intensity increases, in general the magnetic fields of the molecules better align with and contribute more to the applied magnetic field (up until saturation occurs, that is, all magnetic fields are aligned).[14] Magnetization is incorporated into the magnetic flux density as follows:

$$\overline{B} = \mu_o \overline{H} + \mu_o \overline{M} = \mu_o \overline{H} + \mu_o \chi_m \overline{H} = \mu_o (1 + \chi_m) \overline{H} \tag{4.61}$$

Note that $\overline{H}$ is the applied magnetic field intensity, whether the magnetic material is present or not. Thus, the magnetization $\overline{M}$ from a magnetic material increases the total magnetic flux density $\overline{B}$. The **relative permeability** μ_r is defined to be $(1 + \chi_m)$, resulting in the well-known B–H relationship:

$$\overline{B} = \mu \overline{H} = \mu_o \mu_r \overline{H} \tag{4.62}$$

Relative permeability is normally tabulated for materials in standard references. For magnetic materials, μ is not constant and specific magnetic material characterizations must be referenced, such as B–H curves. A broad classification of magnetic materials is given in Table 4.1.

4.5.3 Magnetic Field Boundary Conditions

Motivational Question The magnetic BC normal to a boundary, $B_{n1} = B_{n2}$, was examined in Chapter 3. What is the magnetic BC tangential to a boundary?

Recall from Chapter 3 that a change of materials with different permeability values had no effect on the magnetic flux density that was normal to the boundary because magnetic flux is solenoidal. This boundary condition is $B_{n1} = B_{n2}$. The tangential boundary condition for magnetic flux can now be developed using ACL. We will

Table 4.1 Classification of magnetic materials.

Classification	Basic properties
Diamagnetic $\mu_r \approx 1^-$ (slightly less than one)	• Magnetically neutral • Slight opposing effect to an applied external magnetic field
Paramagnetic $\mu_r \approx 1^+$ (slightly greater than one)	• Weak magnetic effects • Small but measurable effect when placed in an external magnetic field
Ferromagnetic $\mu_r \approx 100$–$200{,}000$	• Strong magnetic effects, form magnetic domains[a] • Generally conductive, exhibit hysteresis; iron, cobalt, nickel, ... • Used in power transformers, motors, generators, solenoids, ...
Ferrimagnetic $\mu_r \approx 10$–$20{,}000$	• Significant magnetic effects • Generally low conductivity, used in RF and microwave components

[a] Magnetic domains are essentially regions within magnetic materials where all of the atomic or molecular magnetizations align.

---

14 Magnetization becomes non-linear when saturation occurs.

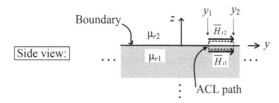

Figure 4.18 Application of ACL to tangential H components (surface current $K = 0$ at the interface).

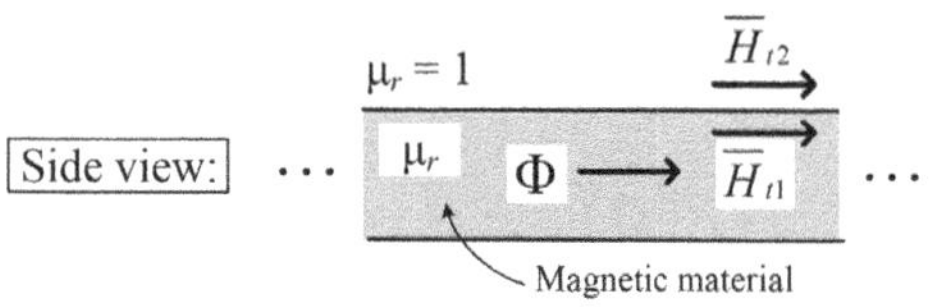

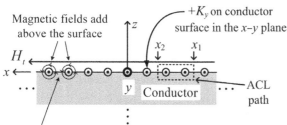

Figure 4.19 Visualization of magnetic flux containment by a magnetic material.

initially examine the special yet important case of zero surface current ($K = 0$), which is often used in magnetic field visualization. Consider the magnetic field intensities on both sides of a material boundary, as sketched in Figure 4.18. The tangential magnetic field intensities are parallel to only two sides of the ACL path. Apply ACL in a clockwise direction around the closed path:

$$\oint \overline{H} \cdot \overline{d\ell} = \int_{y_2}^{y_1} H_{t1}\overline{a}_y \cdot dy\overline{a}_y + \int_{y_1}^{y_2} H_{t2}\overline{a}_y \cdot dy\overline{a}_y = I_{\text{encl}} = 0$$

(4.63)

Assuming a short path length so that H is constant along the path,

$$H_{t1}\int_{y_2}^{y_1} dy + H_{t2}\int_{y_1}^{y_2} dy = 0 \rightarrow H_{t1}(y_1 - y_2) + H_{t2}(y_2 - y_1) = 0$$

(4.64)

$$\boxed{H_{t1} = H_{t2}} \quad (\text{tangential } H \text{ BC}, K = 0) \tag{4.65}$$

The tangential H boundary condition is also a useful tool in visualizing magnetic flux Φ. Consider a bar of magnetic material with Φ running parallel to the axis of the material, as sketched in Figure 4.19. From the tangential magnetic field boundary condition, $H_{t1} = H_{t2}$. The magnetic material permeability $\mu_o\mu_r \gg \mu_o$. Consequently, $B_{t1} \gg B_{t2}$ and most of the magnetic flux is contained within the magnetic material. This result is a primary reason that magnetic materials are used in electronic components that operate magnetically, such as transformers, relays, motors, and generators, and is also why a coil with a magnetic core can often be modeled as an ideal inductor ($L_{\text{coil}} = \mu N^2 A/\ell$) because the flux of each turn is primarily within the magnetic material of the core and consequently links all N turns.

There is another tangential magnetic field boundary condition, and it is useful in visualizing the magnetic field next to the surface of a current-carrying conductor. Assume a surface current density K (A/m) is present at the interface of a perfect conductor and a dielectric or air, as sketched in Figure 4.20, and that the magnetic field does not penetrate into the conductor.[15] Thus, when ACL is evaluated for Figure 4.20 relative to Figure 4.18, only the tangential magnetic field outside of the conductor (H_t) is nonzero around the closed path. However, I_{encl} is no longer zero. Assume K is in the y direction:

$$\oint \overline{H} \cdot \overline{d\ell} = \int_{x_1}^{x_2} H_t\overline{a}_x \cdot dx\overline{a}_x = I_{\text{encl}} = \int_{x_1}^{x_2} K_y dx$$

(4.66)

The first integral reduces to $H_t(x_2 - x_1)$. The surface current density is constant, so the second integral reduces to $K_y(x_2 - x_1)$. Equation (4.66) reduces to

$$H_t(x_2 - x_1) = K_y(x_2 - x_1) \rightarrow H_t = K_y \tag{4.67}$$

Figure 4.20 Tangential magnetic field next to the surface of a current-carrying conductor (note: the coordinate directions differ from those in Figure 4.17).

---

15 Practically, the result of this development, Eq. (4.68), is also applicable to finite conductivity metals at high frequency, where the current is mostly confined to the surface of the metal (a phenomenon called "skin effect"). For DC and lower frequency AC, a volume current density approach must be utilized to determine H at the surface of the conductor.

Note that the direction of K, $\overline{a}_y$, is perpendicular to the direction of H, $\overline{a}_x$. The following vector equation accounts for the directions, as well as the magnitudes:

$$\boxed{\overline{H} = \overline{K} \times \overline{a}_n} \tag{4.68}$$

where $\overline{a}_n$ is the unit vector pointing outward from the conducting surface. This is the tangential magnetic field boundary condition adjacent to a "perfect" conductor–dielectric interface. In Figure 4.20, $\overline{H} = \overline{K} \times \overline{a}_n = K_y\overline{a}_y \times \overline{a}_z = H_t\overline{a}_x$ and $H_t = K_y$, as previously developed.

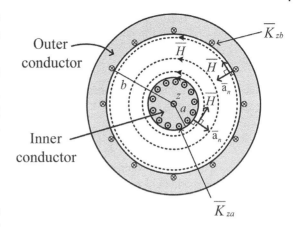

Figure 4.21 End-view sketch of coax.

Example 4.5.3

Show that the direction of the magnetic field pattern inside coax satisfies the conductor-dielectric tangential magnetic boundary condition.

Answer

See the sketch in Figure 4.21. At the surface of the inner conductor, $\overline{K} = K_{za}\overline{a}_z$ and $\overline{a}_n = \overline{a}_\rho$.

$\overline{H} = \overline{K} \times \overline{a}_n = K_{za}\overline{a}_z \times \overline{a}_\rho = K_{za}\overline{a}_\phi$, so $\overline{H} = H_\phi\overline{a}_\phi$, as expected. Also, $H_\phi|_{\rho=a} = K_{za}$.

At the surface of the outer conductor, $\overline{K} = K_{zb}(-\overline{a}_z)$ and $\overline{a}_n = (-\overline{a}_\rho)$.

$\overline{H} = \overline{K} \times \overline{a}_n = K_{zb}(-\overline{a}_z) \times (-\overline{a}_\rho) = K_{zb}\overline{a}_\phi$, so $\overline{H} = H_\phi\overline{a}_\phi$, as expected. Also, $H_\phi|_{\rho=b} = K_{zb}$.

Thus, the tangential magnetic boundary conditions at both the inner and outer conductors of coax are satisfied.

4.5.4   Energy Storage in a Magnetic Field

Motivational Question   Inductors store energy in magnetic fields. How is the amount of energy stored in an H field determined?

Recall that a capacitor stores energy in an electric field. Similarly, an inductor stores energy in a magnetic field. We will use the inductor to illustrate the energy stored in a magnetic field in this section. Let $w(t)$ be the energy stored in an inductor as a function of time. From circuit theory:

$$w(t) = \int_0^t p(\tau)d\tau = \int_0^t v(\tau)i(\tau)d\tau = \int_0^t L\frac{di(\tau)}{d\tau}i(\tau)d\tau = L \int_{i(0)}^{i(t)} i(\tau)di(\tau) = \frac{1}{2}Li^2\Big|_{i(0)}^{i(t)} \tag{4.69}$$

where τ is again being used as a dummy variable for time. Complete the integration:

$$w(t) = \frac{1}{2}Li^2(t) - \frac{1}{2}Li^2(0) = \frac{1}{2}Li^2(t) - w(0) = \frac{1}{2}Li^2(t) \tag{4.70}$$

where $w(0) = 0$, the initial energy stored in the inductor is zero. The equation $\frac{1}{2}Li^2(t)$ expresses the energy stored (accumulated) from a zero energy state. Once the current is nonzero, energy is accumulating. For DC steady state, $w(t)$ is constant, W, and

$$w(t)|_{DC} = W = \frac{1}{2}LI_{DC}^2 \tag{4.71}$$

The energy stored in the magnetic field of an inductor can be generalized (this development is not a proof, just an illustration of a concept). Substitute the expression for the inductance of an ideal coil-type inductor:

$$W = \frac{1}{2}LI^2 = \frac{1}{2}\left(\frac{\mu N^2 A}{\ell}\right)I^2 = \frac{1}{2}\mu\left(\frac{NI}{\ell}\right)^2 A\ell \tag{4.72}$$

Recognize that (NI/ℓ) is the magnitude of the uniform magnetic field intensity H in the ideal long coil (per Example 4.4.2) and that $A\ell$ is the interior volume of the coil:

$$W = \frac{1}{2}\mu H^2 (\text{volume}) \tag{4.73}$$

Thus, $\frac{1}{2}\mu H^2$ is a magnetic field **energy density** (J/m^3). Under the assumption that the development for the ideal coil inductor is valid in general,

$$W = \frac{1}{2}\mu \int H^2 dV = \frac{1}{2}\mu_o\mu_r \int H^2 dV \tag{4.74}$$

where dV is a differential volume in the region of the magnetic field.[16] The integration is necessary in general because the magnetic field may be nonuniform. Thus, a magnetic field stores energy and is the means through which an inductor stores energy. Note that a magnetic material ($\mu_r > 1$) increases the energy stored.

4.6 Summary of Important Equations

$$V_{BA} = \frac{W}{Q} = -\int_A^B \overline{E} \cdot \overline{d\ell} = V_B - V_A \qquad \oint \overline{E} \cdot \overline{d\ell} = 0 \quad (\text{static } \overline{E}) \qquad \overline{D} = \varepsilon\overline{E} = \varepsilon_o\varepsilon_r\overline{E}$$

Perfect conductor BCs : $\qquad E_t = 0 \quad D_n = \rho_S \qquad\qquad \varepsilon_o = 8.8542 \times 10^{-12} \text{ (F/m)}$

$$C = \frac{Q}{V} \qquad C_{\text{parallel plate}} = \frac{\varepsilon A}{d} \qquad C'_{\text{coax}} = \frac{2\pi\varepsilon}{\ln\left(\frac{b}{a}\right)} \text{ (F/m)} \qquad W = \frac{1}{2}\varepsilon \int E^2 dV$$

$$\overline{J} = \sigma\overline{E} \qquad R = \frac{\rho\ell}{A} \qquad P = \int \sigma E^2 dV \qquad \oint \overline{H} \cdot \overline{d\ell} = I_{\text{encl}} \qquad \overline{H} = \frac{I}{2\pi\rho}\overline{a}_\phi$$

$$L = \frac{\Lambda}{I} = \frac{N\Phi}{I} \qquad L_{\text{coil}} = \frac{\mu N^2 A}{\ell} \qquad L'_{\text{coax}} = \frac{\mu}{2\pi}\ln\left(\frac{b}{a}\right) \text{ (H/m)}$$

$$\overline{B} = \mu\overline{H} = \mu_o\mu_r\overline{H} \qquad \mu_o = 4\pi \times 10^{-7} \text{ (H/m)} \qquad W = \frac{1}{2}\mu \int H^2 dV$$

Magnetic BCs : $\qquad B_{n1} = B_{n2} \quad H_{t1} = H_{t2}(K = 0) \qquad \overline{H} = \overline{K} \times \overline{a}_n(\text{perfect conductor})$

4.7 Appendices

Appendix 4.A Dielectric–Dielectric Electric Field Boundary Conditions

The general dielectric–dielectric electric field boundary conditions can be developed using KVL and Gauss's law. These boundary conditions can be used to obtain special cases such as the boundary conditions next to a perfect conductor, which are discussed in Sections 3.2 and 4.1.3. These derivations also serve as additional examples of KVL and Gauss's law applications.

The tangential electric field boundary condition results from the application of KVL around a closed path where two sides of the path are adjacent to each side of a dielectric–dielectric boundary that lies in the $z = 0$ plane, as shown in Figure 4.22. Assume there are tangential electric field components above and below the boundary: $\overline{E}_{t2} = E_{t2}\overline{a}_y$ and $\overline{E}_{t1} = E_{t1}\overline{a}_y$, respectively. Assume that the KVL path dimensions are very small such that the

---

16 It has been assumed that the permeability is constant in this simplified, intuitive development of the energy stored in a magnetic field.

electric field magnitudes are constant over the path.[17] Set up the KVL integration by breaking up the closed KVL path integral into open path integrals along the top, sides, and bottom of the KVL path:

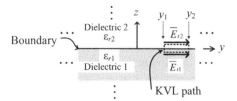

Figure 4.22 Tangential electric fields at a dielectric–dielectric boundary.

$$\oint \overline{E} \cdot \overline{d\ell} = \int_{Top} \overline{E} \cdot \overline{d\ell} + \int_{sides} \overline{E} \cdot \overline{d\ell} + \int_{bottom} \overline{E} \cdot \overline{d\ell} = 0 \qquad (4.75)$$

Note that the path integrals along the sides of the KVL path are zero, since the electric field in both regions is perpendicular to the path (the sides are parallel to the z-axis). The integrals over the top and bottom of the KVL path can then be evaluated. Since the tangential electric fields are constant and parallel to the top and bottom sides of the KVL path, Eq. (4.75) becomes:

$$\int_{Top} \overline{E} \cdot \overline{d\ell} + \int_{bottom} \overline{E} \cdot \overline{d\ell} = \int_{y_1}^{y_2} E_{t2}\overline{a}_y \cdot dy\overline{a}_y + \int_{y_2}^{y_1} E_{t1}\overline{a}_y \cdot dy\overline{a}_y = E_{t2}(y_2 - y_1) + E_{t1}(y_1 - y_2) = 0 \qquad (4.76)$$

which is then manipulated to obtain:

$$E_{t1}(y_2 - y_1) = E_{t2}(y_2 - y_1) \qquad (4.77)$$

The path lengths $(y_2 - y_1)$ cancel and the final boundary condition result is:

$$\boxed{E_{t1} = E_{t2}} \qquad (4.78)$$

Equation (4.78) states that the *tangential* electric fields are continuous across *any* material boundary, that is, the electric field is identical on both sides the boundary regardless of what materials are present! The tangential electric field boundary condition next to a perfect conductor is developed in Section 4.1.3, and it is expressed in Eq. (4.8) as $E_t = 0$. Does this result agree with Eq. (4.78)? Recall that inside a perfect conductor, the electric field is zero. If the region below the boundary is assumed to be filled with a perfect conductor rather than air, then $E_{t1} = 0$. Therefore, according to Eq. (4.78), the tangential component above the interface is identically zero, $E_{t2} = 0$, matching the boundary condition of Eq. (4.8).

The boundary condition for normal electric flux density at the general dielectric–dielectric boundary can be developed using Gauss's law. As shown in Figure 4.23, the boundary lies in the $z = 0$ plane. Assuming there are normally directed electric flux densities in the region above and below the boundary, then $\overline{D}_{n2} = D_{n2}\overline{a}_z$ and

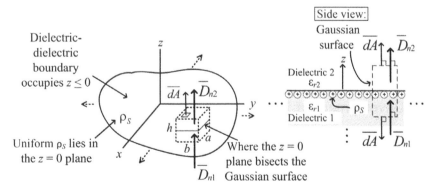

Figure 4.23 Normal electric flux at a dielectric–dielectric boundary.

---

17 To generalize to the time-varying electric fields that will be introduced in Chapter 5, assume the path length is small compared with the operating wavelength, that is, $(y_2 - y_1) \ll \lambda$.

$\overline{D}_{n1} = D_{n1}\overline{a}_z$, respectively. For generality, assume a uniform surface charge density of ρ_S exists at the boundary. To ensure the flux density is everywhere perpendicular or parallel to the surface, a cube is chosen for the GS that extends equally above and below the boundary located in the $z = 0$ plane. The cube is assumed to be very small,[18] such that the electric flux density is constant over the top and bottom surfaces of the cube. The net electric flux integral can then be determined by splitting the closed surface integral into open surface integrals over the top, bottom, and sides of the cube:

$$\Psi_{\text{net}} = \oint \overline{D} \cdot \overline{dA} = \int \overline{D} \cdot \overline{dA}\bigg|_{\text{Top}} + \int \overline{D} \cdot \overline{dA}\bigg|_{\text{Bottom}} + \int \overline{D} \cdot \overline{dA}\bigg|_{\text{sides}} \tag{4.79}$$

The flux density is purely in the z-direction (parallel to the cube sides) and therefore does not pass through the sides of the cube. As a result, the flux integral over the sides is zero, leaving only the top and bottom integrals. Inserting the expressions for the flux densities and differential area vectors results in:

$$\Psi_{\text{net}} = \int \overline{D}_{n2} \cdot \overline{dA}\bigg|_{\text{Top}} + \int \overline{D}_{n1} \cdot \overline{dA}\bigg|_{\text{Bottom}} = \int D_{n2}\overline{a}_z \cdot dA\overline{a}_z + \int D_{n1}\overline{a}_z \cdot dA(-\overline{a}_z) \tag{4.80}$$

Note that the differential surface vector on the bottom of the cube points outward in the $-\overline{a}_z$ direction. The dot products can then be evaluated and since the flux density is constant over the top and bottom surfaces (due to the small cube size), the electric flux densities can be pulled out of the integrals:

$$\Psi_{\text{net}} = D_{n2}\int dA - D_{n1}\int dA = D_{n2}ab - D_{n1}ab = (D_{n2} - D_{n1})ab \tag{4.81}$$

The enclosed charge can then be found by integrating the boundary surface charge contained inside the cube in the $z = 0$ plane:

$$Q_{\text{encl}} = \int \rho_S dA = \rho_S \int dA = \rho_S ab \tag{4.82}$$

Gauss's law states that the net flux of Eq. (4.81) equals the enclosed charge of Eq. (4.82):

$$(D_{n2} - D_{n1})ab = \rho_S ab \tag{4.83}$$

Since the ab term cancels on both sides, the resulting normal boundary condition relationship is given by:

$$\boxed{D_{n2} - D_{n1} = \rho_S} \tag{4.84}$$

Note that if there is no charge density at the boundary, $D_{n2} = D_{n1}$, as expected by Gauss's law: there is zero net flux from a closed surface that encloses the boundary if the enclosed charge is zero. Additionally, the normal electric field boundary condition next to a perfect conductor was developed in Section 3.2 to be $D_n = \rho_S$. Does this result agree with Eq. (4.84)? Assume the space below the boundary ($z < 0$) is filled with a perfect conductor. Recalling that the electric field inside a perfect conductor is equal to zero, $D_{n1} = 0$. As a result, Eq. (4.84) becomes $D_{n2} - 0 = \rho_S \rightarrow D_{n2} = \rho_S$, in agreement with the boundary condition for the normally directed electric flux density next to a perfect conductor.

Appendix 4.B Development of Relative Permittivity

This development starts with Figure 4.8. The electric field intensity due to the bound charge $\overline{E}_{\text{bound }Q}$ is parallel to but *opposite* in direction relative to the electric field intensity due to the free charge $\overline{E}_{\text{free }Q}$. Thus, the total electric field intensity within the dielectric $\overline{E}_{\text{diel}}$ decreases:

$$E_{\text{diel}} = E_{\text{free }Q} - E_{\text{bound }Q} \tag{4.85}$$

---

18 To generalize to the time-varying electric flux that will be introduced in Chapter 5, assume the cube is small compared to the operating wavelength, that is, $a \ll \lambda$ and $b \ll \lambda$.

Only magnitudes are used here because all of the vector components are parallel (or antiparallel). However, $\overline{D} = \varepsilon_o \overline{E}_{\text{free } Q}$, that is, the flux density depends only on the free charge on the plates. Note that the bound charges contribute zero net flux because there are equal amounts of positive and negative charge. Consequently, on a macroscopic scale, D must be the same with or without the dielectric. Substituting from Eq. (4.85):

$$D = \varepsilon_o E_{\text{free } Q} = \varepsilon_o (E_{\text{diel}} + E_{\text{bound } Q}) \tag{4.86}$$

Each of the dipoles in the dielectric is exposed to $\overline{E}_{\text{diel}}$, that is, each molecule "feels" the E due to the free charge on the plates and the E due to the bound charges. This electric field intensity polarizes the charges in each molecule to form a dipole. **Polarization** $\overline{P}$ is a measure of how well a dielectric material can be polarized. *Polarization is defined to be in the direction that corresponds to the negative-to-positive* direction within a dipole. See Figure 4.8e. It is proportional to and parallel with the electric field intensity in the dielectric

$$\overline{P} = \chi_e \varepsilon_o \overline{E}_{\text{diel}} \tag{4.87}$$

where a material-dependent parameter, the **electric susceptibility** χ_e, is defined as a factor in the proportionality constant. Polarization is a flux density-type quantity (Eq. (4.87) has the same form as $\overline{D} = \varepsilon \overline{E}$) that is parallel to the applied electric field intensity by the definition of its direction, but its effect is to *oppose* the applied electric field intensity ($\overline{E}_{\text{free } Q}$ here) because polarization of dipoles creates the oppositely directed $\overline{E}_{\text{bound } Q}$. The stronger the polarization of the dipoles, the stronger the $\overline{E}_{\text{bound } Q}$ is. Hence, the polarization corresponds to the second term in Eq. (4.86), again using magnitudes:

$$P = \varepsilon_o E_{\text{bound } Q} \tag{4.88}$$

Then by comparison of Eqs. (4.87) and (4.88),

$$E_{\text{bound } Q} = \chi_e E_{\text{diel}} \tag{4.89}$$

Substitute $E_{\text{bound } Q} = \chi_e E_{\text{diel}}$ into $D = \varepsilon_o (E_{\text{diel}} + E_{\text{bound } Q})$ and simplify:

$$D = \varepsilon_o E_{\text{diel}} + \varepsilon_o \chi_e E_{\text{diel}} \tag{4.90}$$

$$D = \varepsilon_o (1 + \chi_e) E_{\text{diel}} \tag{4.91}$$

$$D = \varepsilon_o \varepsilon_r E_{\text{diel}} \rightarrow \overline{D} = \varepsilon_o \varepsilon_r \overline{E}_{\text{diel}} \tag{4.92}$$

where the **relative dielectric constant** (also called relative permittivity) ε_r is defined to be $(1 + \chi_e)$. This result is the background of the well-known $\varepsilon = \varepsilon_o \varepsilon_r$ used in capacitance calculations. Relative dielectric constant (relative permittivity) values are tabulated for numerous materials in standard references. It is understood that within a dielectric, $\overline{E} = \overline{E}_{\text{diel}}$, resulting in the well-known relationship:

$$\boxed{\overline{D} = \varepsilon \overline{E} = \varepsilon_o \varepsilon_r \overline{E}} \tag{4.93}$$

Appendix 4.C Development of Resistance

A more analytic development of resistance follows. Visualize charges flowing in a conductor with a finite resistance as sketched in Figure 4.24. The charge per unit volume, charge density ρ_V, that is flowing is related to current by

$$dI = \frac{dQ}{dt} = \frac{d(\rho_V V)}{dt} = \rho_V \frac{dV}{dt} = \rho_V \frac{dAd\ell}{dt} = \rho_V \left(\frac{d\ell}{dt}\right) dA = \rho_V v_d dA \tag{4.94}$$

where dA is a differential area through which the charges flow perpendicularly, dV is a differential volume, and v_d is the **drift velocity**, which is the average velocity of charge flow in a conducting material. It is more generally a

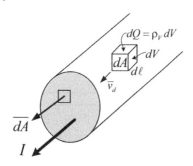

Figure 4.24 Sketch of volume current density flow.

vector quantity: $\bar{v} = v\bar{a}_I$, where $\bar{a}_I$ is the direction of the current flow. A dot product can be used to determine the amount of charge density that crosses a designated surface perpendicularly:

$$dI = \rho_V \bar{v}_d \cdot \overline{dA} \tag{4.95}$$

Integrate Eq.(4.95) for the total current: $\quad I = \int \rho_V \bar{v}_d \cdot \overline{dA}$

$$\tag{4.96}$$

and compare it with volume current density: $I = \int \bar{J} \cdot \overline{dA} \tag{4.97}$

So the relationship between volume current density and charge density is

$$\bar{J} = \rho_V \bar{v}_d \tag{4.98}$$

How can volume current density be related to the electric field? From physics, drift velocity is proportional to electric field intensity and the proportionality constant is **mobility** μ, a material-dependent parameter (unfortunately the same symbol that is used for permeability in magnetics).

$$\bar{v}_d = \mu\overline{E} \tag{4.99}$$

Mobility is a material-dependent parameter that quantifies how easily charges can move through a given material. Mobility is tabulated for numerous materials in standard references. Incorporate the mobility into Eq. (4.98):

$$\bar{J} = \rho_V \mu\overline{E} \tag{4.100}$$

The free charge per unit volume ρ_V is also a material-dependent parameter. Thus, the $\rho_V\mu$ product is a material-dependent parameter and forms the well-known **conductivity** σ (units: S/m, which is also tabulated for numerous materials in standard references:

$$\sigma = \rho_V \mu \tag{4.101}$$

Substitute conductivity into Eq. (4.100):

$$\boxed{\bar{J} = \sigma\overline{E}} \tag{4.102}$$

This relation relates charge flow to the force that causes the charge flow as a function of position. For this reason, it is called the **point form of Ohm's law**, or equivalently, the **microscopic form of Ohm's law**.

The dissipated power in a resistance is now developed. Wentworth[19] employs an insightful approach, which is basically followed here. Let an electric field exist in a conducting material with finite conductivity, as sketched in Figure 4.25. This field will impose a force on the free charge carriers in the material $(\overline{F} = Q\overline{E})$. On a differential charge basis:

$$d\overline{F} = dQ\overline{E} = \rho_V dV\overline{E} \tag{4.103}$$

where ρ_V is the volume charge density in the conductive material and dV is a differential volume. The differential amount of work to move a charge in a differential length $\overline{d\ell}$ is

$$dW = d\overline{F} \cdot \overline{d\ell} = \rho_V dV\overline{E} \cdot \overline{d\ell} \tag{4.104}$$

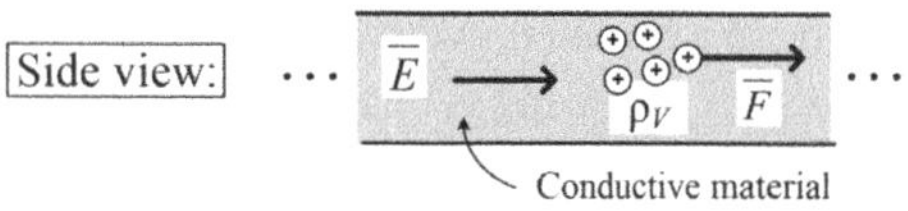

Figure 4.25 Electric force on charges in a conductive material (conventional current).

Power is the time derivative of work. Again, on a differential basis:

$$dP = \frac{dW}{dt} = \rho_V dV\overline{E} \cdot \frac{\overline{d\ell}}{dt} = \rho_V dV\overline{E} \cdot \bar{v}_d \tag{4.105}$$

---

19 Wentworth, S.M. (2005). *EFundamentals of Electromagnetics with Engineering Applications*, 72. New York: McGraw-Hill.

where $\bar{v}_d$ is the average velocity of the charges in the material, named the *drift velocity*. The relationship between volume current density and charge density was previously developed:

$$\bar{J} = \rho_V \bar{v}_d \tag{4.106}$$

$$dP = \bar{E} \cdot \bar{J} dV \tag{4.107}$$

Finally, integrate to obtain power:

$$P = \int \bar{E} \cdot \bar{J} dV \tag{4.108}$$

Normally, $\bar{J}$ and $\bar{E}$ are parallel, and $\bar{J} = \sigma \bar{E}$, so $\bar{E} \cdot \bar{J} = EJ = \sigma E^2$. The power dissipated in a material with finite conductivity is

$$\boxed{P = \int \sigma E^2 dV} \tag{4.109}$$

Appendix 4.D Introduction to Magnetic Circuits

Motivational Question   How is a structure consisting of magnetic material, which is used to guide magnetic flux, modeled for practical, first-order predictions?

Magnetic circuits are magnetic material structures that guide magnetic flux for some intended purpose. For example, the core of an inductor or the iron-core in a power transformer guides magnetic flux through coils. As discussed in magnetic boundary conditions, the use of a magnetic material generally confines the flux to that material if the path of the material is closed (complete). We will analyze a series magnetic circuit consisting of two segments with a different permeability. The objective is to determine a magnetic flux–current relationship for this magnetic circuit. Afterward, the result will be generalized to an N-segment magnetic circuit.

Consider a complete magnetic path with segment lengths of ℓ_1 and ℓ_2, as sketched in Figure 4.26. Each segment may have a different material, a different relative permeability μ_r, and a different magnetic field intensity H. The path encloses a current $I_{\text{encl}} = NI$. Make the following assumptions:

- All magnetic flux Φ is confined to the magnetic core.
- The magnetic flux density B is uniform across the cross-sectional area A of the core.
- Application of ACL along the mean core path (a path "down the middle" of the core) gives valid results.
- The magnetic field intensity H is uniform along the length of each segment, but may differ segment to segment and is parallel in direction to the mean core path.

Apply ACL $\oint \bar{H} \cdot \overline{d\ell} = I_{\text{encl}}$ around the closed magnetic path, where each path segment is assumed to be along the mean core path:

$$\oint \bar{H} \cdot \overline{d\ell} = \int \bar{H_1} \cdot \overline{d\ell} + \int \bar{H_2} \cdot \overline{d\ell} = NI \tag{4.110}$$

Because the magnetic field intensity is constant along each segment and parallel to the path, the integral for segment 1 simplifies to

$$\int \bar{H_1} \cdot \overline{d\ell} = \int H_1 d\ell = H_1 \int d\ell = H_1 \ell_1 \tag{4.111}$$

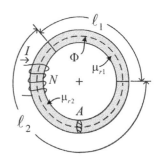

Figure 4.26 Two-segment series magnetic circuit and quantities (Φ is parallel to the dashed mean path).

and similarly for segment 2. Hence, Eq. (4.110) simplifies to

$$H_1\ell_1 + H_2\ell_2 = NI \tag{4.112}$$

The **magnetomotive force** (mmf) is defined for each term in Eq. (4.112). The symbol for mmf is a script F: $\mathfrak{F}$. The units of mmf are amperes (some say ampere-turns due to the N in NI). It represents the source of the magnetic field in a magnetic circuit, NI. It also represents the portion of the magnetic field-length ($H\ell$ from ACL) across a part of the magnetic circuit. An analogy to electric circuits will be shown soon. The sum of the mmfs is the total mmf $\mathfrak{F}_T$ that must equal NI by Eq. (4.112):

$$\mathfrak{F}_1 + \mathfrak{F}_2 = NI = \mathfrak{F}_T \tag{4.113}$$

Given a series path of magnetic material segments, such as those shown in Figure 4.26, the *magnetic flux* Φ *must be the same in each segment*. Why? Magnetic flux is solenoidal, and if it is entirely contained in the core, as assumed, then all of the flux that leaves one segment must enter the next segment. One is reminded of the circuit analogy: the current is constant at any point in a series electric circuit. The magnetic flux is constant at any cross section in a series magnetic circuit. Under the previously stated assumption that the flux is distributed uniformly across the cross-sectional area A of each segment, then the magnetic flux–flux density relationship simplifies to

$$\Phi = \int \overline{B} \cdot \overline{dA} \approx BA \tag{4.114}$$

The magnetic flux density can be expressed in terms of the magnetic field intensity:

$$\Phi \approx BA = \mu_0\mu_r HA \rightarrow H = \frac{\Phi}{\mu_0\mu_r A} \tag{4.115}$$

Insert the previous expression for each H into Eq. (4.112):

$$H_1\ell_1 + H_2\ell_2 = \frac{\Phi}{\mu_0\mu_{r1}A_1}\ell_1 + \frac{\Phi}{\mu_0\mu_{r2}A_2}\ell_2 = NI \tag{4.116}$$

The magnetic flux is the same in each segment of this series magnetic circuit. Factor it out:

$$\Phi\left[\frac{\ell_1}{\mu_0\mu_{r1}A_1} + \frac{\ell_2}{\mu_0\mu_{r2}A_2}\right] = NI = \mathfrak{F}_T \tag{4.117}$$

Note that the mathematical form for each segment $\dfrac{\ell}{\mu_0\mu_r A}$ is similar to the resistance of a conductor $\dfrac{\ell}{\sigma A}$. It is given the name **reluctance** $\mathfrak{R}$ with SI unit of $H^{-1} = \dfrac{A}{Wb}$:

$$\Phi[\mathfrak{R}_1 + \mathfrak{R}_2] = NI = \mathfrak{F}_T \tag{4.118}$$

Hence, the total reluctance of a series magnetic circuit is the sum of the individual series reluctances. The analogies between a series electric circuit, Figure 4.27a, and a series magnetic circuit, Figure 4.27b, are given in Table 4.2, where the magnetic circuit is generalized to P segments. The appropriateness of the term "magnetic circuit" should now be apparent.

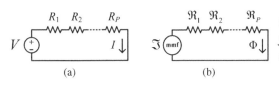

(a)   (b)

Figure 4.27 Analogy between series electric and magnetic circuits. (a) Series electric circuit. (b) Series magnetic circuit.

What is the whole point of these magnetic circuit relations? Given a desired level of magnetic flux, one can *solve for the current in the coil that is required to establish that magnetic flux*:

$$NI = \Phi(\mathfrak{R}_1 + \mathfrak{R}_2 + \cdots + \mathfrak{R}_P)$$
$$\rightarrow I = \frac{\Phi}{N}(\mathfrak{R}_1 + \mathfrak{R}_2 + \cdots + \mathfrak{R}_P) \tag{4.119}$$

Table 4.2 Analogies between series electric and magnetic circuits.

Electric circuit variable	Magnetic circuit variable
I	Φ
V	$\mathfrak{F}$
$R = \dfrac{\ell}{\sigma A}$	$\mathfrak{R} = \dfrac{\ell}{\mu_0 \mu_r A}$
$V = I(R_1 + R_2 + \cdots + R_P)$	$\mathfrak{F} = \Phi(\mathfrak{R}_1 + \mathfrak{R}_2 + \cdots + \mathfrak{R}_P)$

Example 4.7D.1 A ferrite toroid has a mean radius of 2 cm, a cross-sectional area of 0.8 cm², and a relative permeability of 150. A 2000 turn coil wraps a section of the toroid. Assuming the core is not magnetically saturated,[20] what current is required to establish a magnetic flux level of 0.1 mWb in the core?

$$\mathfrak{F} = NI = \Phi\mathfrak{R} = \frac{\Phi\ell}{\mu_0 \mu_r A} \rightarrow I = \frac{\Phi\ell}{N\mu_0 \mu_r A}$$

$$I = \frac{\Phi\ell}{N\mu_0 \mu_r A} = \frac{(0.1 \times 10^{-3}\,\text{Wb})(2\pi 0.02\,\text{m})}{(2000)\left(4\pi \times 10^{-7}\dfrac{\text{H}}{\text{m}}\right)(150)(0.8\,\text{cm}^2)\left(\dfrac{1\,\text{m}}{100\,\text{cm}}\right)^2} = 417\,\text{mA}$$

The previous example has illustrated the usefulness of the magnetic circuit results that were developed from ACL and fundamental magnetic field relations. Magnetic circuits are often analyzed in courses such as Electromechanical Energy Conversion (power, motors, generators, and so forth).

4.8 Homework

1. Starting with the electric field intensity due to a *point* charge at the origin with $Q = 4$ nC, determine the potential at (7, 80°, 55°) with respect to (3, 30°, 10°).

2. Starting with the electric field intensity due to a *line* charge on the z-axis with $\rho_L = 5$ nC/m, determine the potential at (6, 80°, 2) with respect to (3, 30°, 5).

3. A long-charged cylinder of radius 1 m is centered on the z-axis. It has an electric field intensity of
   $$\bar{E} = +\frac{10}{\rho}\bar{a}_\rho \left[\frac{\text{V}}{\text{m}}\right] \text{ for } \rho > 1\,\text{m}.$$
   a) Draw and label a sketch.
   b) Determine the potential (voltage) at $B(6, 70°, 1)$ with respect to $A(2, 40°, 5)$.
   c) Is the potential positive or negative? Physically, why?

4. Two parallel-plate conductors are separated by 2 cm and each have an area of 4 m². The plates are centered about the origin and are parallel to the x–y plane. If the upper plate has a charge of 8 mC and the bottom plate has −8 mC, determine the voltage between $z = -8$ mm and $z = +4$ mm on the z-axis. Hint: Is it reasonable to apply a theoretical result to this practical situation?

5. a) What does $\oint \bar{E} \cdot \overline{d\ell}$ mean?
   b) What does it equal for static electric fields? Why?

---

20 When a material is magnetically saturated, all magnetizations of the molecules are aligned, and B is no longer proportional to H. In this situation, the relation $B = \mu H$ is not valid, which was assumed in the magnetic circuit development.

6. a) What is E_{normal} at the interface of a dielectric–perfect conductor interface? Why?

b) What is $E_{\text{tangential}}$ adjacent to a perfect conductor? Why?

7. A parallel-plate capacitor with a dielectric between the plates is connected to a DC voltage source. Then the dielectric is removed (the source is still connected). Does Q, D, E, V, Ψ, and C increase, decrease, or remain the same relative to the initial situation? Justify *each* of the six answers.

8. Initially, a parallel-plate capacitor *with air* between the plates is connected to a DC voltage source. Then the source is removed. Finally, a dielectric is inserted between the plates. Does Q, D, E, V, Ψ, and C increase, decrease, or remain the same relative to the initial situation? Justify *each* of the six answers.

9. a) Distinguish $C = Q/V$ from $C = \varepsilon A/d$.

b) If the distance between the plates of a parallel plate capacitor were on the order of one dimension of the plate area, would $C = \varepsilon A/d$ still be valid? Why or why not? Hint: consider the assumptions in the capacitance development.

c) In the development of the capacitance of a coaxial capacitor, is the charge on the *outer* conductor utilized in the development? Why or why not?

10. For the two capacitance developments in Section 4.2.1:

a) How does the capacitance depend on each capacitor dimension?

b) Does the capacitance of coax change if the radii are increased but the ratio between them is held constant?

11. a) A development of the capacitance between two concentric spheres is outlined below.

- Sketch and label the geometry: Inner radius is a and outer radius is b (actually the radius to the inner surface of the outer conductor). The inner spherical shell is positive.
- State the assumptions about the electric field pattern symmetry.
- Determine the GS and add it to the sketch.
- Explain and justify each step that follows.

$$\Psi_{\text{net}} = Q_{\text{encl}} \tag{4.120}$$

$$Q = \rho_s 4\pi a^2 \tag{4.121}$$

$$D_r = \frac{\Psi}{4\pi r^2} = \frac{Q}{4\pi r^2} = \frac{\rho_s 4\pi a^2}{4\pi r^2} = \rho_s \frac{a^2}{r^2} \tag{4.122}$$

$$E = \frac{\rho_s a^2}{\varepsilon r^2} \rightarrow \overline{E} = \frac{\rho_s a^2}{\varepsilon r^2} \overline{a}_r \tag{4.123}$$

$$V = -\int_A^B \overline{E} \cdot \overline{d\ell} = -\int_b^a \frac{\rho_s a^2}{\varepsilon r^2} (\overline{a}_r) \cdot dr \overline{a}_r \tag{4.124}$$

$$V = -\frac{\rho_s a^2}{\varepsilon} \int_b^a \frac{dr}{r^2} = -\frac{\rho_s a^2}{\varepsilon} \left(\frac{-1}{r} \right) \Big|_b^a \tag{4.125}$$

$$V = \frac{\rho_s a^2}{\varepsilon} \left(\frac{1}{a} - \frac{1}{b} \right) \tag{4.126}$$

$$C = \frac{Q}{V} = \frac{\rho_s 4\pi a^2}{\dfrac{\rho_s a^2}{\varepsilon} \left(\dfrac{1}{a} - \dfrac{1}{b} \right)} = \frac{4\pi\varepsilon}{\left(\dfrac{1}{a} - \dfrac{1}{b} \right)} \cdot \frac{ab}{ab} \tag{4.127}$$

$$\boxed{C = \frac{Q}{V} = 4\pi\varepsilon \left(\frac{ab}{b-a} \right)} \tag{4.128}$$

12. Interpret the result of the development in Problem 11:
    a) Does the capacitance change if the radii are increased but the ratio between them is held constant?
    b) Is this capacitor subject to fringing?
    c) Is this capacitor favorable to be manufactured?

13. Consider two identical parallel-plate capacitors. Using only the capacitance equation for a parallel-plate capacitor, $C = \dfrac{\varepsilon A}{d}$, show that:
    a) when connected in parallel, the equivalent total capacitance is $C_{total} = 2C$.
    b) when connected in series, the equivalent total capacitance is $C_{total} = \dfrac{C}{2}$.

14. Dielectric breakdown is briefly discussed in Section 4.2.2, where it describes the electric field strength at which an insulator begins to conduct current.
    a) Describe why high-voltage power lines need large separation between conductors in order to avoid arcing (sparks!) between the conductors.
    b) Using concepts involving dielectric and electric fields, for a fixed voltage and separation between two conductors, will a dielectric such as Teflon ($\varepsilon_r = 2.1$) decrease, increase, or have no effect on the maximum power handling of a circuit?

15. Explain why a smaller gauge wire (larger radius) has a lower resistance per unit length than a larger gauge (smaller radius) wire of the same conductivity.

16. Consider two identical, ideal resistors formed from rectangular blocks of conductive material. Assume the fields and currents are uniform throughout the resistors. Using the equation for resistance, Eq. (4.42):
    a) Show that the total equivalent series resistance of the two resistors is twice the resistance of a single resistor.
    b) Show that the total equivalent parallel resistance of the two resistors is half the resistance of a single resistor.

17. A solid cylindrical conductor of radius b centered on the z-axis has a volume current density of $\bar{J} = J_o \rho^3 \bar{a}_z$. Determine the magnetic field intensity in the following regions:
    a) Outside the conductor, and
    b) Within the conductor. Hint: what happens to the total I as radius increases?
    c) What is the field behavior outside the conductor as radius increases? Why?
    d) What is the field behavior within the conductor as radius increases? Why?

18. A hollow cylindrical conductor of inner radius a and outer radius b centered on the z-axis has a volume current density of $\bar{J} = J_o \rho^3 \bar{a}_z$. Determine the magnetic field intensity
    a) In the air region inside the conductor ($\rho < a$), and
    b) Outside the conductor ($\rho > b$).
    c) What current density would be required in this hollow cylindrical conductor to equal the answer of Problem 17 for H outside the solid conductor?
    d) What do you conclude about the relative need for the inner volume of the conductor with a nonuniform current density that increases in magnitude in a radial direction?

19. A closed path does not enclose a current-carrying conductor.
    a) What does ACL equal? Why?
    b) Qualitatively explain the energy expended as a magnetic monopole is moved around that closed path.
    c) Is the magnetic field conservative? Why or why not? Hint: Consider all closed paths.

20. a) What is the magnetic flux *outside* of coaxial cable if the center conductor and shield have equal but oppositely directed currents? Why?

b) What is the practical ramification of this result?

c) Does a magnetic field exist *inside* a current-carrying conductor (assume finite conductivity)? Explain your answer.

21. Use ACL to determine the magnetic field intensity inside coax of inner radius a and outer radius b. Sketch and label the setup including the ACL path. Assume a uniform current on the inner and outer conductors. Also assume that the magnetic field is purely in the $+\overline{a}_\phi$ direction. (See the line current example in the text.)

22. Repeat Problem 21 if the inner conductor is a line current. What is the range of the radius for which this solution is valid?

23. A hypothetical parallel universe contains plenty of magnetic monopoles (particles with isolated magnetic poles, either N or S).

a) Are all magnetic flux lines solenoidal in the parallel universe? Why or why not?

b) Write the integral form of Gauss's law for magnetics in this universe.

24. a) What does $\oint \overline{B} \cdot \overline{dA}$ mean?

b) What does it equal? Why?

c) Does a change in magnetic material change the answer if the magnetic flux density is *normal* to the boundary? Why or why not?

25. a) State the *general* defining equation for inductance.

b) How does the result $L = \dfrac{\mu N^2 A}{\ell}$ relate to the answer in part a)?

c) Why is the inductance of coils generally proportional to A? To N^2? To $1/\ell$?

d) Does a straight wire with a current I have inductance? Why or why not?

e) Explain the unit *henry*. (Analogy: a farad is a coulomb of electric flux per volt.)

26. a) What is a relative advantage of coils over toroids as inductors? Hint: fabrication. What is a relative advantage of toroids over coils as inductors? Hint: where is Φ?
Assume both inductors have a magnetic core.

b) Why is the equation for the inductance of an ideal long coil and a toroid identical if the coil length and mean path lengths are equal (assume equal A, N, and μ)? Hint: consider the assumptions underlying the inductance expressions.

c) Explain why magnetic cores significantly improve adherence to a key assumption (identify that key assumption first).

27. Develop the equation for the inductance of a toroid with a round cross section. State the underlying assumptions.

28. Develop the equation for the inductance of a toroid with a *rectangular* cross section. The inner radius is a, the outer radius is b, and the height is h. Assume that the magnetic field intensity is *not* uniform inside the core. State any other underlying assumptions. Hint: Solve for H using ACL first.

29. Consider the energy stored in the magnetic field of an inductor with a magnetic core. Eventually, the magnetic material begins to *saturate*, that is, an increase in the magnetic field intensity H does not produce a proportionate increase in the magnetic flux density B.
    a) Does increasing the number of turns in the inductor alleviate core saturation? Explain.
    b) How else might the energy stored in the magnetic field be significantly increased?

30. With regards to magnetic field boundary conditions (BC),
    a) State the normal BC and explain how it expresses a solenoidal field.
    b) State the tangential BC and explain how it expresses magnetic flux containment.

31. Surface current $K_\phi \bar{a}_\phi$ flows on both the inner and outer surfaces of a highly conductive thin-walled cylinder.
    a) Sketch and label the geometry.
    b) Based on the appropriate boundary condition, what is the direction of $\overline{H}$ both inside and outside of the cylinder?
    c) Sketch the magnetic field pattern around this current configuration. Reminder: magnetic flux is solenoidal!
    d) To which electrical component does this magnetic field correspond?

32. In general, the magnetic field intensity adjacent to a conductor with *surface* current density $\overline{K}$ is $\overline{H} = \overline{K} \times \bar{a}_n$, where $\bar{a}_n$ is the unit vector normal to and pointing outward from the conductor.
    a) If a flat, thick conductor occupies $z \leq 0$ and $\overline{K}$ flows in the x-direction in the $z = 0$ plane, what is the direction of the magnetic field intensity?
    b) Make a sketch of this geometry, the current, and the magnetic field to visualize this direction. Hint: sketch the surface current as several adjacent parallel currents.

Problems for Appendix 4.D

33. A toroid has relative permeability of 200. The mean path length in the core is 6 cm and the cross-sectional area is $1\,\text{cm}^2$. If the number of turns is 500,
    a) Determine the current to establish a flux density level of 0.01 T.
    b) Would the current be higher, lower, or the same to establish the same magnetic flux level if the toroid had an air gap? Justify your answer with physical reasoning.

34. A series magnetic circuit consists of a toroid with a small air gap. The length of the air gap, ℓ_g, is small relative to the cross-sectional dimensions of the core material. The magnetic material of the core has length ℓ_c and relative permeability μ_c. The cross-sectional area of the magnetic circuit is A. A coil that is wrapped around the core material has N turns. Develop the expression for the magnetic flux which follows.

    State assumptions. Recommendation: Analyze this structure as a magnetic circuit using reluctances.

    Result: $\Phi = \dfrac{\mu_0 \mu_c A}{\ell_c + \mu_c \ell_g} NI$

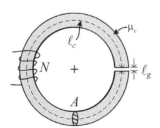

Part II

Time-Changing Electric and Magnetic Fields

5

Maxwell's Equations

Up to this point, we have only considered static electric and magnetic fields, which are fields that are constant with respect to time but may be functions of position. We now generalize this study by also considering **time-changing** fields, which are fields that change as a function of time as well as a function of position. Time-changing fields are also called **dynamic** fields or **time-dependent** fields. When a field is time-changing, new terms appear in the fundamental electric and magnetic fields relationships, and correspondingly, new electromagnetic phenomena appear that lead to many important applications, namely the fundamental voltage–current relationships of electrical components. These applications will be examined in a dynamic fields context in this chapter.

5.1   Introduction to Time-Changing Electromagnetic Fields

Motivational Question   What are the changes to electric and magnetic field laws when the fields are time-dependent?

Electromagnetics and Transmission Lines: Essentials for Electrical Engineering, Second Edition.
Robert A. Strangeway, Steven S. Holland, and James E. Richie.
© 2023 John Wiley & Sons, Inc. Published 2023 by John Wiley & Sons, Inc.
Companion website: www.wiley.com/go/Strangeway/ElectromagneticsandTransmissionLines

So far we have learned a few, very important laws and relationships for electromagnetics. Besides Coulomb's law and the Biot–Savart law, they are:

$$\text{Static electric fields are conservative}: \oint \overline{E} \cdot \overline{d\ell} = 0 \tag{5.1}$$

$$\text{Ampere's circuital law (ACL)}: \oint \overline{H} \cdot \overline{d\ell} = I_{\text{encl}} \tag{5.2}$$

$$\text{Gauss's law}: \oint \overline{D} \cdot \overline{dA} = Q_{\text{encl}} \tag{5.3}$$

$$\text{Gauss's law for magnetic flux}: \oint \overline{B} \cdot \overline{dA} = 0 \;(\text{magnetic flux is solenoidal}) \tag{5.4}$$

It is important to emphasize here that Eqs. (5.1) through (5.4) were developed in Chapters 2 through 4 in the context of static electric and magnetic fields and steady (DC) currents, so these equations are only accurate under these same static conditions. Attempting to apply these equations to more general, time-varying fields will lead to incorrect results. We will see how to modify these equations to a form that accurately describes both general static and dynamic fields.

As a preview of things to come, let's compare the previous four equations to **Maxwell's equations** (Eqs. (5.5) through (5.8) correspond to Eqs. (5.1) through (5.4), respectively):

$$\oint \overline{E} \cdot \overline{d\ell} = -\frac{d}{dt} \int \overline{B} \cdot \overline{dA} \tag{5.5}$$

$$\oint \overline{H} \cdot \overline{d\ell} = I_c + \frac{d}{dt} \int \overline{D} \cdot \overline{dA} \tag{5.6}$$

$$\oint \overline{D} \cdot \overline{dA} = Q_{\text{encl}} \tag{5.7}$$

$$\oint \overline{B} \cdot \overline{dA} = 0 \tag{5.8}$$

where I_c is the conduction current, dQ/dt, which was I_{encl} in ACL up to this point. One observes that Eqs. (5.7) and (5.8) are identical to Eqs. (5.3) and (5.4) respectively, so the electric and magnetic forms of Gauss's law are two of Maxwell's equations. However, there are additional terms in Eqs. (5.5) and (5.6) that are not present in Eqs. (5.1) and (5.2), respectively. Can you identify the macroscopic nature of the new additional terms? They have time derivatives. The addition of these time-dependent terms results in the other two equations that constitute Maxwell's equations. We will discover the meaning and some important, practical applications of these terms in this chapter.

The short discussion in this introductory section establishes the motivation to examine electromagnetic field equations and concepts when *the fields are no longer static*. Let's briefly examine the two laws that are significantly modified when the fields become time-dynamic. Start with the conservative electric field concept (voltage around a closed path):

$$\oint \overline{E} \cdot \overline{d\ell} = 0 \tag{5.9}$$

Why is this result modified with time-dynamic fields? From a previous physics or circuits course, you may recall **Faraday's law**, which states that a time-changing linked magnetic flux induces a voltage v_{ind}:

$$v_{\text{ind}} = -\frac{d\Lambda}{dt} \tag{5.10}$$

where Λ, the linked magnetic flux, is the total magnetic flux that passes through the cross section of a given structure (such as a coil). In circuits, Faraday's law is used to explain the operation of inductors and ideal transformers. Thus, the time-changing magnetic flux will change the voltage, and consequently Eq. (5.9) needs the additional term that is present in Eq. (5.5) to account for this induced potential. This topic is examined in Section 5.2.

Now consider ACL:

$$\oint \overline{H} \cdot \overline{d\ell} = I_c \tag{5.11}$$

where I_c is the conduction current (charge flowing past a point in the conductor per second), that is, the enclosed current. An interesting situation concerning current becomes apparent when one considers a capacitor. If Kirchhoff's current law (KCL) is applied to a node that surrounds only *one* plate of a capacitor, KCL is apparently violated because nonzero current enters the node, but zero current appears to leave the node. However, if current is entering the node, charge is building up on that capacitor plate. Thus, by Gauss's law, the electric flux from that plate is increasing. This action occurs over time. The time-changing electric flux, called **displacement current**, is required to satisfy KCL in the operation of a capacitor. Hence, a displacement current term must be added to Eq. (5.11) – see Eq. (5.6). This topic is examined in Section 5.3.

Maxwell's equations are important and insightful: these four equations, along with the Lorentz force law, describe classical electromagnetics,[1] whether the fields are static or dynamic. Understanding these equations will empower an engineer to explain and advance the state of modern technology. The equations have been shown in integral form.[2] *The content of this chapter is largely conceptual*: relating the mathematical expressions of these electromagnetic field laws to physical meaning. Study this chapter in this spirit. Several practical applications are also examined in this chapter: the connections between the time-dependent i–v circuit relationships and the electromagnetic fields for inductors, capacitors, mutual inductors, and coupled circuits. The electromagnetic plane wave is another major application of time-dynamic electromagnetic fields and opens the door to any future studies of electromagnetic wave propagation, such as in waveguides or with antennas. Plane waves and antennas are examined in Chapter 8.

5.2 Faraday's Law

5.2.1 Lorentz Force Law and Induced Voltage

Motivational Question   How does the Lorentz force law relate to induced voltage as expressed by Faraday's law?

Recall that the Lorentz force law gives the force $\overline{F}$ on a charge Q due to an electric field and motion through a magnetic field ("cutting" magnetic flux) with velocity $\overline{v}$:

$$\overline{F} = Q\left(\overline{E} + \overline{v} \times \overline{B}\right) \tag{5.12}$$

The force due to the electric field has been thoroughly examined in the Coulomb's law context (Chapter 2). Here we concentrate on the force due to the motion of the charge in a magnetic field.[3] Assuming $E = 0$, what does Eq. (5.12) directly tell us?

- The charge must be in motion through the magnetic field ($\overline{v} \neq 0$) for a force to be exerted on the charge.
- The charge must have a vector component of motion in the direction perpendicular to the magnetic flux lines (note the cross product), which is equivalent to saying that the path of the charge motion must "cut" magnetic flux.

---

1 Classical electromagnetics refers to the physics of fields at length scales where quantum effects are negligible and can be ignored. Practical electrical engineering problems are nearly always at size scales where quantum effects are negligible. This text focuses on classical electromagnetics. For very small-size scales, such as particle physics, *quantum* electromagnetics equations must be utilized.
2 There are also differential forms of Maxwell's equations that are useful in solving and visualizing time-dynamic electromagnetic problems – see Appendix 5.B.
3 We define "motion in or through a field" as motion relative to the frame of reference of the field, such as motion relative to the coordinate system of the field.

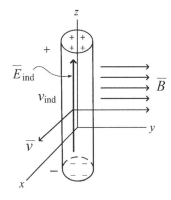

Figure 5.1 Induced voltage.

- The larger the velocity and/or the magnetic flux density, the larger the force on Q.
- The closer the velocity and magnetic flux density vectors are to being perpendicular, the larger the force on Q.

How does the Lorentz force law relate to Faraday's law? Recall that Faraday's law states that a "changing" magnetic field (motion and/or time dependency) induces a voltage into a conductor. A conductor contains free charges. A conductor that moves through a magnetic field automatically has charges moving through the magnetic field. Consequently, these charges will have a force exerted upon them (Lorentz force law) that separates the positive and negative charges, which induces a voltage into the conductor (Faraday's law). One concludes that the Lorentz force law relates to Faraday's law.

The cross product in the Lorentz force law can be used to determine the polarity of the induced voltage. Visualize the scenario sketched in Figure 5.1: $\overline{B}$ is in the $+y$ direction, and a straight wire is aligned parallel to the z-axis and is moving in the x-direction ($\overline{v}$ is in the $+x$ direction). If you perform the $\overline{v} \times \overline{B}$ cross product, the force on positive charges in the conductor is in the $+z$ direction, so the upper end of the wire will be positive and the lower end of the wire will be negative. The polarity of the induced voltage is established.

What is the equation that relates induced voltage to motion through a magnetic field? Consider only the magnetic flux density in the Lorentz force law and divide both sides by Q:

$$\frac{\overline{F}}{Q} = \overline{v} \times \overline{B} \tag{5.13}$$

Hopefully, you recognize the left-hand side of Eq. (5.13) as the definition of electric field intensity. Then,

$$\overline{E}_{\text{ind}} = \overline{v} \times \overline{B} \tag{5.14}$$

which means that motion through a magnetic field induces an electric field $\overline{E}_{\text{ind}}$. Note that *the $\overline{E}$ in Eq. (5.14) is not the $\overline{E}$ in the Lorentz force law* (the latter is an externally applied electric field). The "ind" (induced) subscript will be added to $\overline{E}$ in order to distinguish these two electric field intensities. Recall the relationship between electric field intensity (V/m) and voltage:

$$V = -\int \overline{E} \cdot \overline{d\ell} \tag{5.15}$$

where V is the voltage (as opposed to $\overline{v}$, which is the velocity). Here, the voltage V is time dependent in general. A lower-case v will be used to indicate that the voltage is time dependent, as is usually the convention in circuit analysis. The symbol v_{ind} will be used for induced voltage to distinguish it from velocity $\overline{v}$. Per Eq. (5.15), to determine the voltage induced by motion of a straight wire through a magnetic field, take the dot product of both sides of Eq. (5.14) with $\overline{d\ell}$ and integrate over some path:

$$\int \overline{E}_{\text{ind}} \cdot \overline{d\ell} = \int (\overline{v} \times \overline{B}) \cdot \overline{d\ell} \tag{5.16}$$

The negative sign in Eq. (5.15) must be examined in the context of Eq. (5.16). The induced electric field *inside* the wire creates a charge separation (see Figure 5.1). The $\overline{E}_{\text{ind}}$ *inside* the conductor is *in the direction* that v_{ind} rises (negative to positive). This electric field is in the opposite direction relative to the direction of $\overline{E}$ that is created by separated charge. Hence, the negative sign in Eq. (5.15) must be removed for induced voltages due to motion through a magnetic field. In effect, the voltage induced by motion through a magnetic field is a source. The induced electric field intensity inside a source is in the opposite direction of that expressed in Eq. (5.15):

$$v_{\text{ind}} = +\int \overline{E}_{\text{ind}} \cdot \overline{d\ell} \tag{5.17}$$

Combine Eqs. (5.16) and (5.17):

$$v_{ind} = \int \left(\bar{v} \times \bar{B}\right) \cdot \overline{d\ell} \tag{5.18}$$

which is the voltage induced into a conductor due to relative motion between the conductor and the magnetic field. This voltage is called **motional emf** (a historical term for generated voltage is electromotive force, or emf). Again note that the $\bar{v} \times \bar{B}$ product gives the direction of the force on the charges. The voltage will be positive, that is, it will rise from negative to positive if the integration along the path is in the same direction as $\bar{v} \times \bar{B}$. Thus, Eq. (5.18) answers our earlier question of how the Lorentz force law relates to Faraday's law: a voltage is induced in a conductor moving through a magnetic field (cutting flux).

Example 5.2.1

Given the conductor and magnetic field shown in Figure 5.1,

a) Is a voltage induced if $\bar{v}$ is in the $-x$ direction. If so, which end is positive?
b) Is a voltage induced if $\bar{v}$ is in the $-y$ direction. If so, which end is positive?

Solution

a) $\bar{v} \times \bar{B} = v\left(-\bar{a}_x\right) \times B\left(\bar{a}_y\right) = vB\left(-\bar{a}_z\right)$; yes, a voltage is induced, and the lowermost end of the conductor is positive.
b) $\bar{v} \times \bar{B} = v\left(-\bar{a}_y\right) \times B\left(\bar{a}_y\right) = 0$; no, a voltage is not induced because the conductor is not cutting magnetic flux; $\bar{v}$ is antiparallel to the $\bar{B}$ field.

There is a technicality with Eq. (5.18). A circuit is a closed path. For example, in Figure 5.1 the voltmeter that is used to measure v_{ind} completes the circuit path. There may be some portions of a circuit for which a voltage is not induced, but voltage could be induced around the entire circuit path. Hence, the motional emf integral is shown as a closed path integral with the understanding that some parts of the circuit may not have induced voltage:

$$v_{ind} = \oint \left(\bar{v} \times \bar{B}\right) \cdot \overline{d\ell} \tag{5.19}$$

Notice the contrast between this equation and the KVL equation $\oint \bar{E} \cdot \overline{d\ell} = 0$ from Chapter 4. We will examine the implications of this in Section 5.2.3.

5.2.2 Time-Changing Magnetic Fields

Motivational Question   How does Faraday's law incorporate time-changing magnetic fields?

We have seen that voltage is induced in a conductor by motion of the conductor through a magnetic field. What if the conductor and the magnetic field are stationary relative to each other but $\bar{B}$ is changing with respect to time? Start with Faraday's law:

$$v_{ind} = -\frac{d\Lambda}{dt} \tag{5.20}$$

where Λ is the linked magnetic flux (Λ equals $N\Phi$ for a coil). Thus, Faraday's law states that time-changing linked magnetic flux induces a voltage. The negative sign is an expression of Lenz's law, which indicates that the induced voltage polarity is oriented in an "opposing" manner to the changing linked flux. Specifically, the induced voltage polarity causes a current to flow in the circuit that creates a magnetic field in a direction that counteracts the

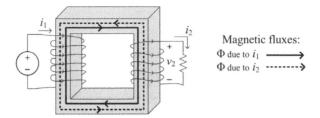

Figure 5.2 Basic transformer operation for time-dependent signals (i_1 increasing).

changing linked magnetic flux. For example, let the primary current i_1 be increasing in the transformer shown in Figure 5.2. The linked magnetic flux also increases. The voltage induced into the secondary causes a current to flow in the secondary that sets up a magnetic field in the core *opposite* in direction to the magnetic flux that originally caused the induced voltage in the secondary. If i_1 is decreasing, then i_2 and the flux due to i_2 would be in the opposite direction.

Given the two ways in which magnetic flux can change over time (motion through a magnetic field and a time-changing magnetic field), the next step is to express Faraday's law in terms of field quantities. Recall that linked flux $\Lambda = \int \overline{B} \cdot \overline{dA}$ and $v_{\text{ind}} = \oint \overline{E} \cdot \overline{d\ell}$ where it is now understood that $\overline{E} = \overline{E}_{\text{ind}}$ in Faraday's law:

$$v_{\text{ind}} = -\frac{d\Lambda}{dt} \rightarrow \oint \overline{E} \cdot \overline{d\ell} = -\frac{d}{dt} \int \overline{B} \cdot \overline{dA} \tag{5.21}$$

Break up Faraday's law to show the two ways that the magnetic flux is changing with respect to time:

$$v_{\text{ind}} = -\frac{d\Lambda}{dt} = -\frac{d\Lambda}{dt}\bigg|_{\substack{\text{time-changing} \\ \text{magnetic flux}}} -\frac{d\Lambda}{dt}\bigg|_{\substack{\text{motion through} \\ \text{magnetic flux}}} \tag{5.22}$$

If $\overline{B}$ is time-changing in Eq. (5.21) but the area of the linked magnetic flux is constant with respect to time, then the time derivative can be moved within the integral and used for the first term on the right-hand-side of Eq. (5.22). If the area in Eq. (5.21) is changing with respect to time, then the motional emf in Eq. (5.18) can be substituted for the second term on the right-hand-side of Eq. (5.22), with the change of the negative sign to be positive as previously discussed. The result is **Faraday's law** in terms of fields:

$$\boxed{\oint \overline{E} \cdot \overline{d\ell} = -\frac{d}{dt} \int \overline{B} \cdot \overline{dA} = -\int \frac{\partial \overline{B}}{\partial t} \cdot \overline{dA} + \oint (\overline{v} \times \overline{B}) \cdot \overline{d\ell}} \tag{5.23}$$

This form of Faraday's law clearly shows the contributions to the induced voltage from the time-changing magnetic flux density and from the motion through the magnetic flux.[4] The whole point of Faraday's law is that *time-changing linked magnetic flux, whether through motion and/or a time-changing magnetic field, induces a voltage.*

Example 5.2.2

A loop of wire in the x–y plane has an area of $1\,\text{cm}^2$ and is immersed in a uniform AC, $+z$-directed sinusoidal magnetic field with a peak magnitude roughly the strength of the Earth's surface magnetic field (about $50\,\mu\text{T}$). It has a series resistor of high resistance compared with the wire resistance. Apply Faraday's law to determine the voltage that develops across the resistor at 60 Hz and at 1 GHz. Form an appropriate conclusion.

Solution

Express $\overline{B}$ as a sinusoidal equation: $\overline{B}(t) = 50 \sin(\omega t) \overline{a}_z \ \mu\text{T}$.

---

4 When the derivative was moved inside the integral, the partial derivative replaced the ordinary derivative because $\overline{B}$ is explicitly a function of multiple independent variables, that is, $\overline{B}(x, y, z, t)$.

The circular loop of wire (with a series resistor) of radius ρ lies in the $z = 0$ plane and the uniform magnetic flux density is in the z-direction, as shown in Figure 5.3. Then $\overline{B}$ crosses the surface area defined by that loop orthogonally $\left(\overline{B} \parallel \overline{dA}\right)$ and is independent of the integration variables because it is uniform. The resultant integral reduces to the product of the magnetic flux density and the area of the loop A:

$$\int \overline{B} \cdot \overline{dA} = \int B_z \overline{a}_z \cdot dA \overline{a}_z = B_z A \text{ where } B_z \text{ is uniform.}$$

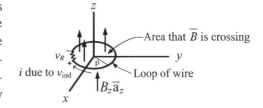

Figure 5.3 Loop of wire immersed in a magnetic field.

Use Faraday's law to determine the induced voltage that drops across R (given $R \gg$ wire resistance, almost all of the induced voltage will drop across R):

$$v_{\text{ind}} = -\frac{d}{dt} \int \overline{B} \cdot \overline{dA} = -(A)\frac{dB}{dt} = -\left(1\,\text{cm}^2\right)\left(\frac{1\,\text{m}^2}{10^4\,\text{cm}^2}\right)\frac{d[50 \sin{(\omega t)}\,\mu\text{T}]}{dt}$$

$$v_{\text{ind}} = -\left(10^{-4}\,\text{m}^2\right)\left[\omega 50 \times 10^{-6} \cos{(\omega t)}\right] = 5\omega \cos{(\omega t)}\,\text{nWb}$$

$$\text{Substitute values}: \quad \text{For 60 Hz,} \quad v_{\text{ind}} = 1.88 \cos{(377t)}\,\mu\text{V}$$
$$\text{For 1 GHz,} \quad v_{\text{ind}} = 31.4 \cos{\left(6.28 \times 10^9 t\right)}\,\text{V}$$

The induced voltage is directly proportional to the AC frequency because the result of the derivative is directly proportional to ω. The magnetic field is changing more quickly as frequency increases, so the induced voltage correspondingly increases.

5.2.3 Another Look at Kirchhoff's Voltage Law

Motivational Question   Given Faraday's law, $\oint \overline{E} \cdot \overline{d\ell} = -\int \frac{\partial \overline{B}}{\partial t} \cdot \overline{dA} + \oint \left(\overline{v} \times \overline{B}\right) \cdot \overline{d\ell}$, why doesn't $\oint \overline{E} \cdot \overline{d\ell} = 0$ per Kirchhoff's voltage law (KVL)?

Recall the application of KVL in Chapter 4 to the circuit shown in Figure 5.4a. For the dashed path shown, KVL obviously holds:

$$\text{Clockwise around the path}: \oint \overline{E} \cdot \overline{d\ell} = -\int \underbrace{\overline{E}_S \cdot \overline{d\ell}}_{\text{antiparallel}} - \int \underbrace{\overline{E}_L \cdot \overline{d\ell}}_{\text{parallel}} = +v_S - v_L = 0 \tag{5.24}$$

where the "S" subscript indicates the source, the "L" subscript indicates the load (v_L is across R), $\overline{E}_L$ is parallel to the path for KVL, and $\overline{E}_S$ is antiparallel to the path. Now consider the circuit shown in Figure 5.4b that has a time-changing linked magnetic flux. Figure 5.4b is redrawn as Figure 5.4c for clarity. Take a closed path for KVL *through the interior* of the conductor. There is an additional electric field present, $\overline{E}_{\text{ind}}$, which is induced by the time-changing linked magnetic flux. Note that the electric field $\overline{E}_S$ exists not only outside of the conductor but also *within* the conductor due to the separated charges created by $\overline{E}_{\text{ind}}$. When the path is taken clockwise through the interior of the conductor,

$$\oint \overline{E} \cdot \overline{d\ell} = \int \overline{E}_{\text{ind}} \cdot \overline{d\ell} \underbrace{- \int \underbrace{\overline{E}_S \cdot \overline{d\ell}}_{\text{anti-parallel}} - \int \underbrace{\overline{E}_L \cdot \overline{d\ell}}_{\text{parallel}}}_{0} = \int \overline{E}_{\text{ind}} \cdot \overline{d\ell} \neq 0 \tag{5.25}$$

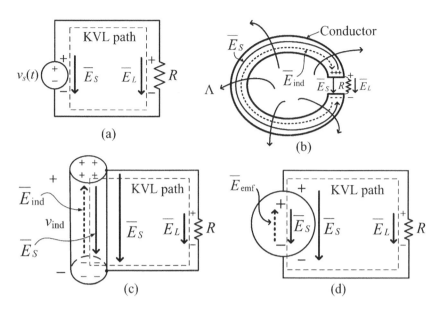

Figure 5.4 KVL applied to a circuit (dotted vectors for $\overline{E}_{ind}$ and $\overline{E}_{emf}$ for clarity). (a) KVL in a circuit. (b) Changing linked flux, $\overline{E}_{ind}$, and $\overline{E}_S$. (c) KVL path through the source with redrawn schematic version of (b). (d) KVL path through a source in general.

The voltages due to the conservative electric field, $\overline{E}_S$ and $\overline{E}_L$, again cancel, leaving only the induced electric field intensity in the closed path:

$$\oint \overline{E}_{ind} \cdot \overline{d\ell} \neq 0 \tag{5.26}$$

This extra electric field intensity term is not canceled, so KVL does not hold. The circuit is not conservative along this path because *net* energy is gained around this closed path!

Why isn't the circuit conservative when a path through the interior of the source is taken? *The energy in the source that separates the charges to form* $\overline{E}_{ind}$ *came from outside* of the circuit. This addition of external energy to a system means that the overall system energy is not conserved when moving along a closed path through the source. In a generator, for example, energy is being injected into the system by the conversion of mechanical energy, motion of the conductor through the magnetic field, into electrical energy, represented by the induced voltage, in motional emf.

This reasoning is applicable for an electrical source in general, as illustrated in Figure 5.4d, where $\overline{E}_{emf}$ represents the electric field within an electrical source that is created by energy conversion.[5] Hence, $\oint \overline{E} \cdot \overline{d\ell} \neq 0$ and KVL does not hold if the path is *through the interior* of the source. The good news is that our circuit measurements are external to the source, not internal, so the path goes around the source. Only the electric field external to the source is in the KVL path for the source and KVL holds. As a result, KVL remains valid for practical AC circuit analysis! This brief discussion has hopefully illuminated fundamental aspects of KVL, emf, conservative fields, and why $\oint \overline{E} \cdot \overline{d\ell} \neq 0$ in the context of Faraday's law and electric sources in general.

5.2.4 Another Look at the Inductor

Motivational Question How does the *i–v* relationship for an inductor, $v = L\,di/dt$, relate to Faraday's law?

We now consider the effects of a time-changing magnetic flux inside an inductor when a time-changing current is flowing through the inductor. Start with the fundamental definition of inductance:

---

5 Reminder: "emf" is an abbreviation for "electromotive force" which is an older term for voltage.

$$L = \frac{\Lambda}{I} \tag{5.27}$$

By Faraday's law, the time rate of change of linked magnetic flux induces a voltage:

$$v = -\frac{d\Lambda}{dt} \tag{5.28}$$

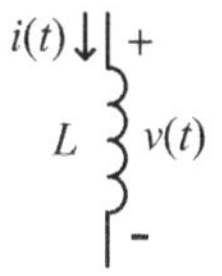

From the definition of inductance, $\Lambda = L\,i$, where $i = i(t)$ is used to explicitly indicate that the current is time-dependent. Substitute Li for Λ in Eq. (5.28) and assume L is a constant inductance:

$$v = \frac{d(Li)}{dt} = L\frac{di}{dt} \tag{5.29}$$

Figure 5.5 Voltage polarity and current direction definitions per the passive sign convention.

which is the familiar circuit i–v relationship for the inductor. Note that the minus sign has been removed to agree with the **passive sign convention** in conventional current flow circuit analysis. See Figure 5.5. The induced voltage opposes the current change (increasing current in this case). This i–v relationship for the inductor is a reminder that electric circuit theory originates from electromagnetic field theory.

5.2.5 The Ideal Transformer

Motivational Question   How does the turns-ratio relationship for an ideal transformer, $n = \dfrac{v_P}{v_S} = \dfrac{N_P}{N_S}$, relate to Faraday's law?

In this section, we examine the ideal transformer (Figure 5.2) further. What are the assumptions behind the ideal transformer circuit relationships?

- All magnetic flux links both coils.
- The coil reactances are significantly greater than the magnitudes of the circuit impedances (load and source).

These conditions are usually well satisfied by transformers with magnetic cores. Given that the magnetic flux Φ is the same in both coils (see Figure 5.2), then the ratio of the voltages induced in each coil, the **turns-ratio** n, is

$$n = \frac{v_P}{v_S} = \frac{\dfrac{d\Lambda_P}{dt}}{\dfrac{d\Lambda_S}{dt}} = \frac{N_P\dfrac{d\Phi_P}{dt}}{N_S\dfrac{d\Phi_S}{dt}} = \frac{N_P\dfrac{d\Phi}{dt}}{N_S\dfrac{d\Phi}{dt}} = \frac{N_P}{N_S}. \tag{5.30}$$

where subscript P is used to indicate the primary side of the transformer, subscript S is used to indicate the secondary side of the transformer, N_P is the number of turns in the primary coil, and N_S is the number of turns in the secondary coil.

Although the flux in each coil is the same, the *linked* flux is *not* the same in both coils, in general, because the number of turns in each coil can differ. The well-known AC ideal transformer phasor relationships result from Eq. (5.30), where the tildes indicate phasors and complex impedances in general:

$$n = \frac{N_P}{N_S} = \frac{\tilde{V}_P}{\tilde{V}_S} = \frac{\tilde{I}_S}{\tilde{I}_P} \text{ and } \tilde{Z}_{\text{in}} = n^2\tilde{Z}_L \tag{5.31}$$

The ideal transformer model is often an acceptable approximation to practical transformers when the coils are wrapped on a highly magnetic core material that ensures most of the magnetic flux links both coils. However, for more accurate modeling of practical transformers, it is important to account for incomplete flux linkage using a mutual inductor analysis, which is discussed next.

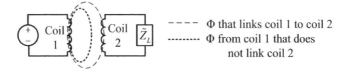

Figure 5.6 Incomplete flux linkage in a mutual inductor.

5.2.6 Mutual Inductors

Motivational Question What is a mutual inductor, and what is its *i*–*v* relationship?

A mutual inductor is basically a transformer where only a portion of the magnetic flux from one coil links the other coil. For example, imagine two coils in close vicinity but no magnetic material path linking them. See Figure 5.6. Some of the flux from coil 1 will link coil 2, but some of the flux will not. This incomplete flux linkage changes the *i*–*v* relationship from that of the ideal transformer. The analysis follows.

In this section, the *i*–*v* relationships for the mutual inductor are developed from the definition of inductance and from Faraday's law. Consider the general case of two coils that link a portion of the magnetic flux from one coil into the other coil. The total linked magnetic flux in each coil consists of the linked flux generated by the current *in that coil* plus the portion of the flux generated *by the other coil* that links into the coil under consideration:

$$\Lambda_1 = N_1\Phi_1 + N_1\Phi_{1\leftarrow 2} \tag{5.32}$$
$$\Lambda_2 = N_2\Phi_{2\leftarrow 1} + N_2\Phi_2 \tag{5.33}$$

where the last term in Eq. (5.32) indicates the flux due to the current in coil 2 that links to coil 1, and the middle term in Eq. (5.33) indicates the flux due to the current in coil 1 that links to coil 2. The linked flux can be substituted with the inductance–current product, from the fundamental definition of inductance in Eq. (5.27). However, a distinction must be made. If the magnetic flux is linking to the coil in which that flux was generated, then the inductance is called **self-inductance** L and $LI = N\Phi$. If the magnetic flux is linking to another coil in which that flux was not generated, then the inductance is called **mutual inductance** M and $MI = N\Phi$. Substituting these relations into Eqs. (5.32)–(5.33) gives:

$$\Lambda_1 = L_1I_1 + M_{1\leftarrow 2}I_2 \tag{5.34}$$
$$\Lambda_2 = M_{2\leftarrow 1}I_1 + L_2I_2 \tag{5.35}$$

where the current directions are defined in Figure 5.7a. Let the currents and magnetic fluxes be time dependent – see Figure 5.7b. Note that the self and mutual inductances are based on the physical layout of conductors and are considered constants here. Take the time derivative of Eqs. (5.34) and (5.35):

$$\frac{d\Lambda_1}{dt} = L_1\frac{di_1}{dt} + M_{1\leftarrow 2}\frac{di_2}{dt} \tag{5.36}$$
$$\frac{d\Lambda_2}{dt} = M_{2\leftarrow 1}\frac{di_1}{dt} + L_2\frac{di_2}{dt} \tag{5.37}$$

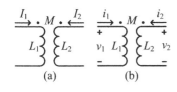

Figure 5.7 Current direction and voltage polarity definitions for a mutual inductor. (a) Current directions defined. (b) Time-dependent voltages and currents.

Does the linkage of magnetic flux from coil 1 to 2 differ from that of coil 2 to 1? A qualitative argument follows. Consider two linked coils, one with many turns and the other with only a few turns. See Figure 5.8a. If coil 1 is energized by an AC current source, it will create a strong magnetic field due to the many turns in coil 1. A voltage will be induced in coil 2.

If the coils are swapped, as in Figure 5.8b, the magnetic field of coil 2 will be significantly smaller than the magnetic field of coil 1 in Figure 5.8 because there are fewer turns in coil 2 relative to coil 1. However, there will be significantly more linked flux in coil 1 due to the larger number of turns in coil 1.

Figure 5.8 Illustration of $M_{2\leftarrow1} = M_{1\leftarrow2}$. (a) Coil 1 excited, $\tilde{V}_{ind}$ in coil 2. (b) Coil 2 excited, $\tilde{V}_{ind}$ in coil 1.

Qualitatively, one could conceive that the induced voltage is the same in both cases. It can be shown to be quantitatively true.[6]

Hence, the mutual inductance M is the same in both equations. Substitute M for $M_{1\leftarrow2}$ and $M_{2\leftarrow1}$ in Eqs. (5.36) and (5.37). By Faraday's law, the time rate of change of linked magnetic flux on the right-hand side of these equations is induced voltage:

$$v_1 = L_1\frac{di_1}{dt} + M\frac{di_2}{dt} \tag{5.38}$$

$$v_2 = M\frac{di_1}{dt} + L_2\frac{di_2}{dt} \tag{5.39}$$

where the voltage polarities and current directions are defined in Figure 5.7b. These equations are the i–v relationships for a mutual inductor. For AC sinusoidal steady-state signals, Eqs. (5.38) and (5.39) simplify into phasor form:

$$\tilde{V}_1 = j\omega L_1\tilde{I}_1 + j\omega M\tilde{I}_2 \tag{5.40}$$

$$\tilde{V}_2 = j\omega M\tilde{I}_1 + j\omega L_2\tilde{I}_2 \tag{5.41}$$

In the complex frequency (Laplace) domain:

$$V_1 = sL_1I_1 + sMI_2 \tag{5.42}$$

$$V_2 = sMI_1 + sL_2I_2 \tag{5.43}$$

Two mutual inductors can be arranged into different configurations to produce new, useful results. If two mutual inductors are connected in **series-aiding** (the linked magnetic fluxes are in the same direction and consequently add), as shown in Figure 5.9a, the total series-aiding inductance L_{Tsa} is:

$$L_{Tsa} = L_1 + L_2 + 2M \tag{5.44}$$

The series aiding is indicated by the **dot convention** (current entering the dot side of one coil induces the positive side of the voltage at the dot side of the other coil), just as it is for transformers. See Figure 5.9a. Likewise, if two mutual inductors are connected in **series-opposing** (the linked magnetic fluxes are in opposite directions and consequently subtract), as shown in Figure 5.9b, the total inductance is:

$$L_{Tso} = L_1 + L_2 - 2M \tag{5.45}$$

The development of the previous two equations is assigned in the chapter homework (problem 9). Note that the total AC impedance increases when the coils are series-aiding and decreases when the coils are series-opposing.

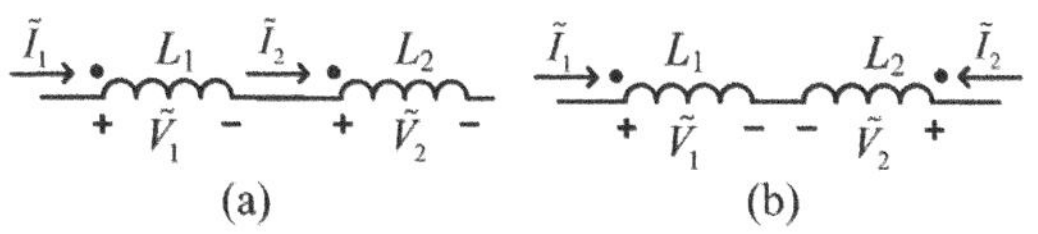

Figure 5.9 Series configurations of mutual inductors. (a) Series-aiding. (b) Series-opposing.

---

6 This argument is under the general circuit analysis topic of *reciprocity*.

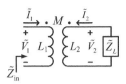

Figure 5.10 Input impedance of a loaded mutual inductor

If Eqs. (5.40) and (5.41) are solved for $\widetilde{V}_1/\widetilde{I}_1$ as defined in Figure 5.10, the input impedance of a mutual inductor results (see homework problem 10):

$$\widetilde{Z}_{\text{in}} = \frac{\widetilde{V}_1}{\widetilde{I}_1} = +j\omega L_1 + \frac{(\omega M)^2}{+j\omega L_2 + \widetilde{Z}_L} \tag{5.46}$$

Note that the input impedance is not the simple $n^2\widetilde{Z}_L$ as it was with the ideal transformer, which is a consequence when the two ideal transformer assumptions do not hold. Refer to a circuit analysis textbook for an introductory discussion, developments, and examples of the analysis of circuits with mutual inductors.[7] The ideal transformer relationships can also be developed from the mutual inductor relationships.

Example 5.2.3

Given a mutual inductor when the load is an open circuit,

a) What is the AC input–output voltage relationship?
b) What is the ideal input impedance?
c) Interpret the results.

Strategy

Set $\widetilde{I}_2 = 0$ in Eqs. (5.40) and (5.41). Solve for $\widetilde{V}_2/\widetilde{V}_1$ and $\widetilde{V}_1/\widetilde{I}_1$.

Solution

a) $\left.\begin{array}{l} \widetilde{V}_2 = j\omega M\widetilde{I}_1 \\ \widetilde{V}_1 = j\omega L_1\widetilde{I}_1 \end{array}\right\}$ $\quad \dfrac{\widetilde{V}_2}{\widetilde{V}_1} = \dfrac{j\omega M\widetilde{I}_1}{j\omega L_1\widetilde{I}_1} = \dfrac{M}{L_1}$

b) $\widetilde{V}_1 = j\omega L_1\widetilde{I}_1 \rightarrow \widetilde{Z}_{\text{in}} = \dfrac{\widetilde{V}_1}{\widetilde{I}_1} = +j\omega L_1$, which is the same result if $\widetilde{Z}_L \rightarrow \infty$ in Eq. (5.46).

c) The voltage ratio result is a real number ratio that is ideally frequency independent. The input impedance is that of the driven coil. Hence, an inductor with a high load impedance can be used to sense the AC magnetic field of another inductor through magnetic coupling without ideally perturbing the circuit with the original inductor.[8] Alternatively, the mutual inductance between two coils can be determined by the ratio of measured AC voltages if the self-inductance of the driven coil is known.

Can circuits couple magnetically, that is, can the time-changing magnetic field in one circuit induce a voltage into another circuit? Yes! See Figure 5.11 for a basic sketch. The time-dynamic magnetic flux that is created by the "driven" circuit (driven by a source) links to the circuit that is labeled "coupled circuit." A voltage is induced into the conductors, which causes a current to flow. The resultant voltage drops across the loads are considered the coupled voltage. The amount of magnetic coupling from one circuit to another is characterized by mutual inductance between the circuits. Note that a signal can be conveyed between two circuits through this magnetic coupling despite no conductive connection between them! Magnetic coupling is one example of **crosstalk** and is a major issue in **signal integrity** (Chapter 9). For example, crosstalk can cause undesired coupling between nearby conductors on a printed circuit

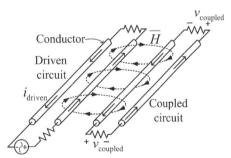

Figure 5.11 Magnetic field coupling results in mutual inductance between two circuits.

---

7 Such as Strangeway, R.A., Petersen, O.G., Gassert, J.D., and Lokken, R.J. (2019). *Electric Circuits*. Wisconsin: RacademicS. Chapter 13.
8 Technically there is always some current, albeit usually very small in an open circuit. The reason for the nonzero current will become apparent in Section 5.3.

board (PCB) and in breadboard circuits. As further examples, crosstalk can occur between wires within a cable or even between different cables!

What is the relationship between the two circuits due to magnetic coupling? Similar to the mutual inductor,

$$v_{\text{coupled}} = M \frac{di_{\text{driven}}}{dt} \tag{5.47}$$

Note the mutual inductance nature ($v = Mdi/dt$) of this i–v relationship: the current is in one circuit (the driven circuit) and the voltage is in the other circuit (the coupled circuit). The derivative has major practical ramifications, such as:

- The coupled voltage increases as frequency increases.
- The coupled waveform shape is generally different from the driven waveform for non-sinusoidal signals because of the derivative relationship.

Since it is the current in the driven circuit that creates the magnetic coupling, this form of coupling is prominent in circuits carrying high currents.

5.3 Displacement Current

Motivational Question How does the current complete its path through a capacitor?

Now that the effects of time-changing magnetic fields have been examined, our attention shifts to time-changing *electric* fields and the associated consequences.

5.3.1 Time-Changing Electric Fields

A capacitor will be used to visualize the effect of a time-changing electric field. Select the upper plate of a parallel plate capacitor as the node for the application of KCL (see Figure 5.12). Recall in Section 5.1 that KCL is satisfied at this node: the conduction current entering the node equals the time-changing electric flux from the node. This concept is now formalized here. At the surface of the upper plate, apply Gauss's law:

$$\Psi_{\text{net}} = Q_{\text{encl}} \tag{5.48}$$

Take the time-derivative:

$$\frac{d\Psi}{dt} = \frac{dQ}{dt} \tag{5.49}$$

The dQ/dt term is recognized as **conduction current** I_c (where it is understood that I, despite the capital letter, may be time-changing); substitute $I_c = dQ/dt$ into Eq. (5.49):

$$I_c = \frac{d\Psi}{dt} \tag{5.50}$$

The time-changing electric flux equals the conduction current! How is the current path completed between the plates of a capacitor? Through the time-changing electric flux. How can a time-changing electric flux look like a current and complete the current path? Consider current flowing onto one plate of the capacitor, building up positive charge. This increasing amount of positive charge creates an increasing electric flux that terminates on the other capacitor conductor and *displaces* an increasing amount of positive charge from this second conductor. These positive charges flow off of the conductor in a current. If a KCL node is placed around the entire capacitor,

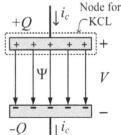

Figure 5.12 KCL applied to one capacitor plate.

the conduction current entering equals the conduction current leaving! Thus, **displacement current** equals the time rate of change of electric flux. In contrast, conduction current is the time rate of charge flowing past a point in the circuit.

The term $\dfrac{d\Psi}{dt}$ is the displacement current and is defined in terms of fields as:

$$I_d = \frac{d\Psi}{dt} = \frac{d}{dt}\int \overline{D} \cdot \overline{dA} \tag{5.51}$$

Notice that the integral is not closed, in general. It is the electric flux through a defined surface.

In a manner similar to how induced voltage modified KVL, we now consider how displacement current is incorporated into ACL $\oint \overline{H} \cdot \overline{d\ell} = I_{\text{encl}}$. ACL states that an enclosed current creates a magnetic field. The enclosed current may be generalized to include *both* conduction current and displacement current contributions, $I_{\text{encl}} = I_c + I_d$. As a result, ACL can then be expressed as:

$$\boxed{\oint \overline{H} \cdot \overline{d\ell} = I_c + I_d = I_c + \frac{d}{dt}\int \overline{D} \cdot \overline{dA}} \tag{5.52}$$

Can a time-changing electric field create a magnetic field? Yes! Consider Eq. (5.52). A conduction current and/or a time-changing electric field create a magnetic field.

Displacement current is the final quantity required to unify electricity and magnetism: a time-changing electric field creates a time-changing magnetic field and Faraday's law states that a time-changing magnetic field creates a time-changing elective field. This interrelationship has profound consequences that will become evident in the discussion of electromagnetic waves in Chapters 6 through 8. Displacement current is important because it is the basis of the i–v relationship of capacitors (the next topic) and is used to explain capacitive coupling between circuits on a PCB, for example (see Chapter 9). We will see that displacement current plays a role in transmission lines (Chapter 6) and antennas (Chapter 8).

> Historical note: James Clerk Maxwell identified the concept of displacement current, which along with Faraday's law, completed the relationship between electric and magnetic fields. He is given credit for this discovery by naming the collection of the four electromagnetic equations in Section 5.4 as Maxwell's equations.

5.3.2 Another Look at the Capacitor

Motivational Question How does the i–v relationship for a capacitor, $i = Cdv/dt$, relate to displacement current?

The i–v relationship for the capacitor is now addressed. Start with the fundamental definition of capacitance:

$$C = \frac{Q}{V} = \frac{\Psi}{V} \tag{5.53}$$

where Ψ is the electric flux. Under time-varying conditions, we will use a lowercase v and write the flux–voltage relationship as $\Psi = Cv$. Now consider that displacement current is the time rate of change of electric flux:

$$i = \frac{d\Psi}{dt} = \frac{d(Cv)}{dt} = C\frac{dv}{dt} \tag{5.54}$$

which gives the familiar circuit i–v relationship for the capacitor (assuming capacitance is a constant). This result is another reminder that electric circuit theory originates from electromagnetic field theory. Note that $i = dQ/dt$, the conduction current due to flowing charges in the conductors connected to the capacitor, equals $i = d\Psi/dt$, the displacement current between the capacitor plates. KCL is satisfied.

The following example illustrates how displacement current can be used to determine a current level at the specified dielectric breakdown strength E_{br} (V/m), which is the electric field intensity at which the dielectric breaks down as an insulator and begins to pass conduction current.

Example 5.3.1

A parallel-plate capacitor has a dielectric with a specified dielectric breakdown strength E_{br} (V/m).

a) Develop the expression for the peak AC current through the capacitor at this breakdown electric field intensity level.

b) Interpret any variable dependence for a given capacitor.

Strategy

This problem requires a setup.

- Assume a parallel plate capacitor with plate area A, plate separation d, and relative dielectric constant ε_r.
- The current is sinusoidal. The peak current value is I_P: $i_C(t) = I_P \cos(\omega t)$.
- The AC electric field intensity between the plates is $E(t) = E_P \sin(\omega t)$, where E_P is the peak electric field intensity. The sine function results from the derivative of the displacement current (based on E) that produces a cosine, which was stated for $i_C(t)$.
- Set the peak electric field intensity to the breakdown level: $E_{max}(t) = E_{br} \sin(\omega t)$.
- KCL: the conduction current equals the displacement current, $i_c(t) = i_d(t)$.
- Insert these expressions into $I_d = \dfrac{d}{dt}\displaystyle\int \overline{D} \cdot \overline{dA}$ and solve for I_P. Here, $I_d = i_d(t)$.

Solution

a) $i_C(t) = I_P \cos(\omega t) = \dfrac{d}{dt}\displaystyle\int \overline{D} \cdot \overline{dA} = \dfrac{d}{dt}\displaystyle\int \varepsilon_o \varepsilon_r \overline{E} \cdot \overline{dA}$

The parallel-plate capacitor has a homogeneous dielectric and uniform electric field, so the integral reduces to a multiplication and the constants can be removed from the derivative:

$$I_P \cos(\omega t) = \dfrac{d}{dt}\left(\varepsilon_o \varepsilon_r E A\right) = \varepsilon_o \varepsilon_r A \dfrac{dE}{dt}$$

Insert $E_{max}(t)$ and solve for I_P:

$$I_P \cos(\omega t) = \varepsilon_o \varepsilon_r A \dfrac{d}{dt}\left[E_{br} \sin(\omega t)\right] = \varepsilon_o \varepsilon_r A \omega E_{br} \cos(\omega t)$$

Result : $\quad I_P = \omega \varepsilon_o \varepsilon_r A E_{br}$

b) Interpretation: Quantities ε_o, ε_r, A, and E_{br} are constant for a given capacitor, so the only variable dependence in I_P is frequency. As the AC frequency increases, the voltage across a capacitor decreases, and so does the electric field intensity between the plates. Hence, a larger peak current can flow to reach the same E_{br} level. (See homework problems 13 and 14 for numerical illustrations.)

Comment

One would not design operation at the edge of breakdown. A safety margin is built in, such as operating at no more than half of the theoretical current that would cause breakdown.

5.3.3 Mutual Capacitance

Motivational Question   Is there an analogy to magnetic field coupling between circuits (mutual inductance) for electric fields?

We observed in Section 5.2.6 that circuits can couple due to time-changing magnetic fields. Can circuits couple electrically, that is, can the time-changing electric field of one circuit induce a current into another circuit? Yes! See Figure 5.13 for a basic sketch. Note how the coupled current would be in the *same* direction in both loads of the

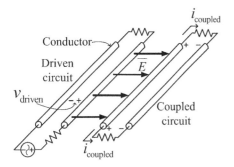

Figure 5.13 Electric field coupling results in mutual capacitance between two circuits.

circuit that is labeled "coupled circuit." The time-changing electric flux (the displacement current), primarily exerts a force on the charges in the conductor of the coupled circuit that is closest to the driven circuit. The force repels (or attracts) charges off of this conductor on both ends of the closest conductor, so the current flows as indicated in Figure 5.13.

The amount of electric field coupling from one circuit to another is characterized by the **mutual capacitance** (C_m) between the circuits:

$$i_{\text{coupled}} = C_m \frac{dv_{\text{driven}}}{dt} \tag{5.55}$$

Note the $i = Cdv/dt$ nature of this i–v relationship differs from that of a capacitor in that the voltage is across one circuit (the driven circuit) and the current is in the other circuit (the coupled circuit). It is a mutual capacitance relationship. The effect of the derivative on frequency dependence and on the waveform shape is the same as it was for the magnetic coupling case. Electric field coupling is another example of **crosstalk** and is also a major issue in **signal integrity** (Chapter 9). Since the driving circuit quantity is a voltage in the mutual capacitance relationship, electric coupling is dominant in high voltage circuits.

5.4   Chapter Summary: Maxwell's Equations in Integral Form

- Faraday's law including both the time-changing magnetic field and motional emf:

$$\oint \overline{E} \cdot \overline{d\ell} = -\frac{d}{dt} \int \overline{B} \cdot \overline{dA} = -\int \frac{\partial \overline{B}}{\partial t} \cdot \overline{dA} + \oint (\overline{v} \times \overline{B}) \cdot \overline{d\ell} \tag{5.56}$$

- ACL and displacement current:

$$\oint \overline{H} \cdot \overline{d\ell} = I_c + \frac{d}{dt} \int \overline{D} \cdot \overline{dA} \tag{5.57}$$

- Gauss's law:

$$\oint \overline{D} \cdot \overline{dA} = Q_{\text{encl}} \tag{5.58}$$

- Gauss's law for magnetic flux:

$$\oint \overline{B} \cdot \overline{dA} = 0 \, (\text{magnetic flux is solenoidal}) \tag{5.59}$$

The first two Maxwell's equations suggest that a time-changing magnetic field generates a time changing electric field that generates a time-changing magnetic field ... that can result in *electromagnetic wave propagation*! This topic is discussed in Chapter 6 and expanded upon in Chapter 8.

5.5 Appendices

Appendix 5.A A Faraday's Law Thought Experiment

Motional emf can be viewed, at least qualitatively, as being equivalent to voltage induced by time-changing magnetic flux. Consider the two cases in the following thought experiment.

A loop of wire in the *x–y* plane is immersed in a uniform, *+z*-directed magnetic field. It has a series resistor. Apply Faraday's law to determine the induced voltage for the following two cases and form an appropriate conclusion.

a) The loop of wire is stationary and the *z*-directed magnetic flux density linking the loop increases.
b) The loop of wire expands in area as the radius increases through a *constant*, uniform *z*-directed magnetic field.

The circular loop of wire (with a series resistor) of radius ρ lies in the $z = 0$ plane and the uniform magnetic flux density is in the *z*-direction, as shown in Figure 5.3. Then $\overline{B}$ crosses the surface area defined by that loop orthogonally $\left(\overline{B} \parallel \overline{dA}\right)$ and is independent of the integration variables because it is uniform. The resultant integral reduces to the product of the magnetic flux density and the area of the loop A.

$$\int \overline{B} \cdot \overline{dA} \Bigg|_{\overline{B} \parallel \overline{dA}} = \int B dA \Bigg|_{B \text{ is uniform}} = B \int dA = BA \tag{5.60}$$

Use Faraday's law to determine the induced voltage:

$$v_{\text{ind}} = -\frac{d}{dt} \int \overline{B} \cdot \overline{dA} = -\frac{d}{dt}(BA) \tag{5.61}$$

Apply the product rule:

$$v_{\text{ind}} = -A\frac{\partial B}{\partial t} - B\frac{\partial A}{\partial t} \tag{5.62}$$

For case a), the derivative of the time-changing magnetic flux density that is linked to the loop is the first term in Eq. (5.62). Thus, the voltage induced by the time-increasing magnetic flux density is

$$v_{\text{ind}} = -A\frac{\partial B_z}{\partial t} \tag{5.63}$$

For case b), the second term in Eq. (5.62) suggests a changing linked flux due to a changing area in a *constant*, uniform magnetic field. Consider the motional emf term of Faraday's law:

$$v_{\text{ind}} = \oint (\overline{v} \times \overline{B}) \cdot \overline{d\ell} \tag{5.64}$$

The differential length along the path is $\overline{d\ell} = \rho d\phi \overline{a}_\phi$. Imagine the conductor is flexible like a rubber band – it can be stretched to cover a larger area. If the circular loop had a time-changing radius, $\rho(t)$, then the velocity of the conductor is $\overline{v} = \dfrac{d\rho}{dt}\overline{a}_\rho$:

$$v_{\text{ind}} = \oint \frac{d\rho}{dt}\overline{a}_\rho \times B_z\overline{a}_z \cdot \rho d\phi\, \overline{a}_\phi \tag{5.65}$$

The B_z factor is uniform and can be pulled out of the integral. The time derivative and ρ do not depend on coordinate ϕ and can also be pulled from the integral.

$$v_{\text{ind}} = \left(\frac{d\rho}{dt}\right)(\rho)B_z \oint (-\overline{a}_\phi) \cdot \overline{a}_\phi d\phi = -\rho\frac{d\rho}{dt}B_z 2\pi \tag{5.66}$$

But $2\rho\dfrac{d\rho}{dt} = \dfrac{d(\rho^2)}{dt}$, so $\rho\dfrac{d\rho}{dt} = \dfrac{1}{2}\dfrac{d(\rho^2)}{dt}$. Make this substitution in Eq. (5.66):

$$v_{\text{ind}} = -B_z\pi\frac{d\rho^2}{dt} = -B_z\frac{d(\pi\rho^2)}{dt} = -B_z\frac{\partial A}{\partial t} \tag{5.67}$$

which matches the second term in Eq. (5.62). The rate of change of the area that links magnetic flux, which was due to the motion of the conductor, is the cause of induced voltage in this case.

The previous discussion illustrates that the motional emf term in the general expression for Faraday's law can be thought of as a time-changing linked flux due to a time-changing area. The whole point of Faraday's law is that *time-changing linked magnetic flux, whether through motion and/or a time-changing magnetic field, induces a voltage.*

Appendix 5.B   Maxwell's Equations in Differential Form

The integral forms of Maxwell's equations are often used to determine and measure practical parameters, such as induced voltage or the magnetic field intensity around a current. Maxwell's equations also exist in differential form (also called "point form"), that is, as differential equations. Differential equations are useful in determining mathematical solutions. Maxwell's equations in differential form are listed below. You may see these forms in the engineering literature.

$$\text{Faraday's law}: \nabla \times \overline{E} = -\frac{\partial \overline{B}}{\partial t} \tag{5.68}$$

$$\text{ACL and displacement current}: \nabla \times \overline{H} = \overline{J}_c + \frac{\partial \overline{D}}{\partial t} \tag{5.69}$$

$$\text{Gauss's law}: \nabla \cdot \overline{D} = \rho_v \tag{5.70}$$

$$\text{Gauss's law for magnetic flux}: \nabla \cdot \overline{B} = 0 \tag{5.71}$$

How are these equation forms related to what we have previously covered, and what does the upside-down delta mean? A non-rigorous approach is taken to answer these questions. First, what is "∇," called "**del**"? It is a symbol for a mathematical operation, one of numerous mathematical operators. It is used frequently in vector calculus and its applications. In Cartesian coordinates, the **del operator** is

$$\nabla \equiv \frac{\partial}{\partial x}\overline{a}_x + \frac{\partial}{\partial y}\overline{a}_y + \frac{\partial}{\partial z}\overline{a}_z. \tag{5.72}$$

The del operator *operates* on a function. For example, one operation of the del operator on the vector function $\overline{D}(x,y,z)$ is

$$\nabla \cdot \overline{D} = \left(\frac{\partial}{\partial x}\overline{a}_x + \frac{\partial}{\partial y}\overline{a}_y + \frac{\partial}{\partial z}\overline{a}_z\right) \cdot \left(D_x\overline{a}_x + D_y\overline{a}_y + D_z\overline{a}_z\right) = \frac{\partial D_x}{\partial x} + \frac{\partial D_y}{\partial y} + \frac{\partial D_z}{\partial z}. \tag{5.73}$$

Notice how the dot product insures that the derivative of each vector component is taken with respect to the position variable *in the direction* of that vector component. The operation $\nabla \cdot \overline{D}$ is called the **divergence** of the electric flux density. There are corresponding expressions for the del operator in the cylindrical and spherical coordinate systems that can be used to determine divergence:

$$\nabla \cdot \overline{D} = \frac{1}{\rho}\frac{\partial(\rho D_\rho)}{\partial\rho} + \frac{1}{\rho}\frac{\partial D_\phi}{\partial\phi} + \frac{\partial D_z}{\partial z} \text{ (cylindrical)} \tag{5.74}$$

$$\nabla \cdot \overline{D} = \frac{1}{r^2}\frac{\partial(r^2 D_r)}{\partial r} + \frac{1}{r\sin\theta}\frac{\partial(D_\theta\sin\theta)}{\partial\theta} + \frac{1}{r\sin\theta}\frac{\partial D_\phi}{\partial\phi} \text{ (spherical)} \tag{5.75}$$

Now that we know what the del operator is, let's take a non-rigorous, intuitive approach to Eqs. (5.70) and (5.71). We start with the following, unproven declaration:

$$\nabla \cdot \overline{D} = \rho_V \tag{5.76}$$

You are already familiar with the volume charge density ρ_V. Integrate both sides of Eq. (5.76) over some volume:

$$\int_V \nabla \cdot \overline{D} dV = \int_V \rho_v dV \tag{5.77}$$

The integral of the charge density is the total charge enclosed in that volume:

$$\int_V \nabla \cdot \overline{D} dV = Q_{\text{encl}} \tag{5.78}$$

The previous expression must equal Gauss's law based on the right-hand-side:

$$\oint_S \overline{D} \cdot \overline{dA} = Q_{\text{encl}} \tag{5.79}$$

Based on this non-rigorous analysis, the original expression, $\nabla \cdot \overline{D} = \rho_V$, is Gauss's law in differential form (also called "point form"):

$$\boxed{\text{div}\overline{D} = \nabla \cdot \overline{D} = \rho_V} \quad \textbf{Gauss's law in point form} \tag{5.80}$$

This form of Gauss's law is common in the literature, and being of differential form, it can be used to form differential equations that can be solved analytically (often used in advanced electromagnetic studies). The overarching idea is that an "inspection" of the field, the derivatives in Eq. (5.73), in a region of interest using the divergence operation reveals whether charge density is present in that region. From the previous developments, we see that divergence is the application of Gauss's law on an infinitesimal volume basis, and it indicates if there is a source of that flux in that infinitesimal volume.

For magnetic fields, $\nabla \cdot \overline{B} = 0$ similarly results, with the result equaling zero because there are no magnetic charges (monopoles), at least practically at the current time. A zero divergence indicates that the field is *solenoidal* in that region and that the flux lines close in on themselves.

$$\boxed{\text{div}\,\overline{B} = \nabla \cdot \overline{B} = 0} \quad \textbf{magnetic Gauss's law in point form} \tag{5.81}$$

One other useful result is obtained by equating the integrals in Eqs. (5.78) and (5.79):

$$\oint_S \overline{D} \cdot \overline{dA} = \int_V \nabla \cdot \overline{D} dV \quad \textit{Divergence theorem} \tag{5.82}$$

The divergence theorem is a mathematical relationship between volume and surface integrals of a vector function. It is used in vector calculus and advanced electromagnetic studies. We will not use it here – it is presented for your awareness.

The vector operation in Eqs. (5.68) and (5.69) is called **curl**. Curl is the cross product between the del operator and a vector field quantity. In Cartesian coordinates,

$$\text{curl}\,\overline{E} = \nabla \times \overline{E} = \left(\frac{\partial}{\partial x}\overline{a}_x + \frac{\partial}{\partial y}\overline{a}_y + \frac{\partial}{\partial z}\overline{a}_z \right) \times \left(E_x \overline{a}_x + E_y \overline{a}_y + E_z \overline{a}_z \right) \tag{5.83}$$

The cross product eliminates any terms with the same unit vector directions between the operation and the vector components, leaving

$$\text{curl}\,\overline{E} = \nabla \times \overline{E} = \left(\frac{\partial E_z}{\partial y} - \frac{\partial E_y}{\partial z} \right)\overline{a}_x + \left(\frac{\partial E_x}{\partial z} - \frac{\partial E_z}{\partial x} \right)\overline{a}_y + \left(\frac{\partial E_y}{\partial x} - \frac{\partial E_x}{\partial y} \right)\overline{a}_z \tag{5.84}$$

Notice how the cross product insures that the derivative of each vector component is taken with respect to the position variable *in a direction perpendicular* to that vector component. There are corresponding expressions for the del operator in the cylindrical and spherical coordinate systems (using $\overline{H}$ instead of $\overline{E}$ in the next two equations):

$$\nabla \times \overline{H} = \left(+ \frac{1}{\rho} \frac{\partial H_z}{\partial \phi} - \frac{\partial H_\phi}{\partial z} \right) \overline{a}_\rho + \left(+ \frac{\partial H_\rho}{\partial z} - \frac{\partial H_z}{\partial \rho} \right) \overline{a}_\phi + \frac{1}{\rho} \left(+ \frac{\partial (\rho H_\phi)}{\partial \rho} - \frac{\partial H_\rho}{\partial \phi} \right) \overline{a}_z \tag{5.85}$$

$$\nabla \times \overline{H} = \frac{1}{r \sin \theta} \left(\frac{\partial (H_\phi \sin \theta)}{\partial \theta} - \frac{\partial H_\theta}{\partial \phi} \right) \overline{a}_r + \frac{1}{r} \left(\frac{1}{\sin \theta} \frac{\partial H_r}{\partial \phi} - \frac{\partial (r H_\phi)}{\partial r} \right) \overline{a}_\theta + \frac{1}{r} \left(\frac{\partial (r H_\theta)}{\partial r} - \frac{\partial H_r}{\partial \theta} \right) \overline{a}_\phi \tag{5.86}$$

We will also use a non-rigorous approach to Eqs. (5.68) and (5.69). First, start with the unproven declaration:

$$\nabla \times \overline{E} = -\frac{\partial \overline{B}}{\partial t} \tag{5.87}$$

Integrate both sides over some surface:

$$\int_S (\nabla \times \overline{E}) \cdot \overline{dA} = -\int_S \frac{\partial \overline{B}}{\partial t} \cdot \overline{dA} \tag{5.88}$$

The time derivative and integration over spatial variables are independent operations and the order of the operations can be switched (without proof):

$$\int_S (\nabla \times \overline{E}) \cdot \overline{dA} = -\frac{d}{dt} \int_S \overline{B} \cdot \overline{dA} \tag{5.89}$$

The previous expression must equal Faraday's law (Eq. 5.56) based on the right-hand side:

$$\oint \overline{E} \cdot \overline{d\ell} = -\frac{d}{dt} \int \overline{B} \cdot \overline{dA} \tag{5.90}$$

Based on this non-rigorous approach, the original expression, $\nabla \times \overline{E} = -\dfrac{\partial \overline{B}}{\partial t}$, is Faraday's law in differential form (also called "point form"):

$$\boxed{\text{curl } \overline{E} = \nabla \times \overline{E} = -\frac{\partial \overline{B}}{\partial t}} \quad \textbf{Faraday's law in point form} \tag{5.91}$$

This form of Faraday's law is common in the literature and being of differential form, it can be used to form differential equations that can be solved analytically (also often used in advanced electromagnetic studies). The over-arching idea is that the curl operation is often used to determine if the field is **conservative**. The curl of the field is zero when the field is conservative.

For magnetic fields, curl $\overline{H} = \nabla \times \overline{H} = \overline{J}_c + \dfrac{\partial \overline{D}}{\partial t}$ similarly results (see homework problem 18). The curl of the magnetic field intensity equals the sum of the conduction current density and the displacement current density. An "inspection" of the magnetic field, the derivatives in Eq. (5.84), in a region of interest using the curl operation reveals whether source currents are present in that region. The curl of the magnetic field intensity can be viewed as the application of ACL on an infinitesimal surface basis, and it indicates if there is a source of that flux crossing that infinitesimal surface.

$$\boxed{\text{curl } \overline{H} = \nabla \times \overline{H} = \overline{J}_c + \frac{\partial \overline{D}}{\partial t}} \quad \textbf{ACL and displacement current} \tag{5.92}$$

One other useful result is obtained by equating the integrals in Eq. (5.89) and (5.90) (using $\overline{H}$ instead of $\overline{E}$):

$$\oint_L \overline{H} \cdot \overline{d\ell} = \int_S (\nabla \times \overline{H}) \cdot \overline{dA} \quad \textit{Stokes' theorem} \tag{5.93}$$

Stokes' theorem is a mathematical relationship between line and surface integrals of a vector function. It is used in vector calculus and advanced electromagnetic studies. We will not use it here – it is presented for your awareness.

Appendix 5.C   Continuity Equation and KCL

Note: The differential operator background from Appendix 5.B is prerequisite to this appendix.

Current is fundamentally defined as the rate of change of electric charge passing a defined surface (often called "a point in a circuit") per unit time:

$$I = \frac{dQ}{dt} \tag{5.94}$$

Imagine a closed surface where current is entering and leaving, but where the net charge enclosed *could* increase or decrease. The net current leaving this closed surface S is related to volume current density by

$$I_{\text{net}} = \oint_S \bar{J} \cdot \overline{dA} \tag{5.95}$$

where positive $\bar{J}$ is defined as leaving S. Apply the divergence theorem:

$$I_{\text{net}} = \oint_S \bar{J} \cdot \overline{dA} = \int_V \nabla \cdot \bar{J} dV \tag{5.96}$$

where V is the volume within S. Then substitute volume charge density into Eq. (5.94) and apply it to V:

$$I_{\text{net}} = \frac{dQ}{dt} = \frac{-\partial \int_V \rho_V dV}{\partial t} = -\int_V \frac{\partial \rho_V}{\partial t} dV \tag{5.97}$$

where the negative sign indicates that if the charge density is increasing, more charge is entering as opposed to leaving S, so the net current is negative (as defined in this development). The time derivative could be moved inside the integral because the surface and associated volume are assumed to be stationary and constant. Combine Eqs. (5.96) and (5.97):

$$\int_V \nabla \cdot \bar{J} dV = -\int_V \frac{\partial \rho_V}{\partial t} dV \tag{5.98}$$

The volumes on both sides of the equation are equal. They can be made arbitrarily small, so that the integrands can be declared equal:

$$\boxed{\nabla \cdot \bar{J} = -\frac{\partial \rho_V}{\partial t}} \tag{5.99}$$

The result is the **continuity equation** (for current). KCL states the sum of the current entering and leaving a node is zero, so there is zero net accumulation of charge within the node. Consider an arbitrarily small closed surface to be the node. If the net charge is constant, then $\partial \rho_V/\partial t = 0$, resulting in KCL as expressed with current density:

$$\nabla \cdot \bar{J} = 0 \tag{5.100}$$

5.6 Homework

1. A straight wire conductor of length 0.3 m is oriented parallel to the z-axis and travels with velocity $\bar{v} = 10\bar{a}_x$ m/s in a uniform magnetic field of density $\bar{B} = 0.05\bar{a}_x$ T.
   a) Is a voltage induced into the wire? Explain why in terms of Maxwell's equations.
   b) Is a voltage induced if $\bar{v} = 10\bar{a}_y$ m/s? Explain why in terms of Maxwell's equations.

2. A straight wire conductor of length 0.3 m is oriented parallel to the x–y plane and lies along the radius at $\phi = 45°$. It travels with velocity $\bar{v} = 10\bar{a}_z$ m/s in a uniform magnetic field of density $\bar{B} = 0.05\bar{a}_x$ T, is a voltage induced into the wire? Explain why or why not.

3. a) Explain how and why a loop of wire rotating in a uniform magnetic field (see the following figure) produces a sinusoidal output, not a constant DC output.
   b) Does this principle relate to any major electrical component? Name it.

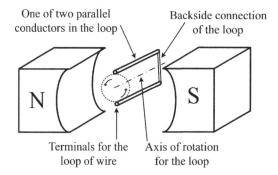

4. A flat disc of copper lies in the x–y plane, centered on a shaft that aligns with the z-axis and can rotate. It is immersed in a uniform magnetic field in the z-direction. A current is fed from the shaft to the outer edge of the disc (assume sliding electrical contacts are present).
   a) Make and label a sketch.
   b) Does the disc rotate? Why or why not?

5. A loop of wire with a resistor in the loop lies in the $z = 0$ plane. Is a voltage induced into the loop if:
   a) $\bar{B} = 0.05 \sin(5000t)\bar{a}_x$ T? Explain why or why not from Maxwell's equations.
   b) $\bar{B} = 0.05 \sin(5000t)\bar{a}_z$ T? Explain why or why not from Maxwell's equations.

6. Repeat Example 5.2.2 if the magnetic field waveform is a triangle wave with the same peak amplitude. What is the voltage waveform shape across the resistor?

7. Explain why an inductor opposes an instantaneous change of current:
   a) From a Faraday's law and Lenz's law viewpoint, and
   b) From a conservation of energy and power viewpoint.

8. a) Explain the difference between self-inductance and mutual inductance.
   b) What is the key difference between an ideal transformer and a mutual inductor?
   c) What is the effect of the difference in b) on the i–v relationships?

9. Develop the results for the total inductance L_T from the phasor voltage–current relationships for a mutual inductor and Kirchhoff's voltage law for:
   a) Two series-aiding mutual inductors. Hint: What is the "series" effect on the currents?
   b) Two series-opposing mutual inductors (caution: be careful with signs).

10. Develop the AC input impedance of a mutual inductor-coupled load from the phasor voltage–current relationships for a mutual inductor. Hint: also use Ohm's law for the load.

11. a) Explain how and why a capacitor completes the current path if a node for KCL surrounds only one of the plates.
   b) Why must the current be time dependent for current to pass through a capacitor?
   c) What is the key condition behind the circuit statement "a capacitor blocks DC"?

12. Explain why a capacitor opposes an instantaneous change of voltage:
   a) From a charge viewpoint, and
   b) From a conservation of energy and power viewpoint.

13. A capacitor has plate area $0.04\,\text{cm}^2$, plate separation $3\,\mu\text{m}$, a relative permittivity of 5, and a dielectric breakdown strength of $115\,\text{MV/m}$ (a mica-type dielectric). At $70\,\text{MHz}$,
   a) Determine the theoretical maximum peak current value for dielectric breakdown.
   b) Determine corresponding peak current for an air dielectric ($3\,\text{MV/m}$) if the plate separation is adjusted to produce the same capacitance as in a). Hint: Does d matter?

14. A capacitor has plate area $0.04\,\text{cm}^2$, plate separation $3\,\mu\text{m}$, and an air dielectric with a dielectric breakdown strength of $3\,\text{MV/m}$. At $70\,\text{MHz}$,
   a) Determine the theoretical maximum peak current value for breakdown.
   b) Determine corresponding peak current for a mica-type dielectric with a relative permittivity of 5 and a dielectric breakdown strength of $115\,\text{MV/m}$ if the plate separation is adjusted to produce the same capacitance as in a). Hint: Does d matter?

15. a) Name and explain two ways that a signal can couple from one circuit to another circuit even if they are not physically connected by conductors.
   b) What is the general name given to both types of coupling between circuits?
   c) Which type of coupling dominates in a given circuit situation? Hint: impedance levels.

16. a) Is there a magnetic field between the plates of a parallel-plate capacitor when the current is time-changing? Explain.
   b) Is there any current in a coax cable with an open-circuit load when the source is time-changing? Explain.

17. a) Identify by name each of the four Maxwell's equations in integral form.
   b) Identify each term in all four Maxwell's equations.
   c) Explain what each of these equations means in electromagnetics.

18. This question pertains to Appendix 5.B.
   a) Write Faraday's law (FL) in both integral and differential form.
   b) Write ACL in both integral and differential form.
   c) What is the mathematical operation in the integral form of FL and ACL that corresponds to the curl operation in the differential forms of these laws?

d) Write Gauss's law (GL) in both integral and differential form.

e) Write Gauss's law for magnetic flux (mag GL) in both integral and differential form.

f) What is the mathematical operation in the integral form of GL that corresponds to the divergence operation in the differential forms of these laws?

g) What is the difference, in general, between the integral form and the differential form of each law? (an open-ended question; think!)

6

Transmission Lines: Waves and Reflections

The previous chapters have focused on building a solid foundation of electromagnetics principles that govern the behavior of static and time-varying electric and magnetic fields. We are now ready to utilize what we have learned to address fundamental questions of electrical engineering: How is electrical energy transferred from point A to point B in any electrical circuit or system? How fast does the electrical energy travel? What does it mean for a coaxial cable to be a "50 Ω" or "75 Ω" cable? Why do practical high-frequency or high-speed digital circuits require specialized design techniques? The answer for these and other questions will become evident as we explore the topic of transmission lines.

What is a **transmission line** (T-line)? It is the set of conductors that connects the source with the load, and this structure forms the path to guide the electrical energy from the source to the load. Three examples are shown in Figure 6.1a–c. You are probably familiar with a pair of parallel wires and with coax. Another example is a trace on a printed circuit board (PCB), called microstrip, where the "hot" conductor is on one side of a dielectric sheet and the ground plane (return path) is on the other side.

This chapter examines phenomena that occur on transmission lines. There will be numerous physical interpretations and insights. We will learn that the voltage is not necessarily constant in a parallel circuit, for example. Several T-line quantities are introduced in the examination of transient waves in DC circuits, including the concept that a wave can reflect from the load (6.1). Then AC sinusoidal steady-state waves are described (6.2). The reflection

Electromagnetics and Transmission Lines: Essentials for Electrical Engineering, Second Edition.
Robert A. Strangeway, Steven S. Holland, and James E. Richie.
© 2023 John Wiley & Sons, Inc. Published 2023 by John Wiley & Sons, Inc.
Companion website: www.wiley.com/go/Strangeway/ElectromagneticsandTransmissionLines

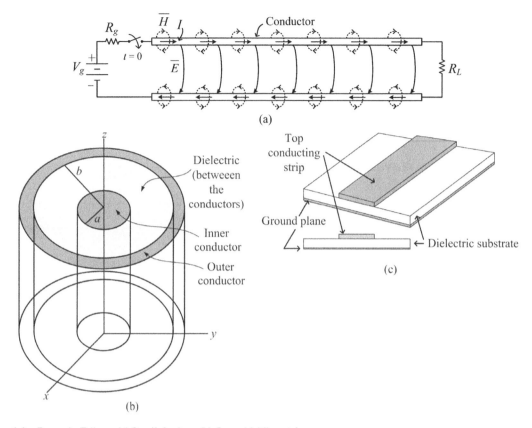

Figure 6.1 Example T-lines. (a) Parallel wires. (b) Coax. (c) Microstrip.

of AC waves in a T-line circuit and practical measures of reflections are then examined (6.3). The power in waves on T-lines leads to the highly useful and practical topics of power, gain, loss, and the scattering parameters, which are used as specifications of gains and reflections in terms of waves (6.4). These parameters are also utilized in high-frequency instrumentation and simulations. The T-line concepts in this chapter also establish a perspective for the theory and applications of T-lines that will be examined in Chapter 7.

6.1 Transient Waves in DC Circuits

This section introduces a key concept: in *any* electric circuit, energy travels as an *electromagnetic wave*! These electromagnetic waves have behavior similar to any other type of waves that you may already be familiar with, such as a finite velocity and reflections. We will initially examine this phenomenon in the context of transients in DC circuits.

6.1.1 Propagation of Waves in DC Circuits

Motivational Question Given that an electromagnetic wave takes time to propagate, what current flows in a T-line *before* the wave reaches the load?

Consider the DC circuit shown in Figure 6.2 with a voltage source V_g that has internal resistance R_g. The parallel "rods" generically represent the conductors in the T-line that connect the source to the load R_L (in practice

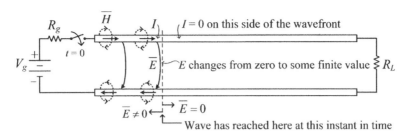

Figure 6.2 Displacement current visualization as the wave propagates.

the T-line could be coax, or traces on a PCB, and so forth). When the switch is closed in a DC circuit, an electromagnetic wave is launched onto the T-line in the circuit. The electromagnetic wave takes time to propagate. It is fast, but not infinitely fast, that is, it has a finite velocity (we will see soon that the wave travels at the speed of light in the medium surrounding the T-line wires). Consequently, the voltage from the source does *not* appear instantaneously across the load. There is a time delay τ_d from the switching event until the wave reaches the load. This is contrary to what you assumed in circuit analysis. The time it takes for the wave to reach the load can be ignored when τ_d is much shorter than time constants and other time effects in the circuit. However, this is not the case in high-speed digital circuits, where rising edges may be in the sub-nanoseconds (ns) time frame. Then the effect of wave propagation in a DC circuit, the topic of this section, must be considered.

When the switch is closed, the charges in the conductors of the circuit move in association with the electromagnetic wave, but the charges move at a far lower velocity.[1] How does the current complete its path just after the switch is closed and the electromagnetic wave has not reached the load yet? Recall how a capacitor operates. Charges accumulating on one plate repel charges off of the opposite plate. Charges have electric fields. Charges that are moving onto/off of plates result in a *time-changing* electric field between the plates. This time-changing electric field, displacement current,[2] is actually the "current" that completes the current path. In the T-line, the displacement current is between the conductors. The electric field changes from zero to some finite value as the electromagnetic wave propagates past each point along the T-line. *Displacement current completes the current path in the wave on a transmission line.*

Before the electromagnetic wave reaches the load, the wave does *not* "know" what R_L is yet (it hasn't reached the load). Hence, how much current flows? In other words, what impedance does the source "see"? A sequence of thoughts follows.

For DC steady-state circuits, impedance is automatically real (a resistance) and is the ratio of the DC voltage to the DC current. Real impedance exists when the voltage and current are in-phase (automatically true in DC). In circuit analysis, the resistance of a load represents the conversion of electrical energy into another form of energy (heat in a resistor, motion in a motor, and so forth). This energy conversion is not what a source directly "sees" when connected to a T-line; if it did, the energy would be converted and less or none of the energy would reach the load.

As the wave front propagates from the source to the load, as visualized in Figure 6.2, the voltage and the current step from zero to some value because E and H have stepped from zero to some value. What is occurring with the energy at the wave front? It is not converting to another form of energy, so it must be *transferring* along the T-line. But there is still a voltage-to-current ratio in this energy transfer. Thus, there is an impedance that represents this

---

1 The average velocity of a charge in a conductor is called the *drift velocity*. See Appendix 4C for a brief discussion.
2 See Section 5.3 for a discussion of displacement current.

energy transfer, which would also be the impedance "seen" by the source for the case of one traveling wave. It is named the **characteristic impedance** Z_0 (Ω). For a DC circuit:

$$\boxed{Z_0 \equiv \frac{V}{I}\Bigg|_{\substack{\text{one wave} \\ \text{(one direction)}}}} \tag{6.1}$$

Characteristic impedance is the voltage-to-current ratio of one traveling wave that represents real power *transfer* on a transmission line.

Characteristic impedance is ideally real (zero phase shift if AC) and practically mostly real. Why? Examine real power: V and I are in-phase (automatic for DC). However, if V and I are in-phase, $Z_0 = V/I$ is also real! It is emphasized that characteristic impedance is *not* the resistance of the conductors, which represents energy conversion to heat (dissipation). Again, Z_0 represents the voltage-to-current ratio of energy transfer on a T-line. For example, you may have heard of the term "50 Ω coax." The "50 Ω" designation refers to the characteristic impedance of the coaxial cable. A wave traveling in the coax will have a ratio of voltage to current equal to 50.

You might be wondering upon what does characteristic impedance depend and how is it determined. As we will examine in Chapter 7, Z_0 depends *only* on the T-line geometrical structure and materials (usually the dielectric).[3]

6.1.2 Reflection of Waves in DC Circuits

Motivational Question   Why doesn't an ohmmeter connected to 50 Ω coax indicate 50 Ω?

The previous section introduced the characteristic impedance of a T-line. Suppose you wished to measure the characteristic impedance of a length of coax with nothing connected to either end (open circuits at each end) by connecting an ohmmeter between the center conductor and outer shield at one end. You will find that the ohmmeter does *not* read 50 Ω; instead it reads as an open circuit. This reading occurs because there is *more than one wave* on the T-line, which is examined in the next section. The voltmeter is not measuring the voltage-to-current ratio of a single wave. However, what would the ohmmeter read if the coax were infinitely long? 50 Ω! The voltage-to-current ratio of that one wave on the infinitely long T-line would equal the characteristic impedance. Now let's investigate why multiple waves may be present in a DC circuit of finite size.

Consider what happens on a T-line when a DC source is switched on. A wave, called the **incident wave**, propagates from the source toward the load, as pictured in Figure 6.3. The voltage-to-current ratio of this wave is Z_0. What if $Z_0 \neq R_L$? When the initial wave from the source reaches the load, the voltage-to-current ratio of the incident wave does not equal the voltage-to-current ratio of the load. Hence, a reflection will occur in general because Ohm's law at the load must always be satisfied. The **reflected wave** (traveling away from the load) voltage and current add to the incident wave (traveling toward the load) voltage and current, respectively, so that Ohm's law at the load is satisfied:

$$R_{\text{Load}} = \frac{V_{\text{Total}}}{I_{\text{Total}}} = \frac{V_{\text{incident}} + V_{\text{reflected}}}{I_{\text{incident}} + I_{\text{reflected}}} \neq Z_0 \tag{6.2}$$

Thus, a key concept is that a wave is reflected to in order to "make up the difference" so that the *total* voltage and current at the load satisfy Ohm's law. You may be familiar with reflections in other contexts, such as waves on a rope or echoes in acoustics. In these other contexts, reflections occur due to similar conditions at the point of reflection. For example, when a wave on a rope reaches the anchored end point of the rope, a reflected wave is generated to satisfy the condition that the rope is stationary at the anchor point (sum of incident and reflected wave amplitudes on the rope cancel out to zero at the anchor point).

---

3 Technical note: This statement holds only for T-lines that carry transverse electromagnetic (TEM) waves, for which E and H are perpendicular to the direction of propagation. The vast majority of practical T-lines carry TEM or approximately TEM waves, and thus we will only consider TEM T-lines in this text.

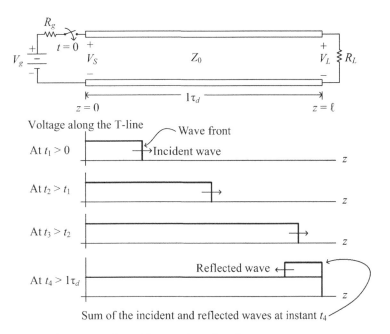

Figure 6.3 DC transient waves (lowercase z is position, not impedance).

The DC transient waves on a T-line can be determined with this concept of reflection in mind. Start with the source. The internal resistance of the source is modeled as R_g in series with V_g, as shown in Figure 6.4. At $t = 0^+$ (just after the switch is closed), the source "sees" Z_0 (it does not "see" the load yet), as previously discussed. By the voltage divider rule, the first incident wave in the circuit, V_1, is:

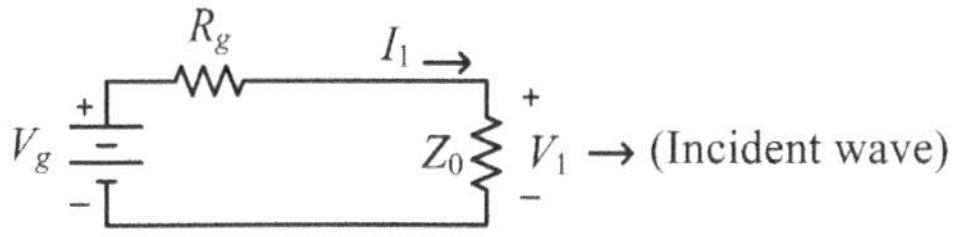

Figure 6.4 T-line circuit at $t = 0^+$.

$$V_1 = V_g \frac{Z_0}{R_g + Z_0} \tag{6.3}$$

When this wave reaches the load, a *reflection* may occur. How much? Use Ohm's law at the load to determine the amplitude of the reflected wave. Let V_1 and I_1 be the voltage and current of the incident DC transient wave (traveling toward the load). Let V_2 and I_2 similarly be the voltage and current of the reflected DC transient wave (traveling away from the load). At the load, the incident and reflected waves superpose, and the voltage and current for each wave are related by the characteristic impedance:

$$V_L = V_1 + V_2; \quad I_1 = \frac{V_1}{Z_0}; \quad I_2 = \frac{-V_2}{Z_0} \tag{6.4}$$

Note that the sign on the current reverses for the reflected wave, relative to the incident wave, because the waves are going in opposite directions. Superpose the currents at the load and incorporate the load resistance:

$$I_L = I_1 + I_2 = \frac{V_1}{Z_0} - \frac{V_2}{Z_0}; \quad R_L = \frac{V_L}{I_L} \tag{6.5}$$

$$R_L = \frac{V_1 + V_2}{\left(\dfrac{V_1 - V_2}{Z_0}\right)} \tag{6.6}$$

Perform the algebra to solve for the V_2/V_1 ratio:

$$\boxed{\frac{V_2}{V_1} = \frac{R_L - Z_0}{R_L + Z_0} \equiv \Gamma_L} \tag{6.7}$$

where Γ (uppercase Greek gamma) is defined as the **reflection coefficient** and the "L" subscript indicates that this reflection coefficient is at the location of the load. The reflection coefficient quantifies how much of the incident wave is reflected. The reflected wave travels back to the source. If the source is not matched, a re-reflection occurs. Why? (Same reasoning except it is the source resistance that is mismatched.) In DC circuits, the time period for which the reflections are significant is called the *transient*.

If $R_L = Z_0$, then $\Gamma_L = 0$, that is, there is *zero reflection* from the load! A load that has zero reflection is called an "impedance matched" load, though in practice is often simply called a "matched" load. The voltage-to-current ratio of the incident wave equals the voltage-to-current ratio of the load, so zero reflection is required to satisfy Ohm's law at the load.

Example 6.1.1

Determine the load and source voltage waveforms for a 100 V DC step input applied to a 50 Ω T-line with a shorted load. The source impedance is 50 Ω.

Solution

A "50 Ω T-line" means that the characteristic impedance of the T-line is 50 Ω. See Figure 6.5. First determine the reflection coefficients of the source and the load.

$$\Gamma_S = \frac{R_g - Z_0}{R_g + Z_0} = \frac{50 - 50}{50 + 50} = 0 \qquad \Gamma_L = \frac{R_L - Z_0}{R_L + Z_0} = \frac{0 - 50}{0 + 50} = -1$$

The source is matched (to the T-line), and the load is mismatched. At $t = 0^+$, determine the initial incident wave V_1. See Figure 6.6.

$$V_1 = \frac{V_g Z_0}{R_g + Z_0} = \frac{(100\,\text{V})50}{50 + 50} = 50\,\text{V}$$

Again, Z_0 is the V/I ratio of one traveling wave and represents real power *transfer*. Note that the wave launched onto a T-line by a matched source is *half* of the source voltage due to the voltage divider formed by its internal impedance and the characteristic impedance of the T-line. When the wave reaches the load, Eq. (6.7) can be used to determine that a wave is reflected with an amplitude of $V_2 = \Gamma_L V_1 = (-1)(50) = -50$ V. Figure 6.7a–c shows snapshots in time as the incident wave propagates toward the load, as the reflected wave travels away from the load, and the sum of these waves at a given instant in time. Note how the reflected wave subtracts from the incident wave as it propagates toward the source. When the reflected wave reaches the source, there is zero re-reflection because the source is matched. This circuit has reached steady state at $t = 2\tau_d$. The transient is over. We observe, however, that during the transient, the circuit did have a nonzero voltage on the T-line despite the shorted load.

Next the voltage levels for the source and load waveforms are determined:

$$\text{For } V_S: \qquad 0 < t < 2\tau_d: \quad V_S = V_1 = 50\,\text{V}$$

$$t > 2\tau_d \quad V_S = V_1 + V_2 = 50 - 50 = 0\,\text{V}$$

$$\text{Zero reflections from the source after } 2\tau_d$$

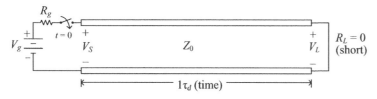

Figure 6.5 T-line circuit for Example 6.1.1 (V_g = 100 V, R_g = 50 Ω, Z_0 = 50 Ω).

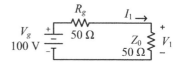

Figure 6.6 T-line circuit for Example 6.1.1 at $t = 0^+$.

Figure 6.7 DC transient waves for Example 6.1.1.
(a) Incident wave traveling toward the load.
(b) Reflected wave traveling toward the source.
(c) Total voltage along the T-line at instant t_6.

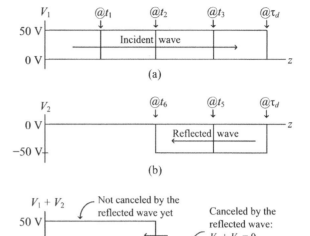

For V_L: $0 < t < 1\tau_d$: $V_L = 0\,\text{V}$ (V_1 has not reached the load yet)

$\quad\quad\quad t > 1\tau_d$: $V_L = V_1 + V_2 = 50 - 50 = 0\,\text{V}$

$\quad\quad\quad$ Zero re-reflections from the load after $1\tau_d$

The final source and load waveforms are sketched in Figure 6.8a. Once the DC transient is over, the DC steady-state voltage on the T-line, V_S and V_L, can be determined using normal circuit theory as a DC steady-state check:

$$V_S = V_L = \frac{V_g R_L}{R_g + R_L} = \frac{(100)(0)}{50 + 0} = 0\,\text{V}$$

The V_S and V_L plots *must always* approach the DC steady-state voltage on the T-line, which they do: 0 V across a short circuit. Notice that the characteristic impedance affects the individual waves in the DC transient but has no effect on steady-state DC.

The next example considers the reflection due to a finite resistance, unmatched load.

Example 6.1.2

Determine the load and source voltage waveforms for a 5 V DC step input applied to a 50 Ω T-line with a load resistance of 30 Ω. Assume the source resistance is 50 Ω.

Solution

First determine the reflection coefficients of the source and the load.

$$\Gamma_S = \frac{R_g - Z_0}{R_g + Z_0} = \frac{50 - 50}{50 + 50} = 0 \quad \Gamma_L = \frac{R_L - Z_0}{R_L + Z_0} = \frac{30 - 50}{30 + 50} = -0.25$$

The source is matched (to the T-line), and the load is mismatched. At $t = 0^+$, determine the initial incident wave V_1:

$$V_1 = \frac{V_g Z_0}{R_g + Z_0} = \frac{(5\,\text{V})50}{50 + 50} = 2.5\,\text{V}$$

Determine the reflected wave: $V_2 = \Gamma_L V_1 = (-0.25)(2.5) = -0.625\,\text{V}$

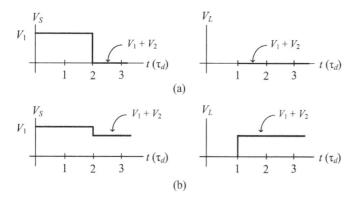

Figure 6.8 Voltage waveforms vs. t at *fixed* positions (load and source); $t(\tau_d)$ is the time in increments of time delays. (a) Example 6.1.1, with $R_L = 0\,\Omega$. (b) Example 6.1.2, with $R_L = 30\,\Omega$.

The source and T-line are matched. There are no more reflected waves. Next determine the voltage levels for the source and load waveforms:

For V_S: $0 < t < 2\tau_d$: $V_S = V_1 = 2.5\,\text{V}$

$t > 2\tau_d$: $V_S = V_1 + V_2 = 2.5 + (-0.625) = 1.875\,\text{V}$

Zero reflections from the source

For V_L: $0 < t < 1\tau_d$: $V_L = 0\,\text{V}$ (V_1 has not reached the load yet)

$t > 1\tau_d$: $V_L = V_1 + V_2 = 2.5 + (-0.625) = 1.875\ \text{V}$

Zero re-reflections from the load after $1\tau_d$

DC steady-state voltage check: $V_S = V_L = \dfrac{V_g R_L}{R_g + R_L} = \dfrac{(5\,\text{V})(30)}{50 + 30} = 1.875\,\text{V}$

The final source and load waveforms are sketched in Figure 6.8b. Note the contrast relative to Figure 6.8a. The steady-state voltage across the T-line is nonzero, as expected with a nonzero resistance load.

These ideas can be generalized into multiple reflections. A **bounce diagram** is often used to keep track of the individual waves (see Figures 6.9 and 6.10). In Example 6.1.1, $V_1 = 50\,\text{V}$, $V_2 = -50\,\text{V}$, and $V_3 = 0\,\text{V}$ (no source re-reflection). In contrast, when the source is not matched to the T-line and the wave that is reflected from the load reaches the source, it will be partially re-reflected and another wave will now travel to the load. This process will repeat until the reflections become negligible. The voltage waveforms at the source and load ends can be plotted by appropriately summing the waves per the bounce diagram. Those plots were shown for Example 6.1.1 in Figure 6.8a, and similarly for Example 6.1.2 in Figure 6.8b. Multiple reflections could lead to significantly longer transient times (much greater than the transient in the matched source cases considered here) and can lead to serious

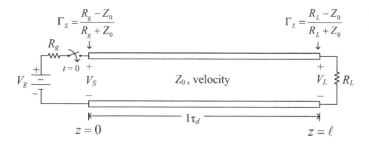

Figure 6.9 T-line notation for bounce diagram analysis.

corruption of signals. Multiple re-reflections will not be pursued in this text but are mentioned here for your awareness that they could exist and the importance of a source impedance that matches the characteristic impedance.

How long does it take for the wave to travel from one end of the T-line to the other end (the time delay τ_d)? An important result from electromagnetics is reported but not developed here. The velocity of an electromagnetic wave[4] is given by

$$\text{vel.} = \frac{\ell}{\tau_d} = \frac{1}{\sqrt{\mu\varepsilon}} = \frac{1}{\sqrt{\mu_o\varepsilon_o\mu_r\varepsilon_r}} = \frac{c}{\sqrt{\mu_r\varepsilon_r}} \qquad (6.8)$$

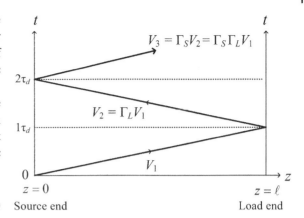

Figure 6.10 The bounce diagram.

where *vel.* is the velocity, ℓ is the distance traveled, and c is the speed of light in vacuum (2.998×10^8 m/s). *The velocity of waves on an ideal transmission line depends on the electrical properties of the insulating materials only*! For a T-line with good conductors, the relevant material properties are the permittivity (dielectric constant) and permeability of the insulating material.

As some final comments to this section,

- Pulse sources are common, such as those in digital circuits. The analysis proceeds in a similar manner to DC step sources, but in the source and load waveforms, the waves don't accumulate as they do with DC steps because the pulse turns off after a set amount of time.
- Instrumentation exists that is based on the principles in this section. The time-domain reflectometer (TDR) is an instrument that is used to locate faults on transmission lines.

6.2 Introduction to AC Wave Phenomena

Now we transition to AC sinusoidal steady-state signals on T-lines. This signal is incredibly important given its role in circuits as well as in signal analysis. The AC sinusoidal wave will be considered to be in *steady state* when the transient of the AC wavefront is over.

6.2.1 Traveling Waves

Motivational Question   How is AC wave propagation visualized in circuits?

Assume that the signal in the circuit varies sinusoidally as a function of position as well as a function of time (the position aspect is related to wavelength, which will be examined soon). Visualize a constant phase point on a traveling sine wave. Two snapshots in time of a traveling sine wave showing propagation of a constant phase point on the sinusoidal wave are illustrated in Figure 6.11.

---

4 Technical note: The velocity expression in Eq. (6.8) holds only for transverse electromagnetic (TEM) waves, for which E and H are perpendicular to the direction of propagation. The vast majority of practical T-lines carry TEM or approximately TEM waves, and thus we will only consider TEM T-lines in this text.

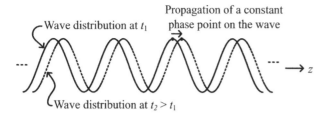

Figure 6.11 Propagation of an AC sinusoidal wave.

As the wave travels, *time goes on!* Thus, there are two independent variables: t, z. What is the mathematical description of a traveling wave? Try a sinusoidal expression that relates t and z in the argument:

$$v(t,z) = A \sin(\omega t\ ?\ \beta z)$$

where A, ω, and β are constants to be examined soon. What mathematical operation should be placed for "?" in the argument? Observe a *constant* phase point on the wave as the wave travels toward increasing z, as marked on Figure 6.11. The argument is constant for this point:

argument $= (\omega t\ ?\ \beta z) =$ total phase $=$ constant.

But z increases as t increases, so the two quantities in the argument must be subtracted in order to remain at a *constant* phase point, that is, for $(\omega t - \beta z) =$ total phase $=$ constant:

$$v(t,z) = A \sin(\omega t - \beta z)$$

which is a traveling wave in the $+z$ direction. What would be the expression for a wave in the opposite direction? $v(t, z) = A \sin(\omega t + \beta z)$

Note that the sine function operates only on phase (angle). Thus ω can be viewed as a time-to-angle converter, so β must be a distance-to-angle converter. The quantity ω is the same **radian frequency** used in circuits and signals. The quantity β is named the **phase constant**. It is a critical parameter in T-line visualization and calculations. Insight into β can be gained by the following viewpoints. Vary one independent variable and keep the other independent variable constant, as plotted in Figure 6.12a for a fixed (constant) position and Figure 6.12b for a fixed time (time is "frozen"). Just as one sinusoidal cycle occurs in one period T of time, one sinusoidal cycle occurs in one **wavelength** λ of distance. The radian frequency ω is the parameter that relates phase shift to time:

$$\omega = 2\pi f = \frac{2\pi}{T} = \frac{2\pi\ \text{rad}}{1\ \text{period}}\ (\text{rad/s}) \tag{6.9}$$

The phase constant β is the parameter that relates phase shift to distance:

$$\beta = \frac{2\pi}{\lambda} = \frac{2\pi\ \text{rad}}{1\ \text{wavelength}}\ (\text{rad/m}) \tag{6.10}$$

The distance viewpoint can be utilized to visualize the wave effects in circuits (the next topic).

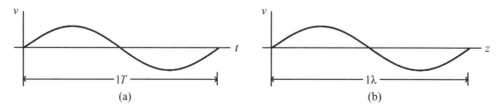

Figure 6.12 Voltage vs. time and position. (a) Fixed position. (b) Fixed time.

6.2.2 Wavelength and Distance Considerations

Motivational Question Why haven't we previously seen traveling wave effects in circuits?

Recall the velocity of a wave, using *vel.* for velocity here to distinguish it from voltage:

$$vel. = \lambda f \tag{6.11}$$

Consider the 15 cm long air-dielectric circuit in Figure 6.13a. At a given instant, what is the distribution of the voltage across the circuit at 60 Hz? At 10 GHz?

$$\text{At 60 Hz, } \lambda = \frac{vel.}{f} = \frac{c}{f} = \frac{2.998 \times 10^8 \text{ m/s}}{60 \text{ cycles/s}} \approx 5000 \text{ km} \approx 3100 \text{ miles}$$

$$\text{At 10 GHz, } \lambda = \frac{vel.}{f} = \frac{c}{f} = \frac{2.998 \times 10^8 \text{ m/s}}{10 \times 10^9 \text{ cycles/s}} \approx 0.03 \text{ m} = 3 \text{ cm}$$

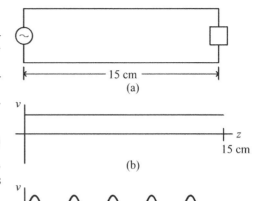

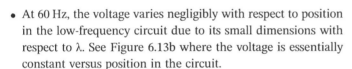

Figure 6.13 Voltage vs. position at different frequencies. (a) The circuit. (b) *v* distribution at 60 Hz. (c) *v* distribution at 10 GHz.

- At 60 Hz, the voltage varies negligibly with respect to position in the low-frequency circuit due to its small dimensions with respect to λ. See Figure 6.13b where the voltage is essentially constant versus position in the circuit.
- At 10 GHz, the voltage varies rapidly with respect to position. The circuit is several wavelengths long. See Figure 6.13c.

Thus, contrary to what you learned in circuits, we see that the voltage may not be constant in a parallel circuit. *Transmission line effects*, defined generally to be the voltage and current dependence on position, must be considered when the dimensions of the circuit are *not* small with respect to one wavelength. Generally, a rule of thumb is to consider T-line effects when the circuit dimensions are 0.10 wavelengths or larger. *All* electric circuits contain electromagnetic waves, and thus all circuits have wave behavior, but the T-line effects are generally negligible when the circuit size is small compared with wavelength. For this reason, you have likely not noticed these effects when working with low-frequency circuits. Note that T-line effects are not just the domain of high-frequency engineering. For example, with modern high-speed digital clock rates and signal rise and fall times, T-line effects must usually be considered in digital circuit design.

Example 6.2.1

Determine whether T-line effects will be negligible or significant for the following scenarios:

a) An audio circuit (20 Hz–20 kHz) on a breadboard (length ~17 cm)
b) A Wi-Fi receiver built into your phone operating at 2.4 GHz (length ~13 cm)
c) A 60 Hz power grid with a length of 10 miles
d) A 20 m length coax cable between your rooftop antenna and your television (assume a maximum frequency of 700 MHz).

Solution

Approach: for each case determine the operating wavelength and compare with the circuit size:

a) $\lambda = \dfrac{vel.}{f} = \dfrac{c}{f} = \dfrac{2.998 \times 10^8 \text{ m/s}}{20 \times 10^3 \text{ Hz}} \approx 15 \text{ km} \rightarrow \dfrac{\lambda}{10} \approx 1.5 \text{ km} \gg 17 \text{ cm: T-line effects are negligible.}$

b) $\lambda = \dfrac{vel.}{f} = \dfrac{c}{f} = \dfrac{2.998 \times 10^8 \text{ m/s}}{2.4 \times 10^9 \text{ Hz}} \approx 0.125 \text{ m} \rightarrow \dfrac{\lambda}{10} \approx 0.125 \text{ m, comparable to 13 cm: T-line effects are significant.}$

c) $\lambda = \dfrac{vel.}{f} = \dfrac{c}{f} = \dfrac{2.998 \times 10^8 \text{ m/s}}{60 \text{ Hz}} \approx 5000 \text{ km} \approx 3100 \text{ miles} \rightarrow \dfrac{\lambda}{10} \approx 310 \text{ miles} \gg 10 \text{ miles: T-line effects are negligible.}$

d) $\lambda = \dfrac{vel.}{f} = \dfrac{c}{f} = \dfrac{2.998 \times 10^8 \text{ m/s}}{700 \times 10^6 \text{ Hz}} \approx 0.428 \text{ m} \rightarrow \dfrac{\lambda}{10} \approx 0.0428 \text{ m} \ll 20 \text{ m: T-line effects are significant.}$

6.2.3 Electromagnetic (EM) Fields on a Transmission Line

Motivational Question   In an electric circuit, how does the electrical energy *transfer* from the source to the load?

As a start to the discussion of how energy is transferred on a T-line, let's first examine the power in a single AC wave. The calculation of the power in an AC traveling wave on a T-line is straightforward in terms of circuit quantities. Assume the voltage and the current in a traveling wave are in-phase.[5] Then using RMS magnitudes:

$$P_{\text{wave}} = V_{\text{wave}}I_{\text{wave}} = I_{\text{wave}}^2 Z_0 = \frac{V_{\text{wave}}^2}{Z_0} \tag{6.12}$$

Using peak values:

$$P_{\text{wave}} = \frac{V_{\text{wave}}I_{\text{wave}}}{2} = \frac{I_{\text{wave}}^2 Z_0}{2} = \frac{V_{\text{wave}}^2}{2Z_0} \tag{6.13}$$

The significance of this power equation is that it represents the power being *transferred* in one wave on a T-line. Notice that Eqs. (6.12) and (6.13) are the same form as AC power that you learned in circuits. While these equations describe the power in terms of the voltage and current associated with a traveling wave on a T-line, the lingering question is *how* does the energy transfer?

Does the energy transfer by the current? If it is by the kinetic energy of the moving charges in the current, then the charge entering the load must have greater energy than the charge leaving the load. By $\frac{1}{2}mv^2$, the velocity of the charges entering the load must be greater than the velocity of the charges leaving the load. But this scenario would lead to an accumulation of charge in the load and a consequent violation of KCL. Hence, the mechanical energy due to the movement of charge is *not* how energy transfers from the source to the load. Then, what is the mechanism for electrical energy transfer in circuits?

In any electric circuit, the energy transfers from the source to the load via the electromagnetic field.

Consequently, electrical energy is transferred in the electromagnetic field in the space around/between the conductors, *not* in the conductors (or the electrons). See Figure 6.14. The purpose of conductors is to *guide* the electromagnetic wave! Furthermore, an example of electrical energy transfer *without* conductors is a signal that propagates between antennas (sometimes called unguided waves) to be discussed in Chapter 8.

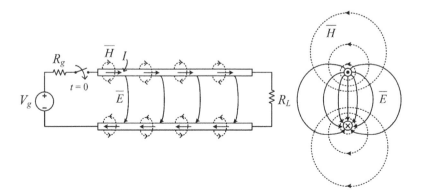

Figure 6.14 EM fields around a T-line (right-hand figure is an end view).

---

5 True for a lossless T-line, and also practically true for a low-loss T-line.

The two parallel long cylinders shown in Figure 6.14 form the *schematic symbol* of a transmission line. The actual T-line may be parallel wires, coax, a PCB trace, and so forth. The T-line symbol is used in circuit schematics to indicate that transmission line effects, such as significant phase shift vs. distance, are present.

The **Poynting vector**[6] $\overline{S}$ gives the direction of power flow in the EM field. See Figure 6.15.

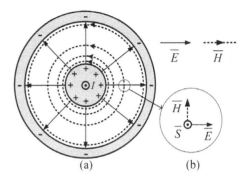

$$\overline{S} = \overline{E} \times \overline{H} \text{ units}: \quad (\text{V/m}) \cdot (\text{A/m}) = \left(\text{W/m}^2\right) \quad (6.14)$$

Note that the Poynting vector is an electromagnetic power *density* (W/m^2). It truly represents the power transfer in the electromagnetic field. Note the Poynting vector form is similar to the voltage-current product in circuit analysis. Consequently, circuit theory is a simplified, convenient representation of electromag-

Figure 6.15 EM fields in coax (end view). (a) Coax, end view. (b) Poynting vector.

netic field theory and is a model that makes reliable predictions under appropriate conditions. At this time, we will not calculate power using the Poynting vector because we will use the voltage and/or current on a T-line for this purpose. However, we will use the Poynting vector to determine the direction of power flow. As a precursor of things to come, the Poynting vector is also used in the power determination from an unguided electromagnetic wave to an antenna, a subject in Chapter 8.

Example 6.2.2
Determine the direction of power flow in the coax in Figure 6.15a.

Answer

Application of the Poynting vector cross product in Figure 6.15b shows that the direction of electromagnetic power flow is out of the page.

Another aspect to consider is how electromagnetic waves are guided. T-lines and waveguides are utilized to guide EM waves. A waveguide is similar to a T-line but only has one conductor or just a dielectric line. Unguided waves propagate between antennas, but antennas generally focus the electromagnetic wave propagation. Lists of common T-lines, waveguides, and antennas follow. You are encouraged to view sketches and pictures from online vendor catalog pages.

T-lines: two or more conductors (conductors are often held in place by dielectrics)

1. Lecher line (two open wire)	5. microstrip
2. twin lead	6. parallel plate
3. coax	7. balanced shielded line
4. stripline	

Waveguides: one conductor or dielectric line

1. rectangular waveguide	4. dielectric waveguide
2. circular waveguide	5. optical fiber
3. elliptical waveguide	

Antennas: no conductor between sending and receiving elements

1. dipole	3. parabolic dish
2. horn	4. PCB antenna

---

6 Not a misspelling of the word "pointing"! This aptly named quantity for the *direction* of power flow in an electromagnetic wave honors its discoverer, John Henry Poynting.

Waveguides will not be addressed in this text, but T-line expressions can often be adapted to waveguides, which are commonly used at higher microwave and millimeter-wave frequencies. Antennas will be examined in Chapter 8.

6.3 Reflections in AC Transmission Line Circuits

We previously developed how and why reflections occur in the context of DC transient waves. This section examines reflections in the context of AC steady-state circuits and introduces several commonly used parameters that quantify the amount of reflections in a circuit. The Smith chart is introduced as an insightful tool in calculating and visualizing reflections.

6.3.1 Reflected Waves and Measures of Reflection

Motivational Question   What is the effect of the reflection of an AC wave in a circuit?

Similar to the DC transient case, if there is a mismatch between the characteristic impedance and the load impedance in an AC circuit, a reflection must occur to satisfy Ohm's law at the load. The voltage-to-current ratio equals the characteristic impedance for both the incoming wave to the load, the **incident wave**, and the outgoing wave from the load, the **reflected wave**. The AC reflection coefficient is the same as for the DC transient case except the voltages are phasors and the load impedance is complex in general:

$$\Gamma_L = \frac{\widetilde{V}_{\text{refl}}}{\widetilde{V}_{\text{inc}}} = \frac{\widetilde{Z}_L - Z_0}{\widetilde{Z}_L + Z_0} \tag{6.15}$$

where the tilde over the phasor voltages and over the load impedance indicates that they are complex numbers in general, as used in AC circuit theory.[7] We will adopt a common AC T-line notation that will be used hereafter: the incident wave will be indicated by a "+" superscript and the reflected wave will be indicated by a "−" superscript:

$$\boxed{\Gamma_L = \frac{\widetilde{V}_L^-}{\widetilde{V}_L^+} = \frac{\widetilde{Z}_L - Z_o}{\widetilde{Z}_L + Z_o}} \tag{6.16}$$

For a passive load connected to a lossless T-line, normally $|\Gamma| \leq 1$.[8] The AC reflection coefficient of the load is complex and usually expressed in polar form. Reflection coefficient is one way to quantify the reflection from a load. Are there other ways that are used to quantify reflections of AC waves? Yes, and they are the **voltage standing wave ratio** (*VSWR*) and the **return loss** (*RL*). The concepts behind each measure are developed next.

What happens to the *total* voltage as a function of position on a T-line when AC incident and reflected waves superpose? Consider generic incident and reflected traveling sinusoidal waves as shown in Figure 6.16. The peak amplitudes are made constant (let $V^+ = V^- = A$), and the initial phase shifts are set to zero for mathematical convenience in this illustration. Start with the sum of the two traveling waves:

$$A \cos(\omega t - \beta z) + A \cos(\omega t + \beta z) \tag{6.17}$$

Apply the trigonometric identity $\frac{1}{2}\cos(x + y) + \frac{1}{2}\cos(x - y) = \cos x \cos y$:

$$A \cos(\omega t - \beta z) + A \cos(\omega t + \beta z) = \underbrace{2A \cos(\beta z)}_{\begin{pmatrix} \text{amplitude is} \\ \text{position} \\ \text{dependent} \end{pmatrix}} \underbrace{\cos(\omega t)}_{\begin{pmatrix} \text{varies} \\ \text{between} \pm 1 \\ \text{vs.time} \end{pmatrix}} \tag{6.18}$$

---

7 Not all complex quantities are marked with a tilde, but these quantities are understood to be complex in general. The reflection coefficient is a prime example of this practice.

8 $|\Gamma|$ can be greater than one, for example in an active circuit or with certain reactive loads on lossy T-lines (see Anderson, E. (1985). *Electric Transmission Lines Fundamentals*, 350–356. Reston).

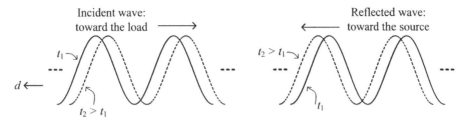

Figure 6.16 Incident and reflected AC sinusoidal waves.

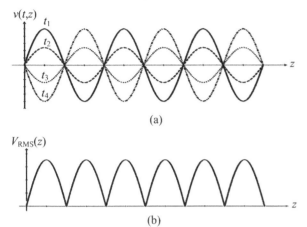

Figure 6.17 Standing wave. (a) Instantaneous wave snapshot at four time instants. (b) RMS standing wave pattern.

The peak amplitude is position dependent and varies "up and down" (increases and decreases) with respect to time, as shown in Figure 6.17a, but the nulls in the wave appear to "stand still." Hence, the sum of two oppositely directed traveling waves (same frequency) is a **standing wave**. Normally, the standing wave pattern is displayed with a root-mean-square (RMS) amplitude, as shown in Figure 6.17b.

The presence of a standing wave suggests there is another measure of reflection because if the reflected wave is absent, then a standing wave does not result. The standing wave patterns for three different loads are shown in Figure 6.18. The incident and reflected waves add at the maxima in the standing wave pattern. Conversely, the incident and reflected waves subtract at the minima in the standing wave pattern. The difference between the amplitudes of the maxima and minima increases as the reflected wave amplitude increases. A shorted load reflects the entire incident signal, so the minima are zero because the incident wave and reflected wave are of equal peak amplitude but 180° out-of-phase. A matched load has zero reflection, so there is zero interference and the standing wave pattern is flat. A general nonreactive load (with finite resistance) reflects less than a shorted load, so the standing wave pattern voltage difference at the minima and maxima is relatively smaller. The voltage levels at the maxima and at the minima for the general load are indicated on Figure 6.18. The specification of reflection that is based on standing waves is the *VSWR*, defined as:

$$VSWR \equiv \frac{V_{max}}{V_{min}} \qquad (6.19)$$

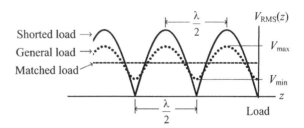

Figure 6.18 Standing wave patterns for three different loads on a T-line. V_{max} and V_{min} are labeled for the general load case.

where $V_{\max}$ is the sum of the peak values of $|\widetilde{V}^{+}|$ and $|\widetilde{V}^{-}|$ because the incident and reflected waves are in-phase at the position of maximum constructive interference, and $V_{\min}$ is the difference of the peak values of $|\widetilde{V}^{+}|$ and $|\widetilde{V}^{-}|$ because they are 180° out-of-phase at the position of maximum destructive interference. The RMS pattern repeats every half wavelength ($\lambda/2$).

The *VSWR* is related to the reflection coefficient:

$$VSWR = \frac{|\widetilde{V}^{+}| + |\widetilde{V}^{-}|}{|\widetilde{V}^{+}| - |\widetilde{V}^{-}|} = \frac{1 + \dfrac{|\widetilde{V}^{-}|}{|\widetilde{V}^{+}|}}{1 - \dfrac{|\widetilde{V}^{-}|}{|\widetilde{V}^{+}|}} = \frac{1 + |\Gamma|}{1 - |\Gamma|} \quad VSWR = \frac{1 + |\Gamma|}{1 - |\Gamma|} \tag{6.20}$$

Equation (6.20) reveals two important consequences about *VSWR*:

- *VSWR* does *not* contain phase information.
- $0 \leq |\Gamma| \leq 1$ corresponds to $1 \leq VSWR \leq \infty$

Another common specification for reflection from a load is return loss, *RL*. It is also related to reflection coefficient and is normally expressed in dB:[9]

$$RL(\mathrm{dB}) \equiv 10 \log\left(\frac{P_{\mathrm{inc}}}{P_{\mathrm{refl}}}\right) = -10 \log\left(\frac{P_{\mathrm{refl}}}{P_{\mathrm{inc}}}\right) = -10 \log\left(\frac{\dfrac{|\widetilde{V}^{-}|^{2}}{Z_0}}{\dfrac{|\widetilde{V}^{+}|^{2}}{Z_0}}\right) \tag{6.21}$$

$$RL(\mathrm{dB}) = -10 \log\left(\left|\frac{\widetilde{V}^{-}}{\widetilde{V}^{+}}\right|^{2}\right) = -20 \log\left|\frac{\widetilde{V}^{-}}{\widetilde{V}^{+}}\right| = -20 \log|\Gamma| \tag{6.22}$$

$$\boxed{RL(\mathrm{dB}) = -20 \log\,|\Gamma|} \tag{6.23}$$

The relationships between the three specifications of reflection that have been examined in this section are summarized in Table 6.1. What is considered to be a good match? This answer depends on the particular component, circuit, or system. An increase in reflection results in an increased percentage of the incident power that is reflected. The impacts of reflection from a mismatched load are (a) some of the incident power is not delivered to the load and (b) the reflected power reenters the source, potentially damaging it. For example, if the *VSWR* = 2.0, then *RL* = 9.54 dB and $|\Gamma|$ = 0.33. About 11% of the incident power is reflected. This level of reflection constitutes a reasonable match that is often a data sheet specification as a worst-case, acceptable match. See Table 6.2 for reflected power levels versus measures of reflection.[10]

Table 6.1 Relationships between specifications of reflection.

Matched condition		Reflection specification		Total reflection
0	$\leq$	$|\Gamma|$	$\leq$	1
1	$\leq$	*VSWR*	$\leq$	∞
∞ dB	$\geq$	*RL* (dB)	$\geq$	0 dB

---

9 One must be careful to distinguish between $RL(\mathrm{dB}) = -20 \log|\Gamma|$ and the reflection coefficient in dB, $\Gamma(\mathrm{dB}) = +20 \log|\Gamma|$; thus, $RL(\mathrm{dB}) = -\Gamma(\mathrm{dB})$.
10 $P_{\mathrm{refl}} = P_{\mathrm{inc}}|\Gamma|^{2}$; note that the power transmitted to the load, assuming no other losses, would be $P_{\mathrm{inc}}(1 - |\Gamma|^{2})$.

Table 6.2 Percent of incident power that is reflected vs. measures of reflection.

VSWR	$|\Gamma|$	RL (dB)	% of P_{inc} reflected
1	0	∞	0
1.5	0.20	14.0	4
2	0.33	9.5	11
3	0.50	6.0	25
4	0.60	4.4	36
6	0.71	2.9	51
8	0.78	2.2	60.5
10	0.82	1.7	67
20	0.90	0.9	82
100	0.98	0.2	96

Example 6.3.1

A load has a *VSWR* of 2.3. Determine the reflection coefficient and *RL* of this load.

Solution

Solve $VSWR = \dfrac{1 + |\Gamma|}{1 - |\Gamma|}$ for $|\Gamma| \rightarrow |\Gamma| = \dfrac{VSWR - 1}{VSWR + 1} = \dfrac{2.3 - 1}{2.3 + 1} = 0.394$

Check: $|\Gamma| \leq 1$ (good)

The angle of the reflection coefficient cannot be determined because there is no phase information in *VSWR*. *RL* is normally reported in dB:

$$RL(\mathrm{dB}) = -20 \log |\Gamma| = -20 \log (0.394) = 8.09 \, \mathrm{dB}$$

Check: $RL(\mathrm{dB}) \geq 0 \, \mathrm{dB}$ (good)

6.3.2 Smith Chart: Impedance and Measures of Reflection

Motivational Question   How are impedances and reflections plotted and related graphically?

Consider plotting impedances on a rectangular chart as shown in Figure 6.19a. It would be difficult to plot both large and small impedances on the same grid. The grid has no limits in three directions (positive real axis and positive and negative imaginary axes). A **Smith chart** is an impedance plot where these three axes "curve around" to meet up at infinity on the right-hand side. See Figure 6.19b. Thus, all impedances with a real part greater than or equal to zero are plotted on a finite-size graph. The grid obviously has nonuniform spacing.

An important convention used in engineering is **normalization**: to standardize by dividing the quantities of interest by a selected reference quantity. The normalization used with Smith charts is to divide all impedances by the characteristic impedance (Z_0) of the T-line and to divide all admittances by the characteristic admittance (Y_0) of the T-line:

$$z_L = \frac{\widetilde{Z}_L}{Z_0} \qquad y_L = \frac{\widetilde{Y}_L}{Y_0} = \widetilde{Y}_L Z_0 \qquad Y_0 = \frac{1}{Z_0} \tag{6.24}$$

Note that normalized impedances and admittances (no tilde shown) are complex in general. Also, be sure to distinguish normalized impedance z vs. distance variable z. The normalization of impedances allows a single, normalized Smith chart to be used regardless of the system characteristic impedance.

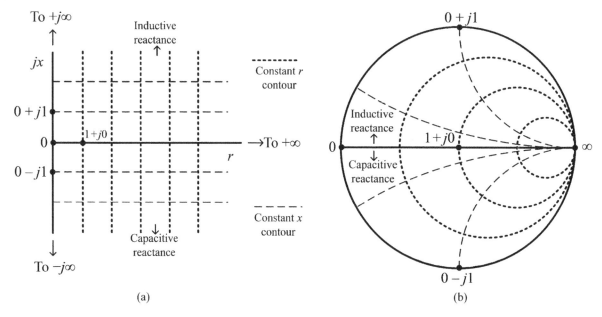

Figure 6.19 Two types of impedance plots. (a) Rectangular impedance plot. (b) Smith chart.

The tools required for manual Smith chart use are a Smith chart, a compass, a ruler, a pencil, and a competent operator. Historically, the Smith chart was developed in the late 1930s as a "graphical calculator" to simplify the numerous complex number computations needed in AC T-line analysis in a time before digital computers. Why is the Smith chart used in the computer age? The Smith chart is often used as follows:

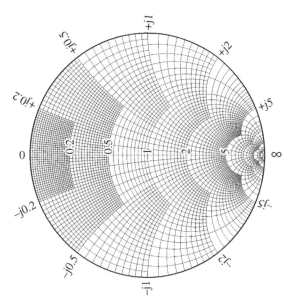

Figure 6.20 Finer Smith chart grid on commercial Smith charts.

- The Smith chart is a visual representation of an impedance "graph" with widespread understanding.
- Simulation software output is often plotted on Smith charts.
- Measurement instrumentation output on a Smith chart is standard industry practice (see the vector network analyzer [VNA] discussion later in this chapter).
- Certain high-frequency design techniques, such as impedance matching, rely on the Smith chart.

Commercial versions of the Smith chart are readily available. The resolution between divisions on a commercial Smith chart grid changes as one goes from low impedances to higher impedances. See Figure 6.20 for a Smith chart grid that was generated using software. A few uses of the Smith chart that are related to measures of reflection are examined in this section:

- Γ determination, both magnitude and phase
- *VSWR* determination
- *RL* determination

There are many more uses of the Smith chart in high-frequency electrical engineering, and some of these are examined in a continuation of the Smith chart discussion in Chapter 7. The remainder of this section is organized as follows:

- Smith chart conventions ("geography")
- How to read normalized impedances and admittances on a Smith chart
- Determining measures of reflection on a Smith chart (Γ, *VSWR*, and *RL*)
- An example

Some of the important Smith chart conventions (Smith chart "geography") are identified on Figure 6.21. The radially scaled parameters, which will be examined later in this section, utilize the following common abbreviations on the Smith chart:

- SWR is *VSWR*.
- RTN. LOSS is *RL*.
- RFL. COEFF, P is the fraction of the incident power that is reflected, $|\Gamma|^2$.
- RFL. COEFF, E or I is $|\Gamma|$.

How are impedances and admittances read on the Smith chart? Normalized impedance, $z = r + jx$, is plotted on a nonlinear grid consisting of circles or arcs of circles (contours). Similarly, normalized admittance, $y = g + jb$, can be plotted on the same grid. The circles are listed below:

a) Constant normalized resistance (r) contours are shown in Figure 6.22a.
b) Constant normalized reactance (x) contours are shown in Figure 6.22b.
c) Constant normalized conductance (g) contours are shown in Figure 6.22a.
d) Constant normalized susceptance (b) contours are shown in Figure 6.22b.

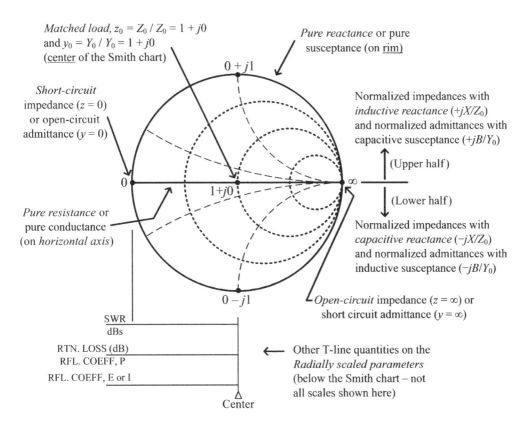

Figure 6.21 The Smith chart with some important normalized impedances and admittances labeled.

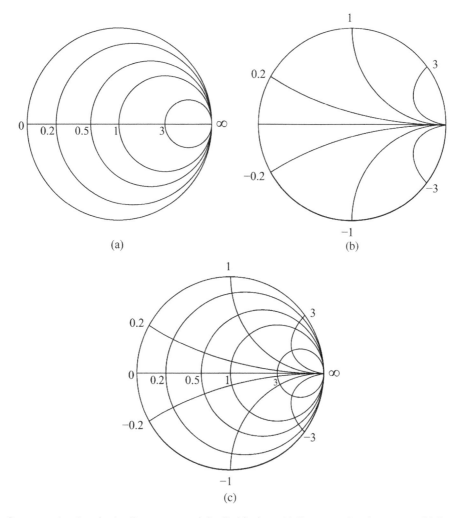

Figure 6.22 Constant *r* (or *g*) and *x* (or *b*) contours and the Smith chart. (a) Constant *r* (or *g*) contours. (b) Constant *x* (or *b*) contours (the contours are arcs of circles). (c) The Smith chart.

These circles are superposed to form a grid, as shown in Figure 6.22c.

Now let's consider how measures of reflection are accommodated on a Smith chart. Reflection coefficient has been related to impedance by $\Gamma_L = \dfrac{\widetilde{V}_L^-}{\widetilde{V}_L^+} = \dfrac{\widetilde{Z}_L - Z_0}{\widetilde{Z}_L + Z_0}$. Is there a way to plot this relationship? Both z_L and Γ_L are complex. A standard rectangular plot is insufficient. However, the Smith chart will accommodate both z_L and Γ_L. In fact, the Smith chart is also a polar chart of reflection coefficient as shown in Figure 6.23a. For example, the following reflection coefficients relate to the corresponding Smith chart normalized impedances (see Figure 6.23a and b):

$\Gamma = 1\angle 180° \rightarrow z = 0$ (far left on the Smith chart)

$\Gamma = 0 \rightarrow z = 1 + j0$ (center of the Smith chart)

$\Gamma = 1\angle 0° \rightarrow z = \infty$ (far right on the Smith chart)

One could superpose the reflection coefficient and Smith chart grids, but the grids, even in different colors, become difficult to read (note the fine grid on a commercial Smith chart). Instead, the magnitude of the reflection coefficient is plotted on the Smith chart separately from the angle of the reflection coefficient:

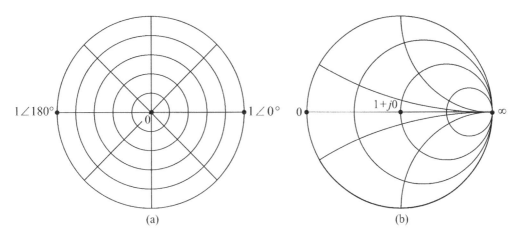

Figure 6.23   Reflection coefficient and Smith charts. (a) Polar chart of Γ ($|\Gamma| \propto$ radius, $\angle \Gamma$ = angle). (b) Smith chart.

Figure 6.24   Measures of reflection on the Smith chart.

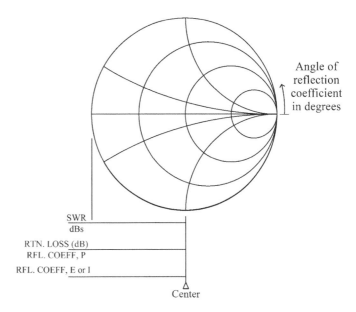

- $|\Gamma|$, *VSWR*, and *RL* (dB) are plotted on the **RADIALLY SCALED PARAMETERS** below the Smith chart. The radius is measured from the **CENTER**.
- $\angle \Gamma$ is plotted on one of the four scales just outside the rim of the Smith chart.

A generic plot of these four scales is shown in Figure 6.24. You should locate these scales on a commercial Smith chart. The radially scaled parameters may be in different places below the Smith chart. Often, the quantities are labeled on the Smith chart as follows (bold):

$|\Gamma|$ = **RFL. COEFF, E or I** [note: **RFL. COEFF, P** = $|\Gamma|^2$]
VSWR = **SWR** [note: **dBs** = 20 log (*VSWR*) = *SWR* (dB)]
RL (dB) = **RTN. LOSS (dB)**

One further note on *VSWR*: the *VSWR* can also be read off of the right-hand real axis if a circle that has a radius from the center of the Smith chart to the impedance of concern is drawn, centered on the Smith chart center. This

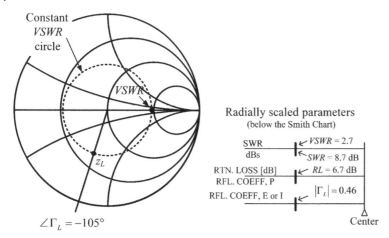

Constant
VSWR
circle

VSWR

z_L

$\angle\Gamma_L = -105°$

Figure 6.25   Smith chart for Example 6.3.2.

Radially scaled parameters
(below the Smith Chart)

SWR / dBs	← *VSWR* = 2.7
	← *SWR* = 8.7 dB
RTN. LOSS [dB]	← *RL* = 6.7 dB
RFL. COEFF, P	
RFL. COEFF, E or I	$|\Gamma_L| = 0.46$

Center

circle is called a **constant VSWR circle** because any normalized impedance on this circle has the same *VSWR* (a constant *VSWR* has a constant reflection coefficient magnitude). The intersection of the constant *VSWR* circle with the right-hand real axis equals the *VSWR*.[11] The use of these scales is illustrated in the next example.

Example 6.3.2

If the load impedance is $28 - j32\,\Omega$ and the characteristic impedance of a lossless T-line is $52\,\Omega$, determine the (a) reflection coefficient, (b) *VSWR*, (c) *SWR* (dB), and (d) *RL* (dB) for the load.

Solution

Plot z_L : $\widetilde{Z}_L = 28 - j\,32\,\Omega, \ Z_0 = 52\,\Omega \rightarrow z_L = \dfrac{\widetilde{Z}_L}{Z_0} = \dfrac{28 - j32}{52} = 0.538 - j0.615$

See Figure 6.25.

- Use a compass to draw a *constant VSWR circle* through z_L centered on the Smith chart center with a radius from the center of the Smith chart to the normalized load impedance.
- Use the radius of the constant *VSWR* circle to mark $|\Gamma_L|$, *VSWR*, *SWR* (dB), and *RL* (dB) on the radially scaled parameters, from the CENTER outward.
- Extend the line drawn from the Smith chart center to z_L through the scales on the outer rim of the Smith chart. The angle of Γ_L is on one of the Smith chart rim scales.
- The *VSWR* is also at the intersection of the constant *VSWR* circle and the right-hand side of the horizontal axis.

Results: $\Gamma_L = 0.46\,\angle -105°$ *VSWR* = 2.7   *SWR* = 8.7 dB   *RL* = 6.7 dB.

6.4   Scattering Parameters (*S*-parameters)

The **scattering parameters** (*S*-parameters) are a set of quantities (parameters) that specify transmissions through and reflections from a component or system. The *S*-parameters are important in measurements and design because they specify quantitative behavior (gains and reflection coefficients, both magnitude and phase, for example) *in terms of waves*. In this section, power, gain, and loss are introduced first because the magnitudes of the *S*-parameters are often related to these quantities. Then the *S*-parameters are defined and illustrated by several examples. Finally, the VNA is examined because it displays measurements primarily in terms of *S*-parameters.

---
11 Ramo, S., Whinnery, J.R., and Van Duzer, T. (1994). *Fields and Waves in Communication Electronics*, 3e, 241. New York: Wiley.

6.4.1 Power, Gain, and Loss

Figure 6.26 P_{in} and P_{out}.

Motivational Question How are gain and loss practically utilized in circuits with waves?

It was noted previously that the power in a traveling wave on a T-line is V^2/Z_0, where V is the RMS value of the traveling wave voltage. Power is a useful quantity at high frequencies because it can be easily measured and is crucial in assigning power levels at various points within a system (system power budget). Specifications and measurements often involve ratios of powers. Some of these topics are examined in this section.
Note: The terms **loss**, **attenuation**, and **insertion loss** (*IL*) are used interchangeably in this section.[12]

Recall **power gain** is fundamentally a power ratio: $G = $ Gain, $G = \dfrac{P_{out}}{P_{in}}$. See Figure 6.26. The gain in decibels equals 10 times the log of gain as a power ratio: G (dB) $= 10 \log (G)$. **Attenuation** (loss) is also a power ratio: $A \equiv \dfrac{P_{in}}{P_{out}} = \dfrac{1}{G}$, so $A(\text{dB}) = 10 \log \left(\dfrac{P_{in}}{P_{out}} \right) = -G(\text{dB})$. The attenuation in dB is a positive number for components with a power gain ratio less than one.

Example 6.4.1
What is (a) the gain and (b) the loss in dB of a component that dissipates 10% of the input power?
See Figure 6.27.

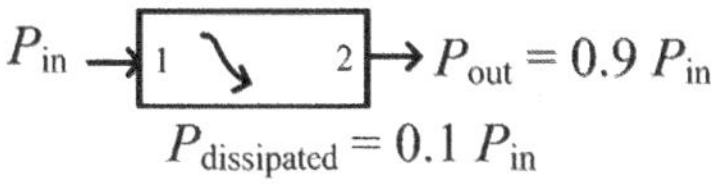

Figure 6.27 Diagram for Example 6.4.1.

Solution

Determine the gain and loss power ratios within the corresponding dB expressions.

$$G = 10 \log \left(\frac{P_{out}}{P_{in}} \right) = 10 \log \left(\frac{0.9 P_{in}}{P_{in}} \right) = 10 \log (0.9) = -0.46 \, \text{dB}$$

$$A = 10 \log \left(\frac{P_{in}}{P_{out}} \right) = 10 \log \left(\frac{1}{0.9} \right) = +0.46 \, \text{dB}$$

Losses give negative dB values for gain! To keep numbers "nice," one usually specifies gain when $P_{out} > P_{in}$ and loss when $P_{in} > P_{out}$.

An observation from a basic property of logarithms follows:

$$G(\text{dB}) = 10 \log \left(\frac{P_{out}}{P_{in}} \right) = 10 \log (P_{out}) - 10 \log (P_{in}) \tag{6.25}$$

How is a power level represented in dB form? The power in dBm, P (dBm), is pronounced "the power in d b m" where the "m" reminds us that the power is referenced to 1 milli-watt (mW):

Definition: $P(\text{dBm}) \equiv 10 \log \left[\dfrac{P(\text{W})}{1 \, \text{mW}} \right] = 10 \log \left[\dfrac{P(\text{W})}{0.001 \, \text{W}} \right]$

Practically: Take 10 log of the power in mW: $P(\text{dBm}) = 10 \log [P \, (\text{mW})]$
The power in dBm is considered to be an absolute quantity (unlike the dB, which is used for gain and attenuation).

---

12 There can be a difference between attenuation and insertion loss if the source and/or the load are not well-matched. See Lance, A.L. (1964). *Introduction to Microwave Theory and Measurements*, 212–213. New York: McGraw-Hill.

Example 6.4.2

Solution

Express 0.0182 W in dBm.

Formally: $P(\text{dBm}) = 10 \log \left[\dfrac{0.0182\,\text{W}}{0.001\,\text{W}} \right] = 10 \log (18.2) = 12.6$ dBm

Practically: $P(\text{dBm}) = 10 \log \left[0.0182\,\text{W} \left(\dfrac{1000\,\text{mW}}{\text{W}} \right) \right] = 10 \log (18.2) = 12.6$ dBm

Example 6.4.3 Express 12.6 dBm in mW.

Solution

Invert the logarithm in the P (dBm) expression:

$$P(\text{mW}) = 10^{[P(\text{dBm})/10]} = 10^{(12.6/10)} = 10^{1.26} = 18.2\,\text{mW} = 0.0182\,\text{W}$$

The schematic symbols for commonly used high-frequency components are given in Table 6.3.[13] These symbols will be used in subsequent figures.

Why should one bother with P (dBm)? Consider the implications of the next example.

Table 6.3 Schematic symbols for common RF/microwave components.

Component name	Symbol	Function
AC source		Generates an RF/microwave signal
Fixed attenuator		Provides a fixed amount of loss to the signal
Variable attenuator		Provides a variable amount of loss to the signal
Amplifier		Provides gain to the signal
Directional coupler		Samples signal (see text for elaboration)
Phase shifter		Shifts the phase of the signal as it passes through
Termination		Load equal to Z_0, absorbs $\approx$ all incident power

---

13 See Laverghetta, T.S. (2005). *Microwave and Wireless Simplified*, 2e. Massachusetts: Artech House, Chapter 4, for enhanced descriptions. Several RF/microwave component vendors provide component descriptions in their website literature. There are numerous RF/microwave/wireless textbooks with detailed analysis and/or design content. See Pozar, D.M. (2012). *Microwave Engineering*, 4e. New Jersey: Wiley; Ludwig, R. and Bretchko, P. (2000). *RF Circuit Design – Theory and Applications*. New York: Pearson; and Rizzi, P.R. (1988). *Microwave Engineering – Passive Circuits*. New York: Pearson, for example.

Example 6.4.4

Express the relationship between P_{in} and P_{out} of the circuit in Figure 6.28 in dB form.

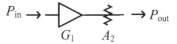

Solution

$$P_{in}G_1G_2 = P_{out}$$

$$P_{in}G_1 \left(\frac{1}{A_2}\right) = P_{out}$$

$$10\log\left(P_{in}\frac{G_1}{A_2}\right) = 10\log\left(P_{out}\right)$$

$$10\log\left(P_{in}\right) + 10\log\left(G_1\right) - 10\log\left(A_2\right) = 10\log\left(P_{out}\right)$$

$$P_{in}(\text{dBm}) + G_1(\text{dB}) - A_2(\text{dB}) = P_{out}(\text{dBm})$$

Figure 6.28 Circuit for Example 6.4.4.

The implication of the previous example is that using the dB form converts a multiplication/division calculation into an addition/subtraction calculation (convenient!). RF, microwave, and optical lab and test equipment readouts and simulations are often in dB or dBm.

Example 6.4.5

If the power meter reading is -2.2 dBm at the output of an attenuator, determine the power at the input in mW if the attenuator is set to 5 dB.

Solution

See Figure 6.29.

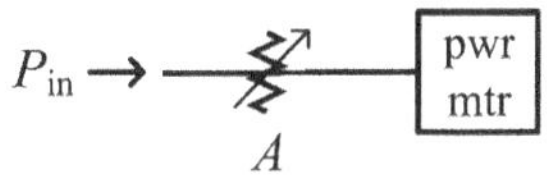

$$P_{in}(\text{dBm}) - A(\text{dB}) = P_{out}(\text{dBm}) \rightarrow P_{in}(\text{dBm}) = P_{out}(\text{dBm}) + A(\text{dB})$$

$$P_{in}(\text{dBm}) = P_{out}(\text{dBm}) + A(\text{dB}) = -2.2 \text{ dBm} + 5 \text{ dB} = +2.8 \text{ dBm}$$

$$P_{in}(\text{mW}) = 10^{[P(\text{dBm})/10]} = 10^{0.28} = 1.9 \text{ mW}$$

Figure 6.29 Circuit for Example 6.4.5.

Note that a "dB" power ratio modifies a "dBm" power level to a new "dBm" power level. The reader is referred to Section 6.6 for elaboration on dBm "Dos" and dBm "Don'ts." Also note that two powers in dBm cannot be added because that would result in multiplication of power!

For higher power levels, or in communication link calculations, the dBW is often used.

$$\text{Definition:} \quad P(\text{dBW}) \equiv 10\log\left[\frac{P(\text{W})}{1\text{ W}}\right]; \quad \text{Practically:} \quad P(\text{dBW}) = 10\log\left[P(\text{W})\right] \tag{6.26}$$

All calculations are analogous to those using the dBm. Note that a dB scale for power "compresses" the number range, as shown in Table 6.4.

Example 6.4.6

A ground transmitter in a satellite link has a power output of 35 dBW. What is the power in W and in kW?

Solution

$$P(\text{W}) = 10^{[P(\text{dBW})/10]} = 10^{(35/10)} = 10^{3.5} = 3162 \text{ W} = 3.162 \text{ kW}$$

Table 6.4 Power in dBm and dBW.

P (mW)	P (dBm)	P (W)	P (dBW)
$0.001 = 1\,\mu W$	-30	10^{-6}	-60
$0.01 = 10\,\mu W$	-20	10^{-5}	-50
$0.1 = 100\,\mu W$	-10	10^{-4}	-40
1	0	10^{-3}	-30
2	$+3$	2×10^{-3}	-27
5	$+7$	5×10^{-3}	-23
10	$+10$	10^{-2}	-20
100	$+20$	10^{-1}	-10
1000	$+30$	1	0
10^4	$+40$	10	$+10$
10^5	$+50$	100	$+20$
10^6	$+60$	$1000 = 1\,kW$	$+30$

6.4.2 *S*-parameter Definitions

Motivational Question   What are the quantities used and displayed in high-frequency measurements and simulations?

At low frequencies, voltages and currents can be measured in circuits. Measurements of voltages and currents at high frequencies become difficult. Instead, traveling waves are commonly utilized in modern high-frequency measurements and simulations. Ratios of these waves are often expressed as the **scattering parameters**, usually abbreviated as *S-parameters*. VNAs give direct readouts in terms of *S*-parameters. Other specifications, such as reflection coefficient, *RL*, power gain, and insertion loss are easily related to *S*-parameters. The Smith chart is a common plot for reflection *S*-parameters. The objectives of this section are to:

- Express *S*-parameters in terms of waves on a transmission line
- Explain why the *S*-parameters are useful at high frequencies
- Relate *S*-parameters to reflection coefficient, *RL*, power gain and insertion loss

A set of two terminals in the circuit is called a *port*. The basis for *S*-parameters begins with the idea that a component or circuit can be viewed from a system viewpoint as having multiple ports. A *two-port* network is illustrated in Figure 6.30. For example, recall that an amplifier has an input set and an output set of terminals, that is, two ports. The internal details of the two-port network are not important in this viewpoint. The variables at each of the ports, and the relationships between those variables, are important. For example, in an audio amplifier, the voltages and currents at the input and output ports are the variables of interest. The voltage and current gains and the input and output impedances are the relationships of interest between the variables. For high-frequency analysis, the variables that are usually of interest are the incident and reflected *waves* at each port, labeled in Figure 6.30a, b, respectively. The relationships of interest are the reflection coefficients of ports and the transmission coefficients between pairs of ports, that is, the *S*-parameters.

Note: Although there are *n*-port networks in general, we will restrict the following *S*-parameters discussion to two-port networks. The concepts are easily extended to *n*-ports.[14]

---

14 See Ghannouchi, F.M. and Mohammadi, A. (2009). *The Six-Port Technique with Microwave and Wireless Applications*. Massachusetts: Artech House. section 1.1.2, for example.

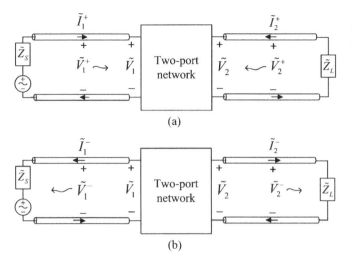

(a)

(b)

Figure 6.30 Waves for *S*-parameters setup.
(a) Incident waves. (b) Reflected (outgoing) waves.

The two-port schematic for the *S*-parameter setup is shown in Figure 6.30. The wave that is outgoing from a port consists of two components, one due to a reflection (refl) from the port and one due to a transmission (x-mit) from another port. For port 1:

$$\tilde{V}_1^- = \tilde{V}_1^-\Big|_{\text{(due to refl from port 1)}} + \tilde{V}_1^-\Big|_{\text{(due to x-mit from port 2)}} \tag{6.27}$$

and for port 2:

$$\tilde{V}_2^- = \tilde{V}_2^-\Big|_{\text{(due to refl from port 2)}} + \tilde{V}_2^-\Big|_{\text{(due to x-mit from port 1)}} \tag{6.28}$$

It is important to distinguish between the *total* port voltage at each port (such as $\tilde{V}_1$) from the *incident* and *reflected* waves at that port (such as $\tilde{V}_1^-$ and $\tilde{V}_1^+$).

So how are the reflections from a port separated from the transmission from another port? The key realization is that if the "other" port is terminated with the characteristic impedance Z_0, then a transmission from the other port will not occur. For example, if port 2 is terminated in Z_0 as shown in Figure 6.31(a) $(\tilde{Z}_L = Z_0)$, then there is zero signal input into port 2, and $\tilde{V}_1^-$ is due only to a reflection from port 1. Then the reflection coefficient can be formed.

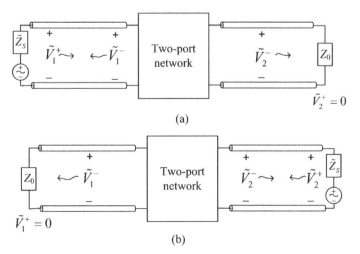

(a)

(b)

Figure 6.31 Waves for the individual *S*-parameters. (a) waves for S_{11} and S_{21}; note that port 2 is matched. (b) waves for S_{22} and S_{12}; note that port 1 is matched.

A **reflection coefficient** is designated by S_{ii}, where i is the port number (the two numbers in the subscript are identical), assuming *the other port is terminated in* Z_0. For port 1:

$$S_{11} = \Gamma_1 = \left.\frac{\widetilde{V}_1^-}{\widetilde{V}_1^+}\right|_{\text{port 2 is terminated in } Z_0} \tag{6.29}$$

Notice in Figure 6.31a if port 2 is terminated in Z_0, then $\widetilde{V}_2^+ = 0$ and the only contribution to $\widetilde{V}_1^-$ is the reflection of $\widetilde{V}_1^+$. Hence, it means that *the other port of the network is terminated in the characteristic impedance of the T-line connected to that port*. This concept is why the S-parameters are so applicable at higher frequencies. Matching a port is often (but not always) straightforward. Thus, in general, S_{ii} is the reflection coefficient of the ith port with all other ports connected to a matched termination (only one other port in our discussion). Similarly, for port 2, Figure 6.31b:

$$S_{22} = \Gamma_2 = \left.\frac{\widetilde{V}_2^-}{\widetilde{V}_2^+}\right|_{\text{port 1 is terminated in } Z_0} \tag{6.30}$$

A **transmission coefficient** is designated by S_{ij}, where the wave transmits from port j to port i (note the *reverse* order of the port numbers in the index).

$$S_{12} = \left.\frac{\widetilde{V}_1^-}{\widetilde{V}_2^+}\right|_{\text{port 1 is terminated in } Z_0} \qquad S_{21} = \left.\frac{\widetilde{V}_2^-}{\widetilde{V}_1^+}\right|_{\text{port 2 is terminated in } Z_0} \tag{6.31}$$

The same concept of matching the other ports of the network is applied to transmission coefficients. For example, consider S_{12}. If port 1 in Figure 6.31b is terminated in Z_0, then $\widetilde{V}_1^+ = 0$ and the only contribution to $\widetilde{V}_1^-$ is the transmission from port 2.

Practically, the magnitudes of S-parameters are often utilized in dB form. For *reflections*, S_{ii} in dB equals the negative of the *RL* for the ith port:

$$\boxed{|S_{ii}|(\text{dB}) = 20 \log\left(|S_{ii}|\right) = 20 \log\left(|\Gamma_i|\right) = -RL_i(\text{dB})} \tag{6.32}$$

Thus, S_{ii} can be determined from the reflection coefficient or, for the *magnitude* of S_{ii}, from any of its related measures, namely *RL* or *VSWR*.

For *transmissions*, $|S_{ij}|$ can be related to power gain G and insertion loss *IL* (using S_{21} to illustrate):

$$|S_{21}| = \frac{|\widetilde{V}_2^-|}{|\widetilde{V}_1^+|} = \sqrt{\frac{P_{\text{out2}}}{P_{\text{in1}}}} = \sqrt{G} = \frac{1}{\sqrt{IL}} \tag{6.33}$$

Squaring both sides: $|S_{21}|^2 = \dfrac{P_{\text{out2}}}{P_{\text{in1}}} = G = \dfrac{1}{IL}$ $\qquad\qquad$ (6.34)

Converting into dB form: $|S_{21}|(\text{dB}) = 20 \log|S_{21}| = G(\text{dB}) = -IL(\text{dB})$ $\qquad\qquad$ (6.35)

The quantity $|S_{21}|^2$ is often called the forward transducer power gain and $|S_{12}|^2$ the reverse transducer power gain. An S-parameter in dB is automatically a magnitude-only quantity. Thus, in general,

- S_{ii} = port reflection coefficient, $|S_{ii}|$ (dB) $= 20 \log(|S_{ii}|) = 20 \log(|\Gamma_i|) = -RL_i(\text{dB})$
- $S_{ij}|_{i \neq j}$ = transmission coefficient, $|S_{ij}|(\text{dB}) = 20 \log|S_{ij}| = G(\text{dB}) = -IL(\text{dB})$
- $|S|^2$ = a power ratio in general

An important property of many passive circuits and components[15] is **reciprocity**. In general, reciprocity is the equality of the transfer relationships between two ports of a network in both directions. For the *S*-parameters, it is expressed as a straightforward equality for a two-port network:

$$S_{21} = S_{12} \tag{6.36}$$

or, in general:

$$S_{ij} = S_{ji} \quad (i \neq j) \tag{6.37}$$

The indices are not equal because reciprocity concerns a transmission between two distinct ports, not reflections from ports. Thus, a reciprocal network has equal transmission coefficients (magnitude and phase) in both directions for each pair of ports. All passive circuits, such as attenuators and filters, are reciprocal. Active circuits, such as amplifiers, are examples of nonreciprocal two-port components. Note that reciprocity does *not* imply equality of reflection coefficients.

6.4.3 *S*-Parameter Examples

Example 6.4.7

Why is an amplifier not reciprocal?

Answer

An amplifier is not reciprocal because the transfer of power from port 1 (input) to port 2 (output) does not equal the transfer of power from port 2 (output) to port 1 (input). In other words, the forward gain of an amplifier, typically 10–30 dB, does not equal the reverse **isolation** (insertion loss in the undesired direction through a component), typically 20 or more dB for an amplifier. Thus, $S_{21} \neq S_{12}$ and the amplifier is not reciprocal.

Example 6.4.8

A practical termination has $R_L \approx Z_0$ but $R_L \neq Z_0$, so there is a small residual reflection. Consider the termination shown in Figure 6.32.

a) Determine the magnitude of S_{11}.
b) Determine the reflection coefficient and the *VSWR*.

Solution

a. $RL_1 \text{ (dB)} = -20 \log |\Gamma_1| = -20 \log |S_{11}| \rightarrow |S_{11}| = 10^{[-RL(\text{dB})/20]} = 10^{[-(50 \text{ dB})/20]} = 0.00316$
b. $|S_{11}| = |\Gamma_1| = 0.00316$

There is not enough information to determine the angle of S_{11}!

$VSWR = \dfrac{1 + |\Gamma|}{1 - |\Gamma|} = 1.006$ This *VSWR* level is generally considered to be very low.

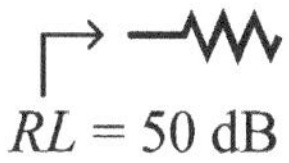

$RL = 50 \text{ dB}$

Figure 6.32 Termination for Example 6.4.8.

Example 6.4.9

a) Determine the magnitude of all four *S*-parameters from the specifications for the attenuator shown in Figure 6.33.
b) Is the attenuator reciprocal? Why or why not?

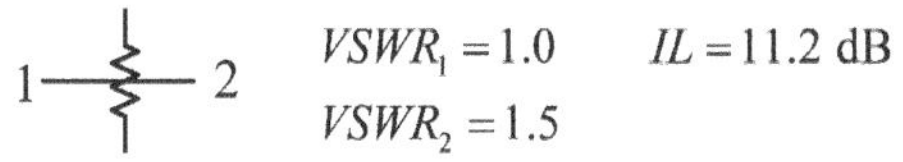

$VSWR_1 = 1.0 \qquad IL = 11.2 \text{ dB}$
$VSWR_2 = 1.5$

Figure 6.33 Fixed attenuator for Example 6.4.9.

Solution

a. $VSWR_1 = 1.0 \rightarrow |S_{11}| = |\Gamma_1| = 0$

---

15 Circuits that consist only of resistors, inductors, capacitors, mutual inductors, and ideal transformers are passive.

$$VSWR = \frac{1 + |\Gamma|}{1 - |\Gamma|} \rightarrow VSWR - VSWR \, (|\Gamma|) = 1 + |\Gamma| \rightarrow$$

$$VSWR - 1 = |\Gamma| \, (VSWR + 1) \rightarrow |\Gamma| = \frac{VSWR - 1}{VSWR + 1}$$

$$VSWR_2 = 1.5: \quad |S_{22}| = |\Gamma_2| = \frac{VSWR - 1}{VSWR + 1} = \frac{1.5 - 1}{1.5 + 1} = 0.200$$

$$|S_{21}|(\text{dB}) = -IL(\text{dB}) = -11.2\,\text{dB} \rightarrow |S_{21}| = |S_{12}| = 10^{[|S_{21}|(\text{dB})/20]} = 10^{(-11.2/20)} = 0.275$$

b. Reciprocal: yes, because an attenuator is a passive component. Note that reciprocity does not relate S_{11} and S_{22}.

Example 6.4.10

a) Determine all four S-parameters (both magnitude and phase) for the phase shifter shown in Figure 6.34.
b) Is the phase shifter reciprocal? Why or why not?

Solution

$VSWR_1 = 1.0 \qquad IL = 0\,\text{dB}$

$VSWR_2 = 1.0 \qquad \phi = 112°$

Figure 6.34 Phase shifter for Example 6.4.10.

a. $VSWR = 1.0$, so $|S_{11}| = |S_{22}| = 0$ (again, has nothing to do with reciprocity.

$IL = 0\,\text{dB} \rightarrow IL = 1$ (ratio); thus $|S_{21}| = |S_{12}| = 1$
A phase shifter shifts phase: $\angle S_{21} = \angle S_{12} = 112°$
Incorporate both the magnitude and phase: $S_{21} = S_{12} = 1 \angle + 112°$

b. A phase shifter is a passive component, so $S_{21} = S_{12}$ and it is reciprocal.

Example 6.4.11

Determine the *S*-parameter magnitudes given the following amplifier specifications: input $RL = 20\,\text{dB}$, output $VSWR = 1.5$, gain $= 30\,\text{dB}$, and isolation (Iso.) $= 25\,\text{dB}$.

Solution

$$|S_{11}|\,(\text{dB}) = -RL\,(\text{dB}) = -20\,\text{dB} = 20\log|S_{11}| \rightarrow \quad |S_{11}| = 10^{(-20/20)} = 0.10$$

$$|S_{22}| = |\Gamma_2| = \frac{VSWR - 1}{VSWR + 1} = \frac{1.5 - 1}{1.5 + 1} = \frac{0.5}{2.5} = 0.200$$

$$|S_{21}|\,(\text{dB}) = G\,(\text{dB}) = 30\,\text{dB} = 20\log|S_{21}| \rightarrow |S_{21}| = 10^{(30/20)} = 31.6$$

$$|S_{12}|\,(\text{dB}) = -\text{Iso.}(\text{dB}) = -25\,\text{dB} = 20\log|S_{12}| \rightarrow |S_{12}| = 10^{(-25/20)} = 0.0562$$

Results : $|S_{11}| = 0.10, \quad |S_{22}| = 0.200, \quad S_{11} \neq S_{22}$
$|S_{12}| = 0.0562, \quad |S_{21}| = 31.6, \quad S_{11} \neq S_{12}$

Note: $S_{21} \neq S_{12}$, so the amplifier is *not* reciprocal (comment: most active components are nonreciprocal).

6.4.4   Vector Network Analyzer

Motivational Question   How does instrumentation measure waves and *S*-parameters?

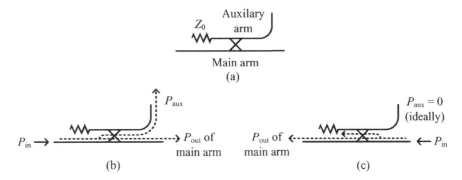

Figure 6.35 Operation of a directional coupler. (a) Coupler arm names. (b) Forward coupling direction. (c) Reverse coupling direction.

How can incident and reflected waves be distinguished in practice? There is a component that separates a wave in one direction from the wave in the other direction. It is called a **directional coupler**, shown in Figure 6.35a. When a wave enters the main arm of a directional coupler, the power splits, with a fraction of the wave power routed into the auxiliary arm and the remaining power exiting the other port of the main arm, as shown in Figure 6.35b. Consequently, a portion of the power that is input into the main arm in the forward coupling direction is routed to the auxiliary arm output:

$$P_{aux}(dBm) = P_{in}(dBm) - C(dB) \tag{6.38}$$

where C is the coupling factor of the directional coupler, usually stated in dB:

$$C(dB) = 10 \log \left(\frac{P_{in}}{P_{aux}} \right) \bigg|_{forward\ direction} = P_{in}(dBm) - P_{aux}(dBm) \tag{6.39}$$

The coupling factor expresses the ratio of the power entering the main arm in the forward direction to the power exiting the auxiliary arm. The use of dB and dBm within the same equation is similar to that used for attenuation and gain (see Section 6.4.1). When the power is input into the main arm in the reverse coupling direction, the portion of the power routed to the auxiliary arm is absorbed by the matched termination, as shown in Figure 6.35c. Ideally, no power appears at the auxiliary arm output[16]. Hence, the directional coupler can separate a wave from another wave that travels in the opposite direction.

The magnitudes of the S-parameters may be determined for a device under test (DUT) using a power meter and a directional coupler. However, this approach is tedious and valid at only one frequency per measurement. The frequency of the source can be *swept* (varied repeatedly between a lower frequency limit and an upper limit), which is often accomplished by stepping a synthesized source in small frequency increments. A swept frequency source and suitable detectors are used to obtain the magnitudes of the S-parameters over a desired frequency band of interest (this is scalar network analysis). However, to obtain the phases as well as the magnitudes of the S-parameters swept over a frequency range, the vector network analyzer (VNA) is used. The basic network analyzer configuration is shown in Figure 6.36a. Directional couplers in the RF/microwave circuit, shown in Figure 6.36b, route and separate the signals for processing and display as S-parameters. The lines between components and to the outputs represent transmission lines.

A practical VNA configuration that routes transmitted and reflected signals from both ports is shown in Figure 6.37. Note the RF/microwave switch that is used to select the port for the incident wave. The directional couplers can function to route the reflection from the incident wave port as well as to route a transmitted wave. The reader is directed to vendor websites for in-depth discussions of modern VNAs.

---

16 The pertinent specification for directional couplers is *directivity*.

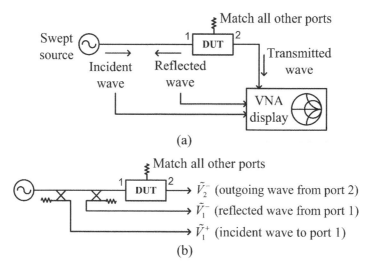

Figure 6.36 Basic VNA operation (DUT = Device Under Test). (a) Overall VNA signal routing. (b) Signal routing with couplers.

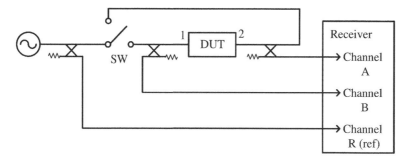

Figure 6.37 Practical configuration of a VNA (SW = Switch).

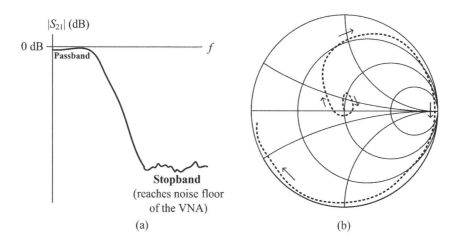

Figure 6.38 Example VNA displays (arrows on the Smith chart indicate the direction of increasing frequency). (a) $|S_{21}|$. (b) S_{11}.

As a practical matter, usually both ports of a VNA are fairly well matched so that the port-match conditions for the *S*-parameters are met. Calibration of the VNA before a measurement provides corrections to any mismatch conditions to further improve the adherence to the port-match conditions.

As an example, two generic VNA displays for a low-pass filter (LPF) are displayed in Figure 6.38. Note that the passband in the $|S_{21}|$ plot in Figure 6.38a corresponds to the trace near the center of the Smith chart in the S_{11} plot, shown in (b), because the signal is not being reflected, instead it is being passed by the LPF. The opposite occurs in the stopband – the signal is reflected, so the trace on the Smith chart is near the rim for these higher frequencies. The VNA is capable of displaying all of the *S*-parameters in various formats (linear and logarithmic rectangular, polar, Smith chart, *VSWR*, and so forth) of the reflection and transmission parameters.

6.5 Summary of Important Equations

$$Z_0 \equiv \left.\frac{V}{I}\right|_{\substack{\text{one wave} \\ \text{(one direction)}}} \qquad \text{DC: } \Gamma_L = \frac{V_{\text{reflected}}}{V_{\text{incident}}} = \frac{R_L - Z_0}{R_L + Z_0} \qquad \text{AC: } \Gamma_L = \frac{\widetilde{V}_L^-}{\widetilde{V}_L^+} = \frac{\widetilde{Z}_L - Z_0}{\widetilde{Z}_L + Z_0}$$

$$\text{vel.} = \frac{\ell}{\tau_d} = \frac{1}{\sqrt{\mu\varepsilon}} = \frac{1}{\sqrt{\mu_o\varepsilon_o\mu_r\varepsilon_r}} = \frac{c}{\sqrt{\mu_r\varepsilon_r}} \qquad \text{vel.} = \lambda f \qquad \beta = \frac{2\pi}{\lambda} \qquad \overline{S} = \overline{E} \times \overline{H}$$

$$VSWR \equiv \frac{|\widetilde{V}_{\max}|}{|\widetilde{V}_{\min}|} = \frac{1 + |\Gamma|}{1 - |\Gamma|} \qquad RL(\text{dB}) = -20\log|\Gamma| \qquad \text{normalization: } z_L = \frac{\widetilde{Z}_L}{Z_0}$$

$$A(\text{dB}) = 10\log\left(\frac{P_{\text{in}}}{P_{\text{out}}}\right) = -G(\text{dB}) \qquad P(\text{dBm}) \equiv 10\log\left[\frac{P(\text{W})}{1\,\text{mW}}\right]$$

$$S_{ii} = \Gamma_i = \left.\frac{\widetilde{V}_i^-}{\widetilde{V}_i^+}\right|_{\text{port } j \text{ terminated in } Z_0} \qquad S_{12} = \left.\frac{\widetilde{V}_1^-}{\widetilde{V}_2^+}\right|_{\text{port 1 terminated in } Z_0} \qquad S_{21} = \left.\frac{\widetilde{V}_2^-}{\widetilde{V}_1^+}\right|_{\text{port 2 terminated in } Z_0}$$

$$S_{ij} = S_{ji} \ (i \neq j) \text{ for reciprocal devices}$$

$$|S_{ii}|(\text{dB}) = 20\log(|S_{ii}|) = 20\log(|\Gamma_i|) = -RL_i(\text{dB}) \qquad |S_{21}|(\text{dB}) = 20\log|S_{21}| = G(\text{dB}) = -IL(\text{dB})$$

6.6 Appendix: dBm "Dos" and dBm "Don'ts"

First, recall the definition of a dBm for guidance: $\boxed{P(\text{dBm}) = 10\log_{10}\left(\frac{P(\text{W})}{1\,\text{mW}}\right)}$.

When you see "__ dBm," what this really means is "this power level is __dB relative to 1 mW." If the dBm value is positive, then the power is greater than 1 mW. If the dBm value is negative, then the power is less than 1 mW.

Do...

1) Subtract two power levels in dBm, if you want to find the *ratio* of the two powers in dB, for example if calculating gain or attenuation: **Subtracting two powers in dBm results in dB.**

$$P_2(\text{dBm}) - P_1(\text{dBm}) = 10\log_{10}\left(\frac{P_2(\text{W})}{1\,\text{mW}}\right) - 10\log_{10}\left(\frac{P_1(\text{W})}{1\,\text{mW}}\right) = 10\log_{10}\left(\frac{P_2(\text{W})}{1\,\text{mW}}\right) + 10\log_{10}\left(\frac{1\,\text{mW}}{P_1(\text{W})}\right)$$

$$= 10\log_{10}\left(\frac{P_2(\text{W})}{1\,\text{mW}} \times \frac{1\,\text{mW}}{P_1(\text{W})}\right) = 10\log_{10}\left(\frac{P_2(\text{W})}{P_1(\text{W})}\right) = \boxed{G(\text{dB})}$$

2) Add gain or attenuation in dB to a power level in dBm if you want to determine another power level: **Adding power in dBm and a gain (or attenuation) in dB results in dBm.**

$$P_1(\text{dBm}) + G(\text{dB}) = 10\log_{10}\left(\frac{P_1(\text{W})}{1\,\text{mW}}\right) + 10\log_{10}\left(\frac{P_2(\text{W})}{P_1(\text{W})}\right) = 10\log_{10}\left(\frac{P_1(\text{W})}{1\,\text{mW}} \times \frac{P_2(\text{W})}{P_1(\text{W})}\right)$$

$$= 10\log_{10}\left(\frac{P_2(\text{W})}{1\,\text{mW}}\right) = \boxed{P_2(\text{dBm})}$$

3) Subtract gain or attenuation in dB from a power in dBm if you want to determine another power level: **Subtracting attenuation or gain in dB from a power in dBm results in dBm.**

$$P_2(\text{dBm}) - G(\text{dB}) = 10\log_{10}\left(\frac{P_2(\text{W})}{1\,\text{mW}}\right) - 10\log_{10}\left(\frac{P_2(\text{W})}{P_1(\text{W})}\right) = 10\log_{10}\left(\frac{P_2(\text{W})}{1\,\text{mW}}\right) + 10\log_{10}\left(\frac{P_1(\text{W})}{P_2(\text{W})}\right)$$

$$= 10\log_{10}\left(\frac{P_2(\text{W})}{1\,\text{mW}} \times \frac{P_1(\text{W})}{P_2(\text{W})}\right) = 10\log_{10}\left(\frac{P_1(\text{W})}{1\,\text{mW}}\right) = \boxed{P_1(\text{dBm})}$$

DON'T...

1) Add two power levels in dBm! This is equivalent to *multiplying two power levels*, something **NEVER** done!

$$P_1(\text{dBm}) + P_2(\text{dBm}) = 10\log_{10}\left(\frac{P_1(\text{W})}{1\,\text{mW}}\right) + 10\log_{10}\left(\frac{P_2(\text{W})}{1\,\text{mW}}\right) = 10\log_{10}\left(\frac{P_1(\text{W})}{1\,\text{mW}} \times \frac{P_2(\text{W})}{1\,\text{mW}}\right)$$

$$= 10\log_{10}\left(\frac{P_1(\text{W})P_2(\text{W})}{(1\,\text{mW})^2}\right) = ????...\text{CAREFUL} - \text{This result is never useful!}$$

To sum power levels, first convert the powers to Watts (or mW), add the power in Watts, and then convert back to dBm:

$$P_{\text{Total}}(\text{W}) = P_1(\text{W}) + P_2(\text{W}) = 1\,\text{mW} \times 10^{[P_1(\text{dBm})/10]} + 1\,\text{mW} \times 10^{[P_2(\text{dBm})/10]}$$

$$P_{\text{Total}}(\text{dBm}) = 10\log_{10}\left(\frac{P_{\text{Total}}(\text{W})}{1\,\text{mW}}\right)$$

6.7 Homework

Note: Use $c = 2.998 \times 10^8$ m/s when c is required in calculations.

1. If the clock speed in a computer is 3 GHz, and at least up to the 11th harmonic (a frequency 11 times the given frequency, 33 GHz here) is needed to meet rising and falling edge specifications,
   a) What is the wavelength in the dielectric at 33 GHz if the circuit board material has a dielectric constant of 2.2?
   b) What could be the smallest distance for which T-line effects become significant?

2. What is the physical length of a piece of coax that is a half wavelength long at 950 MHz? The relative permittivity of the dielectric is 2.2.

3. Determine the time delay of a 25.0 m length of RG-58/U coaxial cable. Cite any references.

4. A 3.0 V pulse is launched onto a 75 Ω cable with a Teflon dielectric. The pulse returns to the source 10 μs later. Determine the distance from the source to the load.

5. If an ohmmeter were connected to 75 Ω coax with no load connected at the other end, what would it read if
   a) The length of the coax was infinite? Why?
   b) The length of the coax was a few meters? Why?

6. a) Why do reflections occur in a circuit?
   b) When do multiple reflections occur in a DC circuit?
   c) Why do standing waves occur in an AC circuit?

7. Given the DC circuit below with $V_g = 100$ V, $R_g = 50$ Ω, $Z_0 = 50$ Ω, $R_L = 10$ Ω, and $\tau_d = 4$ μs, sketch and label the source and load transient voltage waveforms for 0–17 μs. Label the DC steady-state level on each plot too.

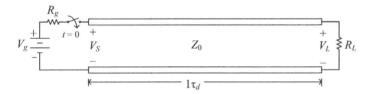

8. Repeat Problem 7 if R_L:
   a) equals 100 Ω,
   b) is an ideal short circuit, and
   c) is an ideal open circuit.

9. a) What is characteristic impedance?
   b) Does characteristic impedance represent power conversion (dissipation here)? If yes, why? If not, then what power does it represent?
   c) How is characteristic impedance used in power calculations? (Identify the equation.)

10. a) Describe what a Smith chart is.
    b) Why is the Smith chart useful when modern computational tools are available?
    c) Why is the *VSWR* the same on a constant *VSWR* circle on a Smith chart?
       Hint: consider the Smith chart relationship with reflection coefficient.

11. a) What is the *VSWR* if the load is a short? An ideal open? An ideal pure reactance?
    b) Explain why from the appropriate equation.
    c) Explain why with physical reasoning (without using equations).

12. a) If the *VSWR* of a load is given, can the reflection coefficient be uniquely determined?
    b) If the reflection coefficient of a load is given, can the *VSWR* be uniquely determined?
    c) Justify both answers.

13. If a given load at a given frequency has a reflection coefficient of 0.328 ∠ + 24°, determine
   a) the *VSWR* and
   b) the return loss.
   c) Develop the expression for and calculate the reflection coefficient from the *VSWR*.
   d) Develop the expression for and calculate the reflection coefficient from the *RL*. Your numerical answers should be consistent.
   e) Use a Smith chart to check all results.

14. If a given load at a given frequency has a reflection coefficient of 0.328 ∠ + 24°,
   a) Develop the expression for and calculate the normalized load impedance from the reflection coefficient.
   b) Use a Smith chart to check the result.

15. a) In any electric circuit, how does the energy transfer from the source to the load?
   b) Is an optical fiber useful in a circuit? Justify your answer based on your answer to a).
   c) What information does the Poynting vector convey?

16. a) Can a quantity in dB be added to a quantity in dB? Explain.
   b) Can a quantity in dB be added to a quantity in dBm? Explain.
   c) Can a quantity in dBm be added to a quantity in dBm? Explain.
   d) What is the difference between dB and dBm?

17. Determine the power from the source in the circuit below given a power meter reading of +3 dBm, $A_1 = 6.3$ dB, and $A_2 = 4.8$ dB. Express answer both in dBm and in mW.

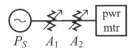

18. Determine the variable attenuator setting in the circuit below given the source power is 16.5 dBm and the power meter reading is +3.5 dBm.

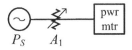

19. Determine P_S and P_1 in the circuit below. The attenuator is set to 12.7 dB, the amplifier has a gain of 13.8 dB, and the power meter reading is +2 dBm.

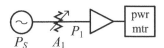

20. Assume a 50 Ω system ($Z_0 = 50$ Ω) for the ideal power combiner (summer) below. Determine the numerical value of the output power in dBm (if possible):
   a) If the input frequencies are *different*,
   b) If the input frequencies are the *same* and the input signals are in-phase, and
   c) If the input frequencies are the *same* and the phase between the input signals is unknown. Hint: recall AC circuit principles.

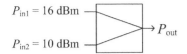

$P_{in1} = 16\,dBm$

$P_{in2} = 10\,dBm$

P_{out}

21. An attenuator has the following specifications: input and output $RL = 25\,dB$, $IL = 10\,dB$.
a) Determine the S-parameters (magnitudes only).
b) Is an attenuator reciprocal? Justify.

22. An amplifier has the following specifications: input $VSWR = 1.10$, output $VSWR = 2.00$, gain $= 20\,dB$, and isolation $= 30.0\,dB$.
a) Determine the S-parameters (magnitudes only).
b) Is an amplifier reciprocal? Justify.

23. a) Do the S-parameters of a component depend on frequency in general? Explain.
b) Give an example of a component that supports your answer to a).
c) Does reciprocity also imply $S_{11} = S_{22}$? Why or why not?

24. Show the switch position and identify the following signal paths (draw the paths, including direction, and label them) on the VNA diagrams that follow (two cases per diagram – you determine which ones go on the same diagram):
a) Reflection from port 1 (for S_{11})
b) Reflection from port 2 (for S_{22})
c) Transmission from port 1 to 2 (for S_{21})
d) Transmission from port 2 to 1 (for S_{12})   Be sure to also label which S-parameters are being measured on each diagram.

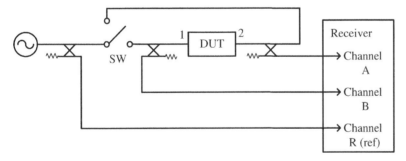

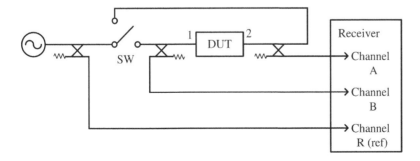

7

Transmission Lines: Theory and Applications

The previous chapter introduced basic transmission line (T-line) concepts, such as the propagation delay, characteristic impedance, and reflections on a T-line in the context of DC transients and AC sinusoidal signals. This chapter continues the examination of T-lines with both theory and applications, especially numerous physical interpretations and insights. The AC steady-state (phasor) equations most commonly used in practical T-line calculations and analyses are developed and interpreted. This chapter, in general, provides the resources and methods to analyze many practical T-line scenarios.

Initially, a key differential equation, the **wave equation**, will be developed from a distributed circuit model of the T-line (7.1). The phasor voltage and current solutions for the waves are then determined from the wave Eq. (7.2). The next section contains several interpretations of these wave solutions to build insights into T-line wave behavior (7.3). The solutions are generalized to incorporate losses in the conductors and the dielectric material of the T-line (7.4). The voltage and current solutions are then used to determine the impedance, standing waves, and the reflection coefficient as functions of position and frequency (7.5). These equations are often the results that are used in practical T-line calculations. The results are then examined on the Smith chart, which again serves as a primary graphical display of T-line results (7.6). Load impedance is determined from input impedance, which is

Electromagnetics and Transmission Lines: Essentials for Electrical Engineering, Second Edition.
Robert A. Strangeway, Steven S. Holland, and James E. Richie.
© 2023 John Wiley & Sons, Inc. Published 2023 by John Wiley & Sons, Inc.
Companion website: www.wiley.com/go/Strangeway/ElectromagneticsandTransmissionLines

often useful and practical, especially from measurements (7.7). The mathematical developments behind some of the key results are developed in the appendices (7.9).

7.1 A Circuit Model for AC Transmission Lines

We have examined waves on a T-line in general in Chapter 6. The next step is the development of the phasor voltage and current wave solutions, which is the initial topic in this chapter. These solutions will then be used to develop numerous practical results and insights later in this chapter. The basic process to obtain these solutions is:

- Build a distributed circuit model for the T-line ("distributed" means spread over position).
- Apply circuit laws to the model to obtain two differential equations, the **telegrapher's equations.**[1]
- Combine the telegrapher's equations into a single second-order differential equation, the **wave equation.**
- Solve the wave equation for the phasor voltage and current wave solutions.
- Insights and applications follow thereafter.

Initially, a lossless T-line is assumed to simplify the analysis and focus on the first-order principles (a T-line with zero attenuation; the effect of losses in a T-line will be incorporated in Section 7.4). The currents in the two conductors of the T-line have an accompanying magnetic field. A magnetic field is associated with inductance ($L = \Lambda/I$). The series nature of the current suggests a series inductance in the T-line model (Figure 7.1). However, recall from the inductance of coax example in Chapter 4 that the inductance is not *lumped* (such as a distinct inductor component), but instead is **distributed**. The inductance per unit length is L' (H/m), where the prime indicates a distributed quantity.

The voltage between (across) the two conductors of the T-line has an accompanying electric field. An electric field is associated with capacitance ($C = \Psi/V$). The parallel nature of the voltage suggests a parallel capacitance in the T-line model (Figure 7.1). However, recall from the capacitance of coax example in Chapter 4 that the capacitance is not *lumped* (such as a distinct capacitor component), but instead is *distributed*. The capacitance per unit length is C' (F/m).

The distributed inductance and capacitance in the T-line model implies that the voltage and the current are functions of position z. Consequently, a *differential length* of T-line with independent variable z should be modeled. Analyzing this model of a differential length of T-line will result in a differential equation. The solution will be a function of position.

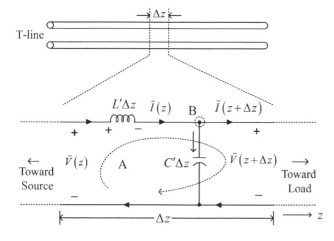

Figure 7.1 Lossless T-line model. The position variable z increases to the right. The KVL path "A" and KCL node "B" used in the analysis are shown with dashed lines.

---

1 Also known as the *telegraphist's equations*. An alternative approach to the telegrapher's equations, starting with Maxwell's equations and using the electromagnetic field equations for coax, is presented in Appendix 7.A.

Thus, the proposed model for T-lines that can be represented by distributed series inductance and shunt capacitance is shown in Figure 7.1. Again, zero losses (lossless) are assumed in this model. The model represents a very small length of T-line that will become a differential length in the limit as Δz approaches zero. Now an AC sinusoidal steady-state (phasor) analysis using Kirchhoff's voltage and current laws will be applied to the model.

Voltage equation: Apply Kirchhoff's voltage law around path "A" shown on the T-line model in Figure 7.1:

$$+ \widetilde{V}(z) - j\omega L' \Delta z\, \widetilde{I}(z) - \widetilde{V}(z + \Delta z) = 0 \tag{7.1}$$

where the tilde over the dependent variables indicates that they are complex number quantities, and $\widetilde{V}$ and $\widetilde{I}$ are **phasors**. Although not always explicitly shown, the phasor voltage and current (the dependent variables) are functions of position: $\widetilde{V} = \widetilde{V}(z)$ and $\widetilde{I} = \widetilde{I}(z)$. The distributed inductance L' (H/m) is multiplied by the small length Δz to give the total inductance in that small length. Then $\Delta \widetilde{V}$ is the voltage drop across the distributed inductance in Δz. Simplifying algebraically:

$$\widetilde{V}(z + \Delta z) - \widetilde{V}(z) = \Delta \widetilde{V} = -j\omega L' \Delta z \widetilde{I}(z) \rightarrow \frac{\Delta \widetilde{V}}{\Delta z} = -j\omega L' \widetilde{I} \tag{7.2}$$

Note that $\Delta \widetilde{V}$ is negative because the voltage drops (across the inductance) as position moves toward the load. To obtain the differential equation, let Δz approach zero:

$$\lim_{\Delta z \to 0} \frac{\Delta \widetilde{V}}{\Delta z} = \frac{d\widetilde{V}}{dz} = -j\omega L' \widetilde{I} \rightarrow \boxed{\frac{d\widetilde{V}}{dz} = -j\omega L' \widetilde{I}} \tag{7.3}$$

Thus, the first of the *telegrapher's equations* results.

Current equation: Apply KCL at node "B" shown on the T-line model in Figure 7.1:

$$+ \widetilde{I}(z) - j\omega C' \Delta z \widetilde{V}(z) - \widetilde{I}(z + \Delta z) = 0 \tag{7.4}$$

The distributed capacitance C' (F/m) is multiplied by the small length Δz to give the total capacitance in that small length. Then $\Delta \widetilde{I}$ is the current through the distributed capacitance in Δz. Simplifying algebraically:

$$\widetilde{I}(z + \Delta z) - \widetilde{I}(z) = \Delta \widetilde{I} = -j\omega C' \Delta z\, \widetilde{V}(z) \rightarrow \frac{\Delta \widetilde{I}}{\Delta z} = -j\omega C' \widetilde{V} \tag{7.5}$$

Note that $\Delta \widetilde{I}$ is negative because the current decreases (due to the current through the shunt capacitance) as position moves toward the load. To obtain the differential equation, let Δz approach zero:

$$\lim_{\Delta z \to 0} \frac{\Delta \widetilde{I}}{\Delta z} = \frac{d\widetilde{I}}{dz} = -j\omega C' \widetilde{V} \rightarrow \boxed{\frac{d\widetilde{I}}{dz} = -j\omega C' \widetilde{V}} \tag{7.6}$$

Thus, the second of the telegrapher's equations results. Equations (7.3) and (7.6) can be combined into a single differential equation. Take the derivative of Eq. (7.3) with respect to z:

$$\frac{d^2\widetilde{V}}{dz^2} = -j\omega L' \frac{d\widetilde{I}}{dz} \tag{7.7}$$

Note that L' and C' are assumed to be constant, that is, they have no z dependence (the uniform T-line assumption), so they are outside of the derivative. Substitute Eq. (7.6) into Eq. (7.7) to eliminate $d\widetilde{I}/dz$:

$$\frac{d^2\widetilde{V}}{dz^2} = -j\omega L' \frac{d\widetilde{I}}{dz} = -j\omega L' \left(-j\omega C' \widetilde{V} \right) = -\omega^2 L' C' \widetilde{V} \tag{7.8}$$

The lossless T-line **wave equation** results:

$$\boxed{\frac{d^2\widetilde{V}}{dz^2} = -\omega^2 L'C'\,\widetilde{V}}$$

(7.9)

where the voltage-dependent variable, $\widetilde{V}(z)$, is a phasor with z dependence. We now have a *differential equation* that can be solved! This differential equation is the one-dimensional *wave equation* for a lossless (zero losses) T-line. The solution will depend on the parameters of the T-line: the inductance per unit length and the capacitance per unit length, the frequency, and the position z, the independent variable. The differential equation must be solved to determine the functional dependence of $\widetilde{V}$ on z (next section). We will discover that the solution will represent the voltage waves that propagate in the plus and minus z directions along a T-line.

7.2 Voltage and Current Solutions for a Lossless Transmission Line

The establishment of the underlying assumptions is an important reason to examine derivations. The assumptions behind this T-line model are:

- AC signals: Sinusoidal steady-state (AC) voltages and currents are functions of position and frequency.
- Lossless T-line: The T-line has zero attenuation.
- Uniform T-line: All distributed parameters L' and C' are constant vs. position.
- Fixed T-line: All distributed parameters are time-independent (this assumption is buried in the AC sinusoidal steady-state assumption that only the voltage and current have a time-dependency, and it is sinusoidal).
- Linear: Sinusoidal voltages and currents of the same frequency can be summed (superposed) to determine the total sinusoidal voltage and current, respectively.
- The voltage polarity, the current direction, and the coordinate directions for position are defined as shown in Figure 7.2.

As will be seen in several T-line equations to be developed, sometimes it is convenient to designate position starting at the source and sometimes it is convenient to designate position starting at the load. The source position is defined to be $z = 0$ and $d = \ell$. The load position is defined to be at $z = \ell$ and $d = 0$. The z variable is defined "to the right" as it normally is, so the $+z$ direction is toward the load. The d variable is the "distance from the load," so the $+d$ direction is to the left toward the source.

<u>Voltage General Solution</u> Start with the second-order differential wave equation:

$$\frac{d^2\widetilde{V}(z)}{dz^2} = -\omega^2 L'C'\widetilde{V}(z)$$

(7.10)

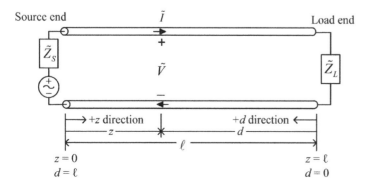

Figure 7.2 T-Line variable definitions ($\ell = z + d$). The parallel set of thick conductors is a schematic symbol for a T-line.

The solution of this differential equation is a function whose second derivative is proportional to itself. These include exponential functions, circular functions (sine and cosine), and hyperbolic functions, the latter two of which can be expressed in terms of exponential functions. If a solution is found that works, we are guaranteed that it is the only solution by the uniqueness theorem. Consider a voltage of the form $\widetilde{V} = \widetilde{V}(z) = Ae^{\gamma z}$. At this point, γ is a multiplicative constant of the exponent to be determined and A is an arbitrary coefficient constant. Then by substitution back into the differential equation,

$$\frac{d^2\widetilde{V}(z)}{dz^2} = A\gamma^2 e^{\gamma z} = -\omega^2 L'C'\widetilde{V}(z) = -\omega^2 L'C'Ae^{\gamma z} \tag{7.11}$$

$$A\gamma^2 e^{\gamma z} = -\omega^2 L'C'A\,e^{\gamma z} \rightarrow \gamma^2 = -\omega^2 L'C' \tag{7.12}$$

$$\gamma = \pm\sqrt{-\omega^2 L'C'} = \pm\sqrt{j^2\omega^2 L'C'} = \pm j\omega\sqrt{L'C'} \tag{7.13}$$

The symbol γ has the dimensions of reciprocal length, since L' and C' are both per unit length quantities, and γ is called the **propagation constant**. It is often expressed as $j\beta$ for lossless T-lines, where β is the **phase constant** (rad/m), the same β that we identified in Chapter 6.

$$\boxed{\gamma = j\omega\sqrt{L'C'} = j\beta \;(\text{rad/m})} \;(\text{lossless T} - \text{line}) \tag{7.14}$$

In general, the solution to the original second-order voltage differential equation in Eq. (7.10) has two terms, one for $+\gamma$ in the exponent and one for $-\gamma$ in the exponent, with different constants A and B for the coefficients. Thus, the general solution is

$$\widetilde{V} = \widetilde{V}(z) = A\,e^{-\gamma z} + Be^{+\gamma z} \tag{7.15}$$

Current General Solution The general solution for the current can be obtained from the general solution for the voltage. Substitute this voltage result into Eq. (7.3) and solve for the current:

$$\frac{d\widetilde{V}}{dz} = -j\omega L'\widetilde{I}$$

$$\frac{d\widetilde{V}}{dz} = -\gamma Ae^{-\gamma z} + \gamma Be^{+\gamma z} = -j\omega L'\widetilde{I} \tag{7.16}$$

$$\widetilde{I} = \frac{\gamma}{j\omega L'}\left(Ae^{-\gamma z} - Be^{+\gamma z}\right) \tag{7.17}$$

Thus, the current solution has the same mathematical form as the solution for the voltage.

Particular Solutions The constants A and B must be determined in order to arrive at the particular solutions. These constants can be determined by applying boundary conditions to the general solutions, Eqs. (7.15) and (7.17). The boundary conditions are the voltage and current at the load, $\widetilde{V}_L$ and $\widetilde{I}_L$, which are assumed to be known or measured. The algebraic details to determine the particular solution are shown in Appendix 7.B. The particular solutions follow:

$$\boxed{\widetilde{V} = \widetilde{V}(d) = \frac{1}{2}\left(\widetilde{V}_L + Z_0\widetilde{I}_L\right)e^{+\gamma d} + \frac{1}{2}\left(\widetilde{V}_L - Z_0\widetilde{I}_L\right)e^{-\gamma d}} \tag{7.18}$$

$$\boxed{\widetilde{I} = \widetilde{I}(d) = \frac{1}{2}\left(\widetilde{I}_L + Y_0\widetilde{V}_L\right)e^{+\gamma d} + \frac{1}{2}\left(\widetilde{I}_L - Y_0\widetilde{V}_L\right)e^{-\gamma d}} \tag{7.19}$$

There are two important details from the development of the particular solution not to ignore:

1) Independent variable z is converted into independent variable d, *the distance from the load*. This conversion is accomplished by substitution of $d = l - z$ (see Figure 7.2).
2) The **characteristic impedance** Z_0 equals[2]:

$$Z_0 \equiv \sqrt{\frac{L'}{C'}} \tag{7.20}$$

and the **characteristic admittance** is defined to be $Y_0 \equiv 1/Z_0$.

The significance of the results in Eqs. (7.18) and (7.19) is that the phasor voltage and current can be predicted at any position on the T-line if the T-line parameters Z_0 and γ, and the load conditions $\widetilde{V}_L$ and $\widetilde{I}_L$ are known. The ratio of $\widetilde{V}_L$ to $\widetilde{I}_L$ is determined by the load at the end of the T-line because $\widetilde{V}_L / \widetilde{I}_L = \widetilde{Z}_L = \widetilde{Z}_{\text{load}}$.

As we examine the solutions, questions come to mind: What is the physical significance of the two terms? What is the meaning of the propagation constant γ? The answers to these questions will be determined in the next section.

7.3 Interpreting the Voltage and Current Solutions

What is the physical significance of the two terms in the phasor voltage and current T-line solutions, Eqs. (7.18) and (7.19)? First define the following notation for the complex constant coefficients, with notational reasons to become apparent soon:

$$\widetilde{V}_L^+ \equiv \frac{1}{2}\left(\widetilde{V}_L + Z_0\widetilde{I}_L\right) \qquad \widetilde{I}_L^+ \equiv \frac{1}{2}\left(\widetilde{I}_L + Y_0\widetilde{V}_L\right) \tag{7.21}$$

$$\widetilde{V}_L^- \equiv \frac{1}{2}\left(\widetilde{V}_L - Z_0\widetilde{I}_L\right) \qquad \widetilde{I}_L^- \equiv \frac{1}{2}\left(\widetilde{I}_L - Y_0\widetilde{V}_L\right) \tag{7.22}$$

Insert these expressions into the phasor voltage and current solutions:

$$\widetilde{V} = \widetilde{V}_L^+ \, e^{+\gamma d} + \widetilde{V}_L^- \, e^{-\gamma d}, \quad \widetilde{I} = \widetilde{I}_L^+ \, e^{+\gamma d} + \widetilde{I}_L^- \, e^{-\gamma d} \tag{7.23}$$

Next determine the *time-domain* voltage expression (or current) from the phasor solution by reinserting the sinusoidal time dependence $e^{j\omega t}$ and taking the real part:

$$v(t, d) = \text{Re}\left[\widetilde{V} e^{j\omega t}\right] \tag{7.24}$$

$$= \text{Re}\left[\left(\widetilde{V}_L^+ \, e^{+\gamma d} + \widetilde{V}_L^- \, e^{-\gamma d}\right) e^{j\omega t}\right] \tag{7.25}$$

where $\widetilde{V}_L^+$ is a phasor, $\widetilde{V}_L^+ = V_L^+ \angle\theta_L^+ = V_L^+ \, e^{j\theta_L^+}$, and similarly for $\widetilde{V}_L^-$. Express the complex coefficients in polar form and substitute $\gamma = j\beta$ because the T-line is assumed to be lossless:

$$v(t, d) = \text{Re}\left[V_L^+ \, e^{j\theta_L^+} e^{+j\beta d} e^{j\omega t} + V_L^- \, e^{j\theta_L^-} e^{-j\beta d} e^{j\omega t}\right] \tag{7.26}$$

$$= \text{Re}\left[V_L^+ \, e^{+j\left(\beta d + \omega t + \theta_L^+\right)} + V_L^- \, e^{j\left(-\beta d + \omega t + \theta_L^-\right)}\right] \tag{7.27}$$

---

2 Equation (7.20) is exactly true for lossless T-lines, but is also a commonly used approximation for practical low-loss T-lines. Lossless or low-loss T-lines will be assumed throughout this text.

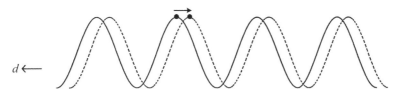

Figure 7.3 Two time snapshots of a traveling sine wave showing propagation of a constant phase point (d decreasing here).

Apply Euler's identity and take the real part:

$$v(t,d) = \text{Re}\left\{ \begin{array}{l} V_L^+ \left[\cos\left(\omega t + \beta d + \theta_L^+\right) + j\sin\left(\omega t + \beta d + \theta_L^+\right)\right] \\ + V_L^- \left[\cos\left(\omega t - \beta d + \theta_L^-\right) + j\sin\left(\omega t - \beta d + \theta_L^-\right)\right] \end{array} \right\} \tag{7.28}$$

Time domain solution: $\boxed{v(t,d) = V_L^+ \cos\left(\omega t + \beta d + \theta_L^+\right) + V_L^- \cos\left(\omega t - \beta d + \theta_L^-\right)}$ (7.29)

Aha! The previous equation contains the traveling waves that were introduced in Chapter 6 (but with position variable d instead of z). What is the nature of each term in the time-domain solution? The key to the meaning is in β. Consider again a sine wave that is traveling through space, as shown in Figure 7.3. As the wave travels, *time goes on*! Thus, there are two independent variables, t and d, which agrees with the independent variables in the time-domain solution.

What is the mathematical description of a traveling AC wave? In generic form,

$$v(t,d) = V\cos\left(\omega t \pm \beta d + \theta\right) \tag{7.30}$$

where V is the peak amplitude, ω is the radian frequency, θ is an initial phase shift, and β is the phase constant. Observe a *constant* phase point on the wave as the wave travels. (Recall a similar discussion in Section 6.2.1.) Note that the total phase ($\omega t + \beta d + \theta$) is a constant if a constant phase point on the wave is tracked. But d decreases as t increases for a wave approaching the load. Consequently, the two quantities in the argument, ωt and βd, must be added to keep the argument constant. Thus, there is a traveling wave in the $-d$ direction (d decreasing, which is toward the load):

$$V\cos\left(\omega t + \beta d + \theta\right) \rightarrow V_L^+ \cos\left(\omega t + \beta d + \theta_L^+\right) \tag{7.31}$$

Note that the sign of $+\beta d$ in this equation for a wave traveling toward the load is different from the $-\beta z$ of Section 6.2.1 because $d = l - z$. What would the expression be for a wave in the opposite direction, which is away from the load ($+d$ direction)? Position increases as time increases, so ωt and βd must be subtracted:

$$V\cos\left(\omega t - \beta d + \theta\right) \rightarrow V_L^- \cos\left(\omega t - \beta d + \theta_L^-\right) \tag{7.32}$$

Thus, each term in the voltage solution shown in Eq. (7.29) is a **traveling wave**, and the waves are moving in opposite directions along the T-line. The wave traveling toward the load ($-d$ direction) is called the **incident wave** (the term with the "+" superscripts) and the wave traveling away from the load ($+d$ direction) is called the **reflected wave** ("$-$" superscript). See Figure 7.4. The notation introduced at the start of this section is now justified.

We can use velocity to also justify the directions of the traveling waves. The incident wave term yields a sinusoid whose phase in the time domain is given by ($\omega t + \beta d + \theta^+$). A constant phase point has a velocity *vel.* given by[3] (distinguish the "d" in the derivative operator from the distance variable d):

$$\frac{d}{dt}\underbrace{\left(\omega t + \beta d + \theta^+\right)}_{\text{argument}} = \frac{d\,(\text{constant})}{dt} = 0 \rightarrow \omega + \beta \frac{d(d)}{dt} = 0 \tag{7.33}$$

---

3 Note that this is the *second* form of velocity for a wave on a T-line we have discussed. The first equation from Chapter 6 was *vel.* $= \dfrac{1}{\sqrt{\mu\varepsilon}}$. Using Eqs. (7.35) and (7.14), one can also show that *vel.* $= \dfrac{\omega}{\beta} = \dfrac{1}{\sqrt{L'C'}}$.

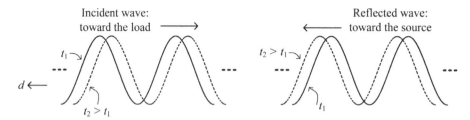

Figure 7.4 Snapshots of incident and reflected traveling waves on a T-line.

$$vel. = \frac{d(d)}{dt} = \frac{-\omega}{\beta} \tag{7.34}$$

Accordingly, the velocity of the incident wave is in the *decreasing* $(-d)$ direction (toward the load). Similarly, the velocity corresponding to the reflected wave term is:

$$vel. = \frac{d(d)}{dt} = \frac{+\omega}{\beta} \tag{7.35}$$

Thus, the velocity of the reflected wave is in the *increasing* $(+d)$ direction (toward the source). The propagation of each wave is illustrated in Figure 7.4. The directions of propagation give rise to the incident (forward) and reflected (backward) names to these components. Wave notation summary:

- A "+" superscript will be used to indicate the wave that is traveling toward the load, which is the *incident* wave that travels in the $-d$ direction (or equivalently, the $+z$ direction), and
- A "−" superscript will be used to indicate the wave that is traveling toward the source, which is the *reflected* wave that travels in the $+d$ direction (or equivalently, the $-z$ direction).

$$\text{Summary:} \qquad v(t,d) = \underbrace{V_L^+ \cos\left(\omega t + \beta d + \theta_L^+\right)}_{\text{incident wave } (d \text{ decr})} + \underbrace{V_L^- \cos\left(\omega t - \beta d + \theta_L^-\right)}_{\text{reflected wave } (d \text{ incr})} \tag{7.36}$$

A significant realization is that if the frequency is changed, β would change proportionately, that is, *the phase constant is frequency dependent*:

$$\beta = \frac{2\pi}{\lambda} = \frac{2\pi f}{vel.} = \frac{2\pi f}{\left(\dfrac{c}{\sqrt{\mu_r \varepsilon_r}}\right)} = \frac{2\pi \sqrt{\mu_r \varepsilon_r}}{c} f = \beta(f) \tag{7.37}$$

This means that voltage and current are both position and frequency dependent on T-lines. Mathematically, one should write $\widetilde{V}[\gamma(f), d]$ because γ is explicitly present in the voltage and current equations, the phase constant β is part of γ, and β is frequency dependent. However, it is frequency that is normally measured or specified, not β or γ. The load impedance is often frequency dependent too. So, although frequency f is often not explicitly shown in voltage and current expressions for T-lines, the argument (f, d) will be used with the understanding that $\widetilde{Z}_L$ (in general) **and γ are frequency dependent**!

$$\widetilde{V} = \widetilde{V}[\gamma(f), d] = \widetilde{V}(f, d) \tag{7.38}$$

- Movement along the T-line (change of position) corresponds to a change in the d variable.
- A change of frequency corresponds to a change in the f variable embedded in $\widetilde{Z}_L$ and γ.

Hereafter, we shall use the argument (f, d) to emphasize both independent variables.

Definition: βd is often called **electrical distance** because it represents the number of radians (phase) in a given length of T-line, and the value of electrical distance depends on both the physical length and the frequency.

Now we address the characteristic impedance Z_0 in the phasor solutions. The voltage-to-current ratio of the incident wave terms of the phasor voltage and current solutions is:

$$\widetilde{Z}^+ = \frac{\widetilde{V}^+}{\widetilde{I}^+} = \frac{\frac{1}{2}\left(\widetilde{V}_L + Z_0\widetilde{I}_L\right)e^{+\gamma d}}{\frac{1}{2}\left(\widetilde{I}_L + Y_0\widetilde{V}_L\right)e^{+\gamma d}} \cdot \frac{Z_0}{Z_0} \rightarrow \widetilde{Z}^+ = \frac{\widetilde{V}^+}{\widetilde{I}^+} = Z_0 \tag{7.39}$$

so *the voltage-to-current ratio of the incident traveling wave is equal to the characteristic impedance.* (Recall that this statement was made but not developed in Chapter 6.) Likewise, for the reflected traveling wave:

$$\widetilde{Z}^- = \frac{\widetilde{V}^-}{\left(-\widetilde{I}^-\right)} = \frac{\frac{1}{2}\left(\widetilde{V}_L - Z_0\widetilde{I}_L\right)e^{-\gamma d}}{-\frac{1}{2}\left(\widetilde{I}_L - Y_0\widetilde{V}_L\right)e^{-\gamma d}} \cdot \frac{Z_0}{Z_0} \rightarrow \widetilde{Z}^- = \frac{\widetilde{V}^-}{\left(-\widetilde{I}^-\right)} = Z_0 \tag{7.40}$$

where the minus sign associated with $\widetilde{I}^-$ accounts for the current direction in the reflected wave relative to $\widetilde{I}^+$. Again, *characteristic impedance, Z_0, is the voltage-to-current ratio of* <u>one</u> *traveling wave.*

How is the characteristic impedance of a T-line determined? Recall Eq. (7.20):

$$Z_0 = \sqrt{\frac{L'}{C'}} \tag{7.41}$$

where a lossless T-line is assumed in this equation. Characteristic impedance can be developed for coax from the inductance per-unit-length L' and the capacitance per-unit-length C' that were developed in Chapter 4, as illustrated in the next example.

Example 7.3.1
Develop the expression for the characteristic impedance of coax.

Solution

The capacitance of a 1 m length of coax was determined in Example 4.2.2 using Gauss's law and the voltage-electric field integral relationship to obtain:

$$C' = \frac{2\pi\varepsilon}{\ln\left(\dfrac{b}{a}\right)} \ \text{(F/m)} \tag{7.42}$$

where $a = $ inner radius and $b = $ outer radius. The inductance of a 1 m length of coax was determined in Example 4.5.2 using Ampere's circuital law and the magnetic flux–flux density relationship to obtain:

$$L' = \frac{\mu}{2\pi}\ln\left(\frac{b}{a}\right) \ \text{(H/m)} \tag{7.43}$$

Substitute L' and C' into Z_0:

$$Z_0 = \sqrt{\frac{L'}{C'}} = \sqrt{\frac{\dfrac{\mu}{2\pi}\ln\left(\dfrac{b}{a}\right)}{\dfrac{2\pi\varepsilon}{\ln\left(\dfrac{b}{a}\right)}}} = \frac{\sqrt{\dfrac{\mu}{\varepsilon}}\ln\left(\dfrac{b}{a}\right)}{2\pi} \tag{7.44}$$

Thus, the characteristic impedance of coax is: $Z_0 = \dfrac{1}{2\pi}\sqrt{\dfrac{\mu}{\varepsilon}}\ln\left(\dfrac{b}{a}\right) \ (\Omega)$ $\tag{7.45}$

As mentioned in Chapter 6, Z_0 depends only on the T-line geometry and the characteristics of the insulating material (μ and ε).

7.4 Lossy Transmission Line Solutions

Every practical T-line has losses. They may or may not be significant, but losses tend to increase as frequency increases, so often they cannot be ignored especially in high-frequency applications. What are the loss mechanisms? Conductors have resistive losses. Dielectrics have loss (which becomes significant as frequency increases). Radiation loss could be present, but will be ignored here – it requires a different model. Losses are incorporated into the T-line model with a distributed series resistance representing conductor losses and a distributed shunt conductance representing dielectric losses. The model is analyzed in the same approach as in Section 7.2 (see Homework Problem 31).

Most often in practice, attenuation is determined from a loss specification, such as from a datasheet, or from measurements. The **attenuation constant** α is converted from the loss specification in dB/unit length into Nepers/unit length using $1\text{ Np} = 8.686\text{ dB}$.[4] T-line loss per unit length is frequency dependent and is usually specified as a function of frequency or at sample frequencies. The effect of loss on each term in the voltage and current solutions is attenuation, that is, the decrease of peak amplitude as a wave propagates, in either direction, as pictured in Figure 7.5. A multiplicative exponential factor $e^{\pm\alpha d}$ with a real exponent is incorporated into each wave in the T-line voltage (and current) solution to account for attenuation:

$$\text{Time domain}: v(t,d) = \underbrace{V_L^+ e^{+\alpha d} \cos\left(\omega t + \beta d + \theta_L^+\right)}_{\text{incident wave } (d\text{ decr})} + \underbrace{V_L^- e^{-\alpha d} \cos\left(\omega t - \beta d + \theta_L^-\right)}_{\text{reflected wave } (d\text{ incr})} \tag{7.46}$$

$$\text{Phasor}: \widetilde{V} = \widetilde{V}(f,d) = \underbrace{\widetilde{V}_L^+ e^{+\gamma d}}_{\substack{\text{incident}\\\text{wave}}} + \underbrace{\widetilde{V}_L^- e^{-\gamma d}}_{\substack{\text{reflected}\\\text{wave}}} = V_L^+ e^{j\theta_L^+} e^{+\alpha d} e^{+j\beta d} + V_L^- e^{j\theta_L^-} e^{-\alpha d} e^{-j\beta d} \tag{7.47}$$

We see from the voltage solution that the **propagation constant** γ consists of two terms for lossy T-lines. The real part is the attenuation constant α and the imaginary part is β, the familiar phase constant:

$$\boxed{\gamma = \alpha + j\beta = \alpha\,(\text{Np/m}) + j\beta\,(\text{rad/m})} \tag{7.48}$$

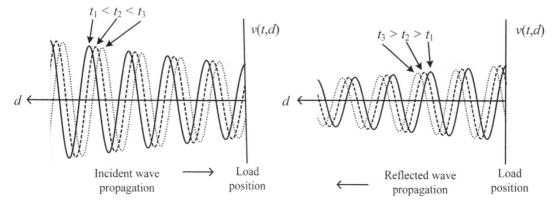

Figure 7.5 Three time snapshots of an incident (on left) traveling wave and a reflected (on right) traveling wave showing attenuation as the waves propagate.

---

4 Nepers are similar to radians and decibels in that they are not physical units. A Neper indicates that the natural logarithm of a voltage ratio has been taken, just as the decibel indicates that 10 log of a power ratio has been taken.

The phase constant β is often practically determined in one of two ways:

- From the ideal relation $\beta = 2\pi/\lambda$ when the dielectric constant is known, where λ is the wavelength in the dielectric of the T-line, or
- From the velocity specification, which is commonly given for manufactured T-lines[5], using $\beta = \omega/vel$.

Attenuation in the T-line does perturb the ideal value of β, but usually only slightly, especially for the low-loss case that is most common in practical T-lines (reminder: this chapter focuses on lossless and low-loss T-lines because they are the most applicable in practice). A typical determination of the propagation constant is illustrated in the next example.

Example 7.4.1

A 50 Ω coax cable with polyethylene (PE) dielectric has a loss of 47 dB/100 ft at 10 GHz. Determine the propagation constant.

Solution

$$\alpha = \left(\frac{47\,\text{dB}}{100\,\text{ft}}\right)\left(\frac{1\,\text{Np}}{8.686\,\text{dB}}\right)\left(\frac{3.2808\,\text{ft}}{\text{m}}\right) = 0.17753\,\text{Np/m}$$

PE : $\varepsilon_r = 2.3,\quad \mu_r = 1$ because the dielectric is nonmagnetic

$$\beta = \frac{2\pi}{\lambda} = \frac{2\pi f}{vel.} = \frac{2\pi f}{\left(\dfrac{c}{\sqrt{\mu_r \varepsilon_r}}\right)} = \frac{2\pi 10 \times 10^9}{\left(\dfrac{2.998 \times 10^8}{\sqrt{(1)(2.3)}}\right)} = 317.843\,\text{rad/m}$$

$$\gamma = \alpha + j\beta = 0.17753\,\text{Np/m} + j317.843\,\text{rad/m}$$

Reminder: The "50 Ω coax" terminology means $Z_0 = 50$ Ω.

7.5 Practical Transmission Line Calculations and Insights

Important T-line quantities (such as characteristic impedance, reflection coefficient, and the *S*-parameters) and the Smith chart were introduced in Chapter 6. These quantities related mostly to the load or to ports on a component. The theory of the T-line was developed in this chapter to produce the expressions for phasor voltage and current as a function of position *d* from the load and of frequency *f*. Now the highly useful impedance expressions are developed and interpreted, especially in terms of standing waves and reflection coefficient as a function of position and frequency. Then the Smith chart is revisited to expand its use for graphical visualization of impedance and reflection coefficient as a function of electrical distance from the load (Section 7.6).

7.5.1 Transmission Line Impedance Expression

What is the impedance at a given position *d* and frequency *f* on the T-line when the load is, in general, not matched? In general, the impedance looking toward the load at a distance *d* from the load is the ratio of the total voltage to the total current:

$$\tilde{V} = \tilde{V}(f, d) = \frac{1}{2}\left(\tilde{V}_L + Z_0\tilde{I}_L\right)e^{+\gamma d} + \frac{1}{2}\left(\tilde{V}_L - Z_0\tilde{I}_L\right)e^{-\gamma d} \tag{7.49}$$

---

5 The velocity specification can be used to determine the effective dielectric constant, which may not be straightforward, for example in mixed air/dielectric insulating transmission media, such as microstrip.

$$\widetilde{I} = \widetilde{I}(f,d) = \frac{1}{2}\left(\widetilde{I}_L + Y_0\widetilde{V}_L\right)e^{+\gamma d} + \frac{1}{2}\left(\widetilde{I}_L - Y_0\widetilde{V}_L\right)e^{-\gamma d} \tag{7.50}$$

$$\widetilde{Z} = \widetilde{Z}(f,d) = \frac{\widetilde{V}(f,d)}{\widetilde{I}(f,d)} = \frac{\frac{1}{2}\left(\widetilde{V}_L + Z_0\widetilde{I}_L\right)e^{+\gamma d} + \frac{1}{2}\left(\widetilde{V}_L - Z_0\widetilde{I}_L\right)e^{-\gamma d}}{\frac{1}{2}\left(\widetilde{I}_L + Y_0\widetilde{V}_L\right)e^{+\gamma d} + \frac{1}{2}\left(\widetilde{I}_L - Y_0\widetilde{V}_L\right)e^{-\gamma d}} \tag{7.51}$$

As a check, at $d = 0$ (at the load), $\widetilde{Z} = \widetilde{Z}(f,0) = \widetilde{V}_L/\widetilde{I}_L = \widetilde{Z}_L$, the load impedance, as expected. As another check, for the infinitely long line, $d \to \infty$, and the terms with $e^{-\gamma d} = e^{-\alpha d}\, e^{-j\beta d}$ approach zero, resulting in $\widetilde{Z} = Z_0$. The impedance is matched because a reflection from the load will never occur on an infinitely long T-line!

The load voltage and current are often unknown in high-frequency circuits. The general $\widetilde{Z}(f,d)$ is manipulated as follows to rearrange $\widetilde{V}_L$ and $\widetilde{I}_L$ into $\widetilde{Z}_L$, which is often known:

$$\widetilde{Z} = \frac{\widetilde{V}}{\widetilde{I}} = \frac{\frac{1}{2}\left(\widetilde{V}_L + Z_0\widetilde{I}_L\right)e^{+\gamma d} + \frac{1}{2}\left(\widetilde{V}_L - Z_0\widetilde{I}_L\right)e^{-\gamma d}}{\frac{1}{2}\left(\widetilde{I}_L + Y_0\widetilde{V}_L\right)e^{+\gamma d} + \frac{1}{2}\left(\widetilde{I}_L - Y_0\widetilde{V}_L\right)e^{-\gamma d}} \cdot \frac{Z_0}{Z_0} \tag{7.52}$$

$$\widetilde{Z} = Z_0\, \frac{\left(\widetilde{V}_L + Z_0\widetilde{I}_L\right)e^{+\gamma d} + \left(\widetilde{V}_L - Z_0\widetilde{I}_L\right)e^{-\gamma d}}{\left(\widetilde{V}_L + Z_0\widetilde{I}_L\right)e^{+\gamma d} - \left(\widetilde{V}_L - Z_0\widetilde{I}_L\right)e^{-\gamma d}} \cdot \frac{\left(\dfrac{1}{\widetilde{I}_L}\right)}{\left(\dfrac{1}{\widetilde{I}_L}\right)} \tag{7.53}$$

$$\widetilde{Z} = Z_0\, \frac{\left(\dfrac{\widetilde{V}_L}{\widetilde{I}_L} + Z_0\right)e^{+\gamma d} + \left(\dfrac{\widetilde{V}_L}{\widetilde{I}_L} - Z_0\right)e^{-\gamma d}}{\left(\dfrac{\widetilde{V}_L}{\widetilde{I}_L} + Z_0\right)e^{+\gamma d} - \left(\dfrac{\widetilde{V}_L}{\widetilde{I}_L} - Z_0\right)e^{-\gamma d}} \tag{7.54}$$

and since $\widetilde{V}_L/\widetilde{I}_L = \widetilde{Z}_L$,

$$\boxed{\widetilde{Z} = \widetilde{Z}(f,d) = Z_0\, \frac{\left(\widetilde{Z}_L + Z_0\right)e^{+\gamma d} + \left(\widetilde{Z}_L - Z_0\right)e^{-\gamma d}}{\left(\widetilde{Z}_L + Z_0\right)e^{+\gamma d} - \left(\widetilde{Z}_L - Z_0\right)e^{-\gamma d}}} \tag{7.55}$$

Again, $\widetilde{Z}_L$ (in general) and γ are frequency dependent. Thus, by knowing the T-line parameters Z_0 and γ, and the load impedance, one can determine the impedance at any position on the T-line (any d) at a given frequency. Alternatively, given a length of T-line, one can determine the impedance as a function of frequency at a given position. The **input impedance** $\widetilde{Z}_{\text{in}}$ is common terminology when the impedance is at the input of the T-line, that is, $\widetilde{Z}_{\text{in}} = \widetilde{Z}(f,\ell)$, where ℓ is the T-line length. See Figure 7.6.

Example 7.5.1

A coax T-line has a characteristic impedance of 75 Ω, a dielectric constant of 2.2, a loss of 88 dB/100 ft, and a length of 0.15 m. Determine the input impedance given a load impedance of $30 + j80$ Ω at 14 GHz.

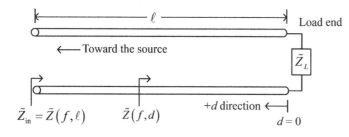

Figure 7.6 Input impedance and impedance as a function of position in general.

Strategy

Determine α in Np/m and β in rad/m, $\gamma = \alpha + j\beta$

Insert the values of γ, $d = \ell$, and $\widetilde{Z}_L$ into $\widetilde{Z}(f, d)$ to calculate $\widetilde{Z}_{\text{in}}$.

Solution

$$\alpha = \left(\frac{88\,\text{dB}}{100\,\text{ft}}\right)\left(\frac{1\,\text{Np}}{8.686\,\text{dB}}\right)\left(\frac{3.2808\,\text{ft}}{\text{m}}\right) = 0.33239\,\text{Np/m}$$

$$\beta = \frac{2\pi}{\lambda} = \frac{2\pi f}{vel.} = \frac{2\pi f \sqrt{\mu_r \varepsilon_r}}{c} = \frac{2\pi 14 \times 10^9 \sqrt{(1)(2.2)}}{2.998 \times 10^8} = 435.19873\,\text{rad/m}$$

$$e^{+\gamma d} = e^{+\alpha d}e^{+j\beta d} = e^{+0.33239(0.15)} \angle [435.19873(0.15)] = 1.051122 \angle + 65.2798\,\text{rad} \text{ (see note 6)}$$

$$e^{-\gamma d} = e^{-\alpha d}e^{-j\beta d} = e^{-0.33239(0.15)} \angle [-435.19873(0.15)] = 0.951364 \angle -65.2798\,\text{rad}$$

Insert the values for $\widetilde{Z}_L, Z_0, e^{+\gamma d}$ and $e^{-\gamma d}$ into $\widetilde{Z}(f, d) = Z_0 \dfrac{\left(\widetilde{Z}_L + Z_0\right)e^{+\gamma d} + \left(\widetilde{Z}_L - Z_0\right)e^{-\gamma d}}{\left(\widetilde{Z}_L + Z_0\right)e^{+\gamma d} - \left(\widetilde{Z}_L - Z_0\right)e^{-\gamma d}}$

Numerator of the fraction in $\widetilde{Z} = -57.04824 - j25.24784$

Denominator of $\widetilde{Z} = -220.2047 + j37.05486$

$$\widetilde{Z}_{\text{in}} = \widetilde{Z}(14\,\text{GHz}, 0.15\,\text{m}) = 17.488 + j11.542\,\Omega \approx 17.5 + j11.5\,\Omega.$$

Concept of a matched termination: An important, general concept is that *impedance varies as a function of position* along a T-line *except when* $\widetilde{Z}_L = Z_0$. If the T-line is terminated in an impedance equal to the characteristic impedance, then Eq. (7.55) reduces to $\widetilde{Z}(f, d) = Z_0$. Thus, *the impedance anywhere on the T-line should equal the characteristic impedance if the load is matched* (another reason why impedance matching is important in practical systems).

Similarly, the matched condition ($\widetilde{Z}_L = Z_0$) removes significant frequency dependence of the impedance at any given position along a T-line: $\widetilde{Z}(f, d) = Z_0$. The reason is because there is no reflected wave to cause interference with the incident wave, so there is no frequency dependency in the standing wave pattern.[7]

7.5.2 Special Case of Lossless Transmission Lines

Often, practical T-lines are low-loss and/or are short enough that the overall attenuation is relatively insignificant. In these cases, setting the attenuation constant to zero is a helpful simplifying approximation. When $\alpha = 0$, the voltage, current, and impedance expressions for the T-line simplify:

$$\gamma = \alpha + j\beta = j\beta \tag{7.56}$$

$$e^{\pm \gamma d} = e^{\pm j\beta d} \tag{7.57}$$

$$\widetilde{V} = \widetilde{V}(f, d) = \frac{1}{2}\left(\widetilde{V}_L + Z_0\widetilde{I}_L\right)e^{+j\beta d} + \frac{1}{2}\left(\widetilde{V}_L - Z_0\widetilde{I}_L\right)e^{-j\beta d} \tag{7.58}$$

$$\widetilde{V} = \widetilde{V}_L \frac{1}{2}\left(e^{+j\beta d} + e^{-j\beta d}\right) + Z_0\widetilde{I}_L \frac{1}{2}\left(e^{+j\beta d} - e^{-j\beta d}\right) \tag{7.59}$$

From Euler's identity: $\cos A = \dfrac{1}{2}\left(e^{+jA} + e^{-jA}\right)$ and $\sin A = \dfrac{1}{2j}\left(e^{+jA} - e^{-jA}\right)$ (7.60)

---

6 Reminder: $e^{j\theta}$ is the angle of a complex number in polar form, so $e^{j\theta} = 1 \angle \theta$.

7 Z_0 has some frequency dependence on lossy T-lines, but usually it can be ignored for low-loss T-lines.

Substitute Eq. (7.60) into Eq. (7.59) to obtain the phasor voltage result on a lossless T-line:

$$\boxed{\widetilde{V} = \widetilde{V}(f, d) = \widetilde{V}_L \cos \beta d + j Z_0 \widetilde{I}_L \sin \beta d} \tag{7.61}$$

Similarly, the phasor current and impedance solutions on a lossless T-line are:

$$\boxed{\widetilde{I} = \widetilde{I}(f, d) = \widetilde{I}_L \cos \beta d + j Y_0 \widetilde{V}_L \sin \beta d} \tag{7.62}$$

$$\widetilde{Z} = \frac{\widetilde{V}}{\widetilde{I}} = \frac{\widetilde{V}_L \cos \beta d + j Z_0 \widetilde{I}_L \sin \beta d}{\widetilde{I}_L \cos \beta d + j Y_0 \widetilde{V}_L \sin \beta d} \cdot \frac{Z_0}{Z_0} \cdot \frac{\left(\dfrac{1}{\widetilde{I}_L}\right)}{\left(\dfrac{1}{\widetilde{I}_L}\right)} \cdot \frac{\dfrac{1}{\cos \beta d}}{\dfrac{1}{\cos \beta d}} = Z_0 \frac{\dfrac{\widetilde{V}_L}{\widetilde{I}_L} + j Z_0 \dfrac{\sin \beta d}{\cos \beta d}}{Z_0 + j Y_0 Z_0 \dfrac{\widetilde{V}_L}{\widetilde{I}_L} \dfrac{\sin \beta d}{\cos \beta d}} \tag{7.63}$$

$$\boxed{\widetilde{Z} = \widetilde{Z}(f, d) = Z_0 \frac{\widetilde{Z}_L + j Z_0 \tan \beta d}{Z_0 + j \widetilde{Z}_L \tan \beta d}} \tag{7.64}$$

This simplified impedance expression is a good approximation for most practical T-lines, and thus it is often the preferred form for T-line analysis and design.

Example 7.5.2

Repeat Example 7.5.1 if the T-line is lossless

Given: $Z_0 = 75\,\Omega$, $\varepsilon_r = 2.2$, $\ell = 0.15$ m, $\widetilde{Z}_L = 30 + j80\,\Omega$, $f = 14$ GHz

Strategy

Determine β in rad/m

Insert the values of β, $d = \ell$, and $\widetilde{Z}_L$ into the lossless $\widetilde{Z}(f, d) = \widetilde{Z}_{\text{in}}$.

Solution

$$\beta = \frac{2\pi}{\lambda} = \frac{2\pi f}{vel.} = \frac{2\pi f \sqrt{\mu_r \varepsilon_r}}{c} = \frac{2\pi 14 \times 10^9 \sqrt{(1)(2.2)}}{2.998 \times 10^8} = 435.19873 \text{ rad/m}$$

$$\widetilde{Z}_{\text{in}} = Z_0 \frac{\widetilde{Z}_L + j Z_0 \tan \beta d}{Z_0 + j \widetilde{Z}_L \tan \beta d}$$

$$\widetilde{Z}_{\text{in}} = 75 \frac{(30 + j80) + j75 \tan [435.19873(0.15)\text{ rad}]}{75 + j(30 + j80) \tan [435.19873(0.15)\text{ rad}]}$$

$$\widetilde{Z}_{\text{in}} = \widetilde{Z}(14\,\text{GHz}, 0.15\,\text{m}) = 13.822 + j11.785\,\Omega \approx 13.8 + j11.8\,\Omega$$

Note the change of the input impedance relative to Example 7.5.1. T-line losses affect voltage, current, and impedances on T-lines!

One can use mathematical software to make magnitude plots of the voltage, current, and impedance standing wave patterns. The standing wave pattern results for two load impedance extremes and a general resistive load are considered next to gain insights on lossless T-lines.

7.5.3 Standing Wave Patterns

Consider the case of a short-circuited line shown in Figure 7.7, where $\widetilde{V}_L = 0$ and $\widetilde{Z}_L = 0$. The voltage, current, and impedance equations on a lossless T-line simplify to:

$$\widetilde{V} = \widetilde{V}(f, d) = j Z_0 \widetilde{I}_L \sin \beta d \rightarrow \left|\widetilde{V}\right| = Z_0 I_L \sin \beta d \tag{7.65}$$

$$\widetilde{I} = \widetilde{I}(f, d) = \widetilde{I}_L \cos \beta d \rightarrow \left|\widetilde{I}\right| = I_L \cos \beta d \tag{7.66}$$

$$\widetilde{Z} = \widetilde{Z}(f, d) = + j Z_0 \tan \beta d \rightarrow \left|\widetilde{Z}\right| = Z_0 \tan \beta d \tag{7.67}$$

Figure 7.7 Short-circuited T-line.

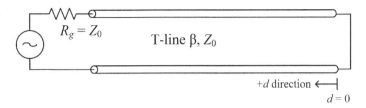

Figure 7.8 Standing wave patterns on a short-circuited T-line.

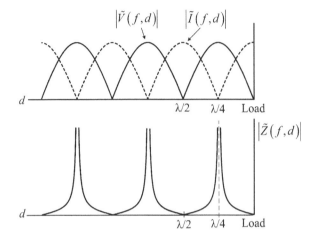

These straightforward trigonometric functions are plotted in Figure 7.8. Observations on the standing wave patterns for a short-circuited T-line:

- $|\widetilde{V}|$, $|\widetilde{I}|$, and $|\widetilde{Z}|$ repeat every half wavelength ($\lambda/2$).
- $|\widetilde{Z}|$ is position-dependent and varies from zero to infinity.
- $\widetilde{Z}(f, n\lambda/2) = \widetilde{Z}_L$, that is, the load impedance repeats every half wavelength.

Consider the case of an open-circuited line shown in Figure 7.9, where $\widetilde{I}_L = 0$ and $\widetilde{Z}_L \to \infty$. The voltage, current, and impedance equations on a lossless T-line simplify to:

$$\widetilde{V} = \widetilde{V}(f, d) = \widetilde{V}_L \cos \beta d \to |\widetilde{V}| = V_L \cos \beta d \tag{7.68}$$

$$\widetilde{I} = \widetilde{I}(f, d) = j Y_0 \widetilde{V}_L \sin \beta d \to |\widetilde{I}| = Y_0 V_L \sin \beta d \tag{7.69}$$

$$\widetilde{Z} = \widetilde{Z}(f, d) = + \frac{Z_0}{j \tan \beta d} \to |\widetilde{Z}| = Z_0 \cot \beta d \tag{7.70}$$

These straightforward trigonometric functions are plotted in Figure 7.10. Observe that the standing wave pattern is shifted $90°$ relative to the short-circuit load case.

Consider a resistive mismatched load where $R_L = 0.5Z_0$. The standing wave patterns change relative to the extreme mismatch of an open or short load. See Figure 7.11. The voltage, current, and impedance minima do not reach zero. The impedance maxima do not approach infinity. The *VSWR* is finite. What would the patterns look like if $R_L = Z_0$ on a lossless T-line? The patterns would be flat horizontal lines for the voltage, current, and impedance plots.

Figure 7.9 Open-circuited T-line.

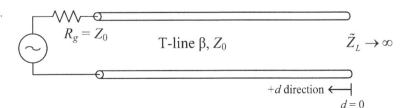

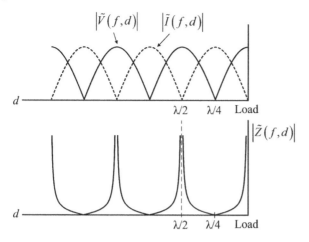

Figure 7.10 Standing wave patterns on an open-circuited T-line.

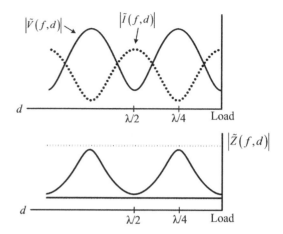

Figure 7.11 Standing wave patterns for a resistive load ($R_L = 0.5Z_0$).

7.5.4 Reflection Coefficient vs. Position

The reflection coefficient at the load ($d = 0$) was previously developed:

$$\Gamma_L = \Gamma_L(f) = \frac{\widetilde{V}_L^-}{\widetilde{V}_L^+} = \frac{-\widetilde{I}_L^-}{\widetilde{I}_L^+} = \frac{\widetilde{Z}_L - Z_0}{\widetilde{Z}_L + Z_0} \qquad (7.71)$$

Now consider the ratio of a reflected wave to an incident wave at *any* position d from the load, that is, the reflection coefficient at position d. The impedance at position d is used instead of the load impedance:[8]

$$\Gamma = \Gamma(f, d) = \frac{\widetilde{Z}(f, d) - Z_0}{\widetilde{Z}(f, d) + Z_0} \qquad (7.72)$$

Like impedance, the reflection coefficient is a function of both frequency and position. See Figure 7.12. This aspect becomes apparent on a Smith chart, as revisited in the next section. The reflection coefficient at the input of the T-line (at $d = \ell$) is

$$\Gamma_{\text{in}} = \Gamma(f, \ell) = \frac{\widetilde{Z}_{\text{in}} - Z_0}{\widetilde{Z}_{\text{in}} + Z_0} \qquad (7.73)$$

Note that on a lossless T-line, the incident and reflected waves are not attenuated and therefore *the reflection coefficient magnitude is constant vs. position along a lossless T-line* (although the phase does change vs. position).

Example 7.5.3

Determine the input reflection coefficient at 0.200 [λ] from the load for the T-line circuit in Example 7.5.2.

Strategy

Insert $\widetilde{Z}_{\text{in}} = 13.822 + j11.785 \, \Omega$ from Example 7.5.2 into $\Gamma_{\text{in}} = \dfrac{\widetilde{Z}_{\text{in}} - Z_0}{\widetilde{Z}_{\text{in}} + Z_0}$

---

8 Alternate forms of $\Gamma(f, d)$ are developed in Appendix 7.C.

Figure 7.12 Input reflection coefficient and reflection coefficient as a function of position in general.

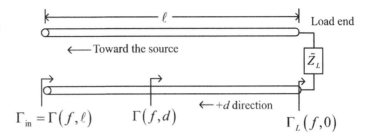

Solution

$$\Gamma_{\text{in}} = \frac{\tilde{Z}_{\text{in}} - Z_0}{\tilde{Z}_{\text{in}} + Z_0} = \frac{(13.822 + j11.785) - 75}{(13.822 + j11.785) + 75} = 0.695\angle 162°$$

This example is expanded in Homework Problems 15 and 16.

7.6 Smith Chart Revisited: Electrical Distance

The Smith chart was introduced in Chapter 6 and was utilized for relating normalized impedance, reflection coefficient, *VSWR*, and *RL*. The Smith chart is also a graphical means for *translating* impedances or admittances and reflection coefficients (finding these quantities as a function of electrical distance) on a T-line. Again, the tools required for manual Smith chart use are a Smith chart, a compass, a ruler, a pencil, and a competent operator. This section will focus on the fundamental concepts of using the Smith chart to analyze lossless T-lines. The Smith chart can be used for lossy T-lines as well (conceptually summarized in Section 7.6.2).

Note that you must distinguish reused symbols in the context that they are used. For example, in this chapter, both normalized impedance and distance use the symbol z. As we have seen, the distance from the load, d, is commonly used in T-line calculations and Smith charts.

Besides the uses of the Smith chart discussed in Chapter 6, the Smith chart is also used for:

- z or y determination with respect to electrical distance on the T-line
- Γ determination with respect to electrical distance on the T-line
- z_{max} and z_{min} along a T-line, both values and positions

and many more uses in high-frequency electrical engineering.

> **Our goal**: Use the Smith chart as a practical engineering graphical tool, especially for visualization of T-line quantities as a function of electrical distance from the load.
>
> **Approach**: Learn the Smith chart using graphical techniques to develop insights and experience with the Smith chart; then, the Smith chart will be a meaningful display.

7.6.1 Rotation on the Smith Chart – an Electrical Distance Perspective

On a commercial Smith chart, note that there are four scales outside the Smith chart rim:

1) Wavelengths toward the generator (electrical distance from the load toward the source)
2) Wavelengths toward the load (electrical distance from a given location on the T-line toward the load)
3) Angle of the reflection coefficient
4) Angle of the transmission coefficient (another use of the Smith chart, but not used here)

The electrical distance in the **Wavelengths** scales is expressed in **number of wavelengths**. If βd is in radians, then it must be converted into the number of wavelengths (notation $[\lambda]$):

$$\beta d \, [\lambda] = \beta d \, (\text{rad}) \left(\frac{1 \, [\lambda]}{2\pi \, (\text{rad})} \right) \tag{7.74}$$

If the actual T-line distance is given, then it must be converted into the number of wavelengths:

$$\beta d \, [\lambda] = d \, (m) \frac{1 \, [\lambda]}{\lambda \, (m)} \tag{7.75}$$

One must distinguish between the length of a wavelength ($\lambda = 2$ m, for example) and the number of wavelengths of electrical distance ($\beta d = 2[\lambda] = 2$ wavelengths, for example). The λ in $\beta = 2\pi/\lambda$ is the length of a wavelength, not the symbol for number of wavelengths. The notation for electrical distance βd in the number of wavelengths $[\lambda]$ is $\beta d \, [\lambda]$ in this text.

Example 7.6.1

A quarter wavelength exists on a T-line of length 2.7 m. Determine (a) $\beta d \, (°)$, (b) $\beta d \, (\text{rad})$, and (c) λ.

Solution

$\beta d \, [\lambda]$ is given. A quarter wavelength means $\beta d = 0.25 \, [\lambda]$. Also given: $d = 2.7$ m.

a) $\beta d \, [°] = \beta d \, [\lambda] \left(\dfrac{360°}{1 \, [\lambda]} \right) = 0.25 \, [\lambda] \left(\dfrac{360°}{1 \, [\lambda]} \right) = 90°$

b) $\beta d \, (\text{rad}) = \beta d \, [\lambda] \left(\dfrac{2\pi \, \text{rad}}{1 \, [\lambda]} \right) = 0.25 \, [\lambda] \left(\dfrac{2\pi \, \text{rad}}{1 \, [\lambda]} \right) = \dfrac{\pi}{2} \, \text{rad} = 1.5708 \, \text{rad}$

c) $\lambda = \dfrac{d \, (m)}{\beta d \, [\lambda]} = \dfrac{2.7 \, \text{m}}{0.25 \, [\lambda]} = 10.8 \, \text{m} \quad \left(= \dfrac{10.8 \, \text{m}}{1 \, [\lambda]} \right)$, that is, 10.8 m in one wavelength,

or $\beta = \dfrac{2\pi}{\lambda} \rightarrow \lambda = \dfrac{2\pi}{\beta} = \dfrac{2\pi d}{\beta d} = \dfrac{2\pi \, (\text{rad})d \, (m)}{\beta d \, (\text{rad})} = \dfrac{(2\pi \, \text{rad})2.7 \, \text{m}}{1.5708 \, \text{rad}} = 10.8 \, \text{m}$

Once the electrical distance in wavelengths is determined, it can be used to translate an impedance, an admittance, or a reflection coefficient on a Smith chart. How? *Rotation!*

A change of electrical distance on the T-line = *rotation* on the Smith chart (centered on $z = 1$)

Recall that on a lossless T-line, the magnitude of the reflection coefficient $|\Gamma|$ is constant versus position on the T-line for a given load at a given frequency; thus, rotation along a circle of constant radius $|\Gamma|$, that is, rotation on a constant *VSWR* circle on the Smith chart is equivalent to movement on a lossless T-line.

One rotation around the Smith chart equals one-half wavelength of electrical distance!

Recall that the phase constant depends on frequency:

$$\beta = \frac{2\pi}{\lambda} = \frac{2\pi f}{vel.} = \frac{2\pi f}{\left(\dfrac{c}{\sqrt{\mu_r \varepsilon_r}} \right)} = \frac{2\pi \sqrt{\mu_r \varepsilon_r}}{c} f \tag{7.76}$$

As previously established, T-line impedance expressions are a function of *both* frequency and position, that is, a function of electrical distance. *If either frequency or position changes, the electrical distance changes* and the location on the Smith chart is *rotated*.

On a vector network analyzer (VNA), the frequency is repeatedly swept, either linearly or logarithmically, over a predetermined frequency range. The impedance locus can be displayed, often on a Smith chart for reflection measurements. The locus will be a circle if the T-line is lossless and the load has negligible reactance (which is usually *not* the case). The Smith chart example from Chapter 6 is continued here to illustrate translation.

Example 7.6.2

If the load impedance is $28 - j32\,\Omega$ and the characteristic impedance of a lossless T-line is $52\,\Omega$, determine

a) The input impedance at 0.200 [λ] from the load, and
b) The minimum and maximum normalized impedances and their locations along the T-line.

Solution

As before, plot the normalized load impedance:

$$\widetilde{Z}_L = 28 - j32\,\Omega, \quad Z_o = 52\,\Omega \rightarrow z_L = \frac{\widetilde{Z}_L}{Z_o} = \frac{28 - j32}{52} = 0.538 - j0.615$$

a) Determine $\widetilde{Z}_{in} = \widetilde{Z}(f, d)$ at 0.200 [λ] from the load: see Figure 7.13a
   - Locate z_L on the wavelengths toward generator scale: z_L is at 0.396 [λ]
   - Move 0.200 [λ] *toward the generator* (clockwise) from the z_L location:

   $$0.396\,[\lambda] + 0.200[\lambda] = 0.596[\lambda] = 0.596[\lambda] - 0.500[\lambda] = 0.096[\lambda]$$

   - The Smith chart scale starts over at 0.500 [λ], so 0.500 [λ] is subtracted
   - Draw a straight line from the Smith chart center to 0.096 [λ]
   - z_{in} is located at the intersection of this straight line and the constant *VSWR* circle

   through z_L: $z_{in} = 0.51 + j0.56$,　$\widetilde{Z}_{in} = z_{in}Z_0 = 27 + j\,29\,\Omega$

b) Determine z_{min}, z_{max}, and their locations along the T-line: see Figure 7.13b.
   - z_{min} is at the intersection of the constant *VSWR* circle with $0.37 + j0$ on the left-hand side of the real (horizontal) axis
   - z_{max} is at the intersection of the constant *VSWR* circle with $2.70 + j0$ on the right-hand side of the real axis; also, notice that $z_{max} = VSWR = 2.70$
   - Use the wavelengths scales to determine the distance in [λ]

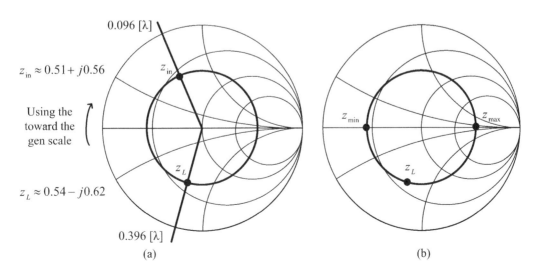

(a)　(b)

Figure 7.13　Smith charts for Example 7.6.2. (a) Determining z_{in}. (b) Determining z_{min} and z_{max}.

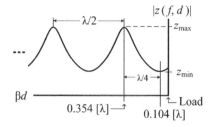

Figure 7.14 Visualization of the standing wave pattern for Example 7.6.2 (the load is at $\beta d = 0$).

- Location of $z_{\min}$: $(0.500 - 0.396) = 0.104[\lambda]$ from the load toward the generator
- Location of $z_{\max}$: $(0.104 + 0.250) = 0.354[\lambda]$ from the load toward the generator
- The standing wave pattern can be visualized: $z_{\min}$ is closest to the load, $z_{\max}$ is a quarter wavelength toward the generator from $z_{\min}$, and the min-max pattern repeats as the distance toward the generator increases. See Figure 7.14. Recall that the standing wave pattern repeats every half-wavelength, just as once around the Smith chart is a half-wavelength.

7.6.2 Lossy Transmission Line Traces on a Smith Chart

While we are not quantitatively analyzing lossy T-lines with the Smith chart in this text, we can use our insights gained in Section 7.6.1 to interpret lossy T-line behavior in the context of a Smith chart display. When the electrical distance is changed on a lossy T-line, the loss causes the radius to change along with the change in angular position. Does the *VSWR* increase or decrease as the load is approached on a lossy T-line? At the load, the reflected wave amplitude is at its maximum because it has not traveled down the T-line back toward the source yet. The incident wave is at its minimum at the load because it has suffered the entire one-way loss of the T-line as it has traveled the entire length of the T-line. Hence, the maximum amount of interference occurs at the load position, and the *VSWR* is maximum at the load. Consequently, the *VSWR* must decrease as the source is approached because the reflected wave amplitude is being attenuated, the incident wave amplitude grows as the source is approached, and the result is that less interference occurs as the source is approached. The impedance locus (trace) spirals out on the Smith chart as the load is approached, or conversely spirals in toward the center as the source is approached. Generic sketches of an impedance locus and standing wave pattern for a lossy T-line are shown in Figure 7.15. As the source is approached, the impedance approaches Z_0, as shown in the rectangular plot. On the Smith chart, the center is being approached. Thus, T-line loss improves input match. An attenuator or lossy T-line is a practical technique to improve the match of an unmatched load, but at the expense of power.

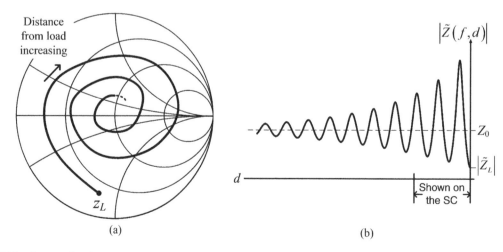

Figure 7.15 Generic sketches of impedance for a lossy T-line (*d* increasing as the source is approached). (a) Impedance locus. (b) Impedance standing wave pattern.

7.7 Determining Load Impedance from Input Impedance

Often, the impedance at the input of the T-line, $\tilde{Z}_{\text{in}} = \tilde{Z}(f, \ell)$, is known and the load impedance $\tilde{Z}_L$ is to be determined. This is especially true in measurements, where the VNA is used to measure the input impedance and then the load impedance is determined. This mathematical translation is often a feature of modern VNAs! If one rearranges the equations for the input impedance to solve for the load impedance, the lossy T-line result is:

$$\tilde{Z}_{\text{in}} = \tilde{Z}(f, \ell) = Z_0 \frac{\left(\tilde{Z}_L + Z_0\right) e^{+\gamma\ell} + \left(\tilde{Z}_L - Z_0\right) e^{-\gamma\ell}}{\left(\tilde{Z}_L + Z_0\right) e^{+\gamma\ell} - \left(\tilde{Z}_L - Z_0\right) e^{-\gamma\ell}} \rightarrow \boxed{\tilde{Z}_L = Z_0 \frac{\left(\tilde{Z}_{\text{in}} + Z_0\right) e^{-\gamma\ell} + \left(\tilde{Z}_{\text{in}} - Z_0\right) e^{+\gamma\ell}}{\left(\tilde{Z}_{\text{in}} + Z_0\right) e^{-\gamma\ell} - \left(\tilde{Z}_{\text{in}} - Z_0\right) e^{+\gamma\ell}}}$$

$$(7.77)$$

and the lossless T-line result is:

$$\tilde{Z}_{\text{in}} = \tilde{Z}(f, \ell) = Z_0 \frac{\tilde{Z}_L + jZ_0 \tan \beta\ell}{Z_0 + j\tilde{Z}_L \tan \beta\ell} \rightarrow \boxed{\tilde{Z}_L = Z_0 \frac{\tilde{Z}_{\text{in}} - jZ_0 \tan \beta\ell}{Z_0 - j\tilde{Z}_{\text{in}} \tan \beta\ell}} \qquad (7.78)$$

The load impedances from Examples 7.5.1 and 7.6.2 are determined from the input impedances in Examples 7.7.1 and 7.7.2, respectively, to illustrate the use of these relations.

Example 7.7.1
A coax T-line has a characteristic impedance of 75 Ω, a dielectric constant of 2.2, a loss of 88 dB/100 ft, and a length of 0.15 m. Determine the load impedance given an input impedance of $17.488 + j11.542\ \Omega$ at 14 GHz.

Strategy
Determine α in Np/m and β in rad/m, $\gamma = \alpha + j\beta$

Insert the values of γ, $d = \ell$, Z_0, and $\tilde{Z}_{\text{in}}$ to calculate $\tilde{Z}_L$.

Solution

$$\alpha = \left(\frac{88\,\text{dB}}{100\,\text{ft}}\right)\left(\frac{1\,\text{Np}}{8.686\,\text{dB}}\right)\left(\frac{3.2808\,\text{ft}}{\text{m}}\right) = 0.33239\,\text{Np/m}$$

$$\beta = \frac{2\pi}{\lambda} = \frac{2\pi f}{vel.} = \frac{2\pi f \sqrt{\mu_r \varepsilon_r}}{c} = \frac{2\pi 14 \times 10^9 \sqrt{(1)(2.2)}}{2.998 \times 10^8} = 435.19873\,\text{rad/m}$$

$$e^{+\gamma\ell} = e^{+\alpha\ell} e^{+j\beta\ell} = e^{+0.33239(0.15)} \angle[435.19873(0.15)] = 1.051122\angle + 65.2798\,\text{rad}$$

$$e^{-\gamma\ell} = e^{-\alpha\ell} e^{-j\beta\ell} = e^{-0.33239(0.15)} \angle[-435.19873(0.15)] = 0.951364\angle - 65.2798\,\text{rad}$$

Insert the values for $\tilde{Z}_{\text{in}}, Z_0, e^{+\gamma\ell}$, and $e^{-\gamma\ell}$ into $\tilde{Z}_L = Z_0 \dfrac{\left(\tilde{Z}_{\text{in}} + Z_0\right) e^{-\gamma\ell} + \left(\tilde{Z}_{\text{in}} - Z_0\right) e^{+\gamma\ell}}{\left(\tilde{Z}_{\text{in}} + Z_0\right) e^{-\gamma\ell} - \left(\tilde{Z}_{\text{in}} - Z_0\right) e^{+\gamma\ell}}$

$$\tilde{Z}_L = 30.000 + j80.000\,\Omega \approx 30.0 + j80.0\,\Omega \quad \text{(matches } \tilde{Z}_L \text{ in Example 7.5.1)}.$$

Example 7.7.2
If the input impedance of a lossless T-line is $26.366 + j29.146\ \Omega$ and the characteristic impedance is 52 Ω, determine the load impedance at 0.200 [λ] from the input (toward the load) (a) analytically, and (b) using a Smith chart.

Solution

a) Convert $\beta\ell$ from wavelengths to radians and insert values into the $\tilde{Z}_L$ relation for lossless T-lines:

$$\beta\ell\,(\text{rad}) = \beta\ell\,[\lambda]\left(\frac{2\pi\,\text{rad}}{1\,[\lambda]}\right) = 0.200\,[\lambda]\left(\frac{2\pi\,\text{rad}}{1\,[\lambda]}\right) = 1.25664\,\text{rad}$$

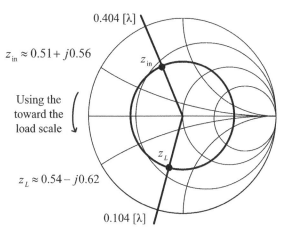

Figure 7.16 Smith chart for Example 7.7.2.

$$\widetilde{Z}_L = Z_0 \frac{\widetilde{Z}_{\text{in}} - jZ_0 \tan \beta\ell}{Z_0 - j\widetilde{Z}_{\text{in}} \tan \beta\ell} = 52 \frac{(26.366 + j29.146) - j52 \tan (1.25664 \text{ rad})}{52 - j (26.366 + j29.146) \tan (1.25664 \text{ rad})}$$

$\widetilde{Z}_L = 28.000 - j31.9995 \approx 28.0 - j32.0 \, \Omega$ (matches $\widetilde{Z}_L$ in Example 7.6.2)

b) Plot the normalized input impedance:

$$z_{\text{in}} = \frac{\widetilde{Z}_{\text{in}}}{Z_0} = \frac{26.366 + j29.146}{52} = 0.507 + j0.561 \approx 0.51 + j0.56$$

Determine z_L at 0.200 [λ] toward the load from z_{in}: see Figure 7.16.

- Locate z_{in} on the *wavelengths toward the load scale*: z_{in} is at 0.404 [λ]
- Move 0.200 [λ] *toward the load* (counterclockwise) from the z_{in} location:

$$0.404[\lambda] + 0.200[\lambda] = 0.604[\lambda] = 0.604[\lambda] - 0.500[\lambda] = 0.104[\lambda]$$

- The Smith chart scale starts over at 0.500 [λ], so 0.500 [λ] was subtracted
- Draw a straight line from the Smith chart center to 0.104 [λ]
- z_L is located at the intersection of this straight line and the constant *VSWR* circle through z_L:
$z_L \approx 0.54 - j0.62, \quad \widetilde{Z}_L = z_L Z_0 \approx 28 - j\,32 \, \Omega$ [matches a)]

7.8 Summary of Important Equations

Telegrapher's eqns.: $\dfrac{d\widetilde{V}}{dz} = -j\omega L'\widetilde{I}$ and $\dfrac{d\widetilde{I}}{dz} = -j\omega C'\widetilde{V}$ Wave eqn.: $\dfrac{d^2\widetilde{V}}{dz^2} = -\omega^2 L'C'\widetilde{V}$

$z =$ distance from the source toward the load $d =$ distance from the load toward the source

$$\widetilde{V} = \widetilde{V}(f,d) = \frac{1}{2}\left(\widetilde{V}_L + Z_0\widetilde{I}_L\right)e^{+\gamma d} + \frac{1}{2}\left(\widetilde{V}_L - Z_0\widetilde{I}_L\right)e^{-\gamma d} \qquad \gamma = j\omega\sqrt{L'C'} = j\beta \text{ (lossless T-line)}$$

$$v(t,d) = V_L^+ \cos\left(\omega t + \beta d + \theta_L^+\right) + V_L^- \cos\left(\omega t - \beta d + \theta_L^-\right) \qquad Z_0 \equiv \sqrt{\frac{L'}{C'}} \qquad Z_0 = \frac{\widetilde{V}^+}{\widetilde{I}^+} = \frac{\widetilde{V}^-}{\left(-\widetilde{I}^-\right)}$$

$$\beta = \frac{2\pi}{\lambda} = \frac{\omega}{vel.} = \frac{2\pi\sqrt{\mu_r\varepsilon_r}}{c}f \qquad \gamma = \alpha + j\beta = \alpha\,(\mathrm{Np/m}) + j\beta\,(\mathrm{rad/m})(\text{lossy T}-\text{line}) \quad 1\,\mathrm{Np} = 8.686\,\mathrm{dB}$$

$$\widetilde{Z} = \widetilde{Z}(f,d) = Z_0 \frac{\left(\widetilde{Z}_L + Z_0\right)e^{+\gamma d} + \left(\widetilde{Z}_L - Z_0\right)e^{-\gamma d}}{\left(\widetilde{Z}_L + Z_0\right)e^{+\gamma d} - \left(\widetilde{Z}_L - Z_0\right)e^{-\gamma d}} \quad \text{(general)}$$

$$\widetilde{Z} = \widetilde{Z}(f,d) = Z_0 \frac{\widetilde{Z}_L + jZ_0\,\tan\beta d}{Z_0 + j\widetilde{Z}_L\,\tan\beta d} \quad \text{(lossless T}-\text{line only)}$$

$$\widetilde{Z}_L = Z_0 \frac{\left(\widetilde{Z}_{\mathrm{in}} + Z_0\right)e^{-\gamma\ell} + \left(\widetilde{Z}_{\mathrm{in}} - Z_0\right)e^{+\gamma\ell}}{\left(\widetilde{Z}_{\mathrm{in}} + Z_0\right)e^{-\gamma\ell} - \left(\widetilde{Z}_{\mathrm{in}} - Z_0\right)e^{+\gamma\ell}} \qquad \widetilde{Z}_L = Z_0 \frac{\widetilde{Z}_{\mathrm{in}} - jZ_0\,\tan\beta\ell}{Z_0 - j\widetilde{Z}_{\mathrm{in}}\,\tan\beta\ell}$$

$$\Gamma_{\mathrm{in}} = \Gamma(f,\ell) = \frac{\widetilde{Z}_{\mathrm{in}} - Z_0}{\widetilde{Z}_{\mathrm{in}} + Z_0} \quad \Gamma = \Gamma(f,d) = \frac{\widetilde{Z}(f,d) - Z_0}{\widetilde{Z}(f,d) + Z_0} \quad \Gamma_L = \Gamma_L(f) = \frac{\widetilde{V}_L^-}{\widetilde{V}_L^+} = \frac{-\widetilde{I}_L^-}{\widetilde{I}_L^+} = \frac{\widetilde{Z}_L - Z_0}{\widetilde{Z}_L + Z_0}$$

$$\beta d\,[\lambda] = \beta d\,(\mathrm{rad})\left(\frac{1\,[\lambda]}{2\pi\,(\mathrm{rad})}\right) \quad \beta d\;[\lambda] = d\,(\mathrm{m})\frac{1\,[\lambda]}{\lambda\,(\mathrm{m})}$$

7.9 Appendices

Appendix 7.A Conversion of Maxwell's Equations into the Telegrapher's Equations[9]

A time-changing magnetic field generates a time-changing electric field which generates a time-changing magnetic field ..., and *electromagnetic wave propagation* may result. The analytic solution of propagating electromagnetic waves is, in general, a three-dimensional problem that can be quite complex to solve in closed form (again, in general). A very useful, practical simplification of waves in circuits is obtained when Maxwell's equations are simplified into circuit variables on a one-dimensional T-line. The circuit variable voltage V corresponds to the fields variable $\overline{E}$ and I to $\overline{H}$. This appendix presents the transition from the well-known coax electric and magnetic fields solutions (that you have already developed in previous chapters) to the *telegrapher's equations*. Then a differential equation is developed. This development is not a general proof for all T-lines, but coax is convenient to illustrate how the T-line equations relate to the corresponding electromagnetic field equations.

The coaxial T-line set-up is shown in Figure 7.17. The radius of the inner conductor is a, and radius of the inner surface of the outer conductor is b. The z direction is out of the page. Assume:

1) perfect conductors,
2) lossless dielectric, and
3) AC sinusoidal signals.

The currents in the inner and outer conductors (technically surface currents) are equal in magnitude but opposite in direction. The dielectric mechanically aligns the centers of the conductors.

Start with Maxwell's curl equations in a lossless medium:

$$\nabla \times \overline{E} = -\frac{\partial\overline{B}}{\partial t} \qquad (7.79)$$

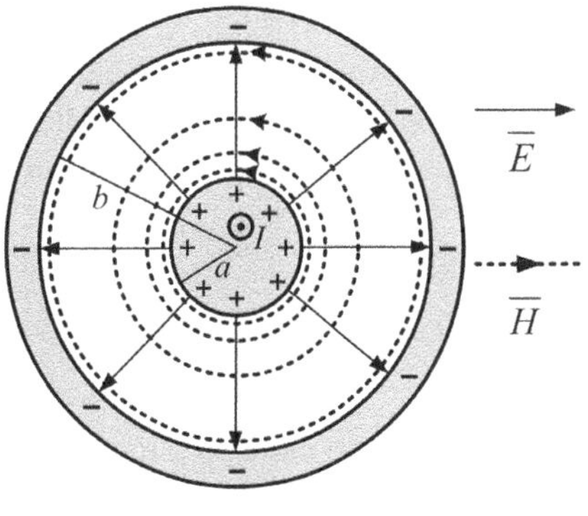

Figure 7.17 Coax end view.

---

9 Appendix 5.B Maxwell's Equations in Differential Form is a prerequisite to this appendix.

$$\nabla \times \overline{H} = \overline{J}_c + \frac{\partial \overline{D}}{\partial t} \rightarrow \nabla \times \overline{H} = \frac{\partial \overline{D}}{\partial t} \tag{7.80}$$

where the del operators will be expressed in *cylindrical coordinates* (ρ, ϕ, z) for coax. Note that the conduction current density *between* the conductors (in the dielectric) is zero, so the $\overline{J}_c$ term drops out. Convert the two previous equations into phasor forms (all time derivatives become multiplication by $j\omega$, just as they do in circuit theory):

$$\nabla \times \overline{\mathbf{E}} = -j\omega\overline{\mathbf{B}} = -j\omega\mu\overline{\mathbf{H}} \tag{7.81}$$

$$\nabla \times \overline{\mathbf{H}} = +j\omega\overline{\mathbf{D}} = +j\omega\,\varepsilon\overline{\mathbf{E}} \tag{7.82}$$

Note the field vector variables are shown in non-italicized **bold** font to clearly distinguish them as phasors. Only the E_ρ and H_ϕ components are present in coax, so the curl expression (in cylindrical coordinates) in Eq. (7.81) simplifies:

$$\nabla \times \overline{\mathbf{E}} = \left(+ \frac{1}{\rho}\frac{\partial \cancel{E_z}}{\partial \phi} - \frac{\partial \cancel{E_\phi}}{\partial z} \right)\overline{a}_\rho + \left(+ \frac{\partial E_\rho}{\partial z} - \frac{\partial \cancel{E_z}}{\partial \rho} \right)\overline{a}_\phi + \frac{1}{\rho}\left(+ \frac{\partial(\rho\cancel{E_\phi})}{\partial \rho} - \frac{\partial E_\rho}{\partial \phi} \right)\overline{a}_z$$
$$= -j\omega\mu\left(\cancel{H_\rho}\overline{a}_\rho + H_\phi\overline{a}_\phi + \cancel{H_z}\overline{a}_z \right) \tag{7.83}$$

and reduces to

$$+ \frac{\partial E_\rho}{\partial z}\,\overline{a}_\phi - \frac{1}{\rho}\frac{\partial E_\rho}{\partial \phi}\overline{a}_z = -j\omega\mu H_\phi\overline{a}_\phi \tag{7.84}$$

Similarly, Eq. (7.82) simplifies to

$$- \frac{\partial H_\phi}{\partial z}\,\overline{a}_\rho + \frac{1}{\rho}\frac{\partial(\rho H_\phi)}{\partial \rho}\,\overline{a}_z = +j\omega\varepsilon E_\rho\overline{a}_\rho \tag{7.85}$$

Note that E_ρ is uniform versus ϕ, so the second term in Eq. (7.84) is zero. The derivative of (ρH_ϕ) is zero because H_ϕ has a $1/\rho$ dependence (recall the magnetic field intensity solution for a straight conductor is $H_\phi = I/(2\pi\rho)$). Thus, the ρ in the numerator and in the denominator cancels in the (ρH_ϕ) product and the second term in Eq. (7.85) is zero. Note that these z-component terms must also equal zero because there are no corresponding vector terms on the other side of each equation. Consequently, the z-direction terms drop out and the curl equations reduce to:

$$+ \frac{\partial E_\rho}{\partial z} = -j\omega\mu H_\phi \tag{7.86}$$

$$- \frac{\partial H_\phi}{\partial z} = +j\omega\varepsilon E_\rho \tag{7.87}$$

Now the conversion from the field variables to the circuit variables begins. Given that this is an AC development, tilde notation will be used for the voltage and current. The electric field intensity and voltage solutions in coax for $a < \rho < b$ were previously obtained using Gauss's law and potential:

$$E_\rho = \frac{\rho_s a}{\varepsilon\rho} \tag{7.88}$$

$$\widetilde{V} = \frac{\rho_s a}{\varepsilon}\ln\left(\frac{b}{a}\right) \tag{7.89}$$

Insert Eq. (7.89) into (7.88) to eliminate ρ_S, which is generally unknown in value:

$$E_\rho = \frac{\widetilde{V}}{\rho\ln\left(\frac{b}{a}\right)} \tag{7.90}$$

Observe that the electric and magnetic fields here must include a z-dependent factor given the derivative with respect to z in Eqs. (7.86) and (7.87); otherwise, the solution is trivial – the fields equal zero. The only variable in Eq. (7.90) that can have z dependence is $\widetilde{V}$:

$$E_\rho = \frac{\widetilde{V}(z)}{\rho \ln\left(\dfrac{b}{a}\right)} \tag{7.91}$$

The static magnetic field intensity solution in coax for $a < \rho < b$ was obtained previously using ACL:

$$H_\phi = \frac{\widetilde{I}}{2\pi\rho} \tag{7.92}$$

The only variable in Eq. (7.92) that can have z dependence is $\widetilde{I}$:

$$H_\phi = \frac{\widetilde{I}(z)}{2\pi\rho} \tag{7.93}$$

Insert the field intensity solutions in Eqs. (7.91) and (7.93) into Eqs. (7.86) and (7.87), respectively:

$$+ \frac{1}{\rho \ln\left(\dfrac{b}{a}\right)} \frac{\partial\left[\widetilde{V}(z)\right]}{\partial z} = -j\omega\mu \frac{\widetilde{I}(z)}{2\pi\rho} \tag{7.94}$$

$$- \frac{1}{2\pi\rho} \frac{\partial\left[\widetilde{I}(z)\right]}{\partial z} = +j\omega\varepsilon \frac{\widetilde{V}(z)}{\rho\ln\left(\frac{b}{a}\right)} \tag{7.95}$$

and simplify:

$$\frac{\partial\left[\widetilde{V}(z)\right]}{\partial z} = -j\omega\left[\frac{\mu}{2\pi}\ln\left(\frac{b}{a}\right)\right]\widetilde{I}(z) \tag{7.96}$$

$$\frac{\partial\left[\widetilde{I}(z)\right]}{\partial z} = -j\omega\left[\frac{2\pi\varepsilon}{\ln\left(\dfrac{b}{a}\right)}\right]\widetilde{V}(z) \tag{7.97}$$

You might recall that the inductance and capacitance *per unit length* (one meter), indicated by the *prime* symbol, of coax were previously developed to be:

$$L' = \frac{\mu}{2\pi}\ln\left(\frac{b}{a}\right) \text{ (H/m)} \quad C' = \frac{2\pi\varepsilon}{\ln\left(\dfrac{b}{a}\right)} \text{ (F/m)} \tag{7.98}$$

Note that L' and C' are assumed to be constant, that is, they have no z dependence. Hence, the electromagnetic field equations for coax convert to the following circuit-variable equations (the partial derivatives have been replaced with ordinary derivatives given that there is only one independent variable, z, in the one-dimensional T-line):

$$\frac{d\left[\widetilde{V}(z)\right]}{dz} = -j\omega L'\widetilde{I}(z) \rightarrow \frac{d\widetilde{V}}{dz} = -j\omega L'\widetilde{I} \tag{7.99}$$

$$\frac{d\left[\widetilde{I}(z)\right]}{dz} = -j\omega C'\widetilde{V}(z) \rightarrow \frac{d\widetilde{I}}{dz} = -j\omega C'\widetilde{V} \tag{7.100}$$

where it is understood that the phasor voltage and current are functions of position z. These two T-line differential equations, the *telegrapher's equations*, are the starting point of developing the T-line wave equation (Section 7.1).

Appendix 7.B Development of the Particular Solutions for T-line Waves

This appendix presents the details in the development of the phasor voltage and current particular solutions starting with the general T-line solutions:

$$\widetilde{V} = \widetilde{V}(z) = A\,e^{-\gamma z} + B e^{+\gamma z} \tag{7.101}$$

$$\widetilde{I} = \frac{\gamma}{j\omega L'}\left(Ae^{-\gamma z} - Be^{+\gamma z}\right) \tag{7.102}$$

The constants A and B must be determined in order to arrive at the particular solution. These constants can be determined from the boundary conditions using the general solutions. More specifically, the $\widetilde{V}$ and $\widetilde{I}$ expressions as a function of position along the T-line depend on the $\widetilde{V}$ and $\widetilde{I}$ at the end (load) of the T-line, $\widetilde{V}_L$ and $\widetilde{I}_L$, which are assumed to be known or measured. However, the *load end* of the T-line is located at $z = \ell$:

$$\widetilde{V}(\ell) = \widetilde{V}_L = A\,e^{-\gamma \ell} + B e^{+\gamma \ell} \tag{7.103}$$

$$\widetilde{I}(\ell) = \widetilde{I}_L = \frac{\gamma}{j\omega L'}\left(Ae^{-\gamma \ell} - Be^{+\gamma \ell}\right) \tag{7.104}$$

Solving for coefficients A and B can be algebraically simplified if the *distance from the load d*, as shown in Figure 7.2, instead of z, is utilized, and the use of d is often of greater practical use as seen in this chapter. Substitute $\ell - d$ for z:

$$\widetilde{V} = \widetilde{V}(d) = A\,e^{-\gamma(\ell-d)} + B e^{+\gamma(\ell-d)} = Ae^{-\gamma \ell}e^{+\gamma d} + B e^{+\gamma \ell}e^{-\gamma d} \tag{7.105}$$

$$\widetilde{I} = \widetilde{I}(d) = \frac{\gamma}{j\omega L'}\left(Ae^{-\gamma(\ell-d)} - Be^{+\gamma(\ell-d)}\right) = \frac{\gamma}{j\omega L'}\left(Ae^{-\gamma \ell}e^{+\gamma d} - Be^{+\gamma \ell}e^{-\gamma d}\right) \tag{7.106}$$

The constants A and B are arbitrary, so let them absorb the $e^{-\gamma \ell}$ and $e^{+\gamma \ell}$ constants, respectively:

$$\widetilde{V} = \widetilde{V}(d) = Ae^{+\gamma d} + Be^{-\gamma d} \tag{7.107}$$

$$\widetilde{I} = \widetilde{I}(d) = \frac{\gamma}{j\omega L'}\left(Ae^{+\gamma d} - Be^{-\gamma d}\right) \tag{7.108}$$

Note: From this point forward in the voltage and current solutions development, the explicitly written independent variable is d (f is also an independent variable, but not shown here). Then the voltage and current at the load are designated as follows:

$$\widetilde{V}(0) = \widetilde{V}_L \quad \text{and} \quad \widetilde{I}(0) = \widetilde{I}_L \tag{7.109}$$

Insert these boundary conditions into the general solutions along with $d = 0$:

$$\widetilde{V}_L = Ae^{+\gamma 0} + Be^{-\gamma 0} \tag{7.110}$$

$$\widetilde{I}_L = \frac{\gamma}{j\omega L'}\left(Ae^{+\gamma 0} - Be^{-\gamma 0}\right) \tag{7.111}$$

Simplify:

$$\widetilde{V}_L = A + B \tag{7.112}$$

$$\widetilde{I}_L = \frac{\gamma}{j\omega L'} (A - B) \tag{7.113}$$

which are mathematically simpler than Eqs. (7.103) and (7.104) because $e^{\pm \gamma \ell} \neq 1$. Define the following ratio to be the *characteristic impedance Z_0*:

$$\frac{j\omega L'}{\gamma} = \frac{j\omega L'}{j\omega\sqrt{L'C'}} = \sqrt{\frac{L'}{C'}} \equiv Z_0 \tag{7.114}$$

Eq.(7.114) is substituted into Eq.(7.113) to give : $\widetilde{I}_L = \dfrac{A - B}{Z_0}$ \hfill (7.115)

Solve the simultaneous Eqs. (7.112) and (7.115) for A and B:

$$A = \frac{1}{2}\left(\widetilde{V}_L + Z_0\widetilde{I}_L\right) \tag{7.116}$$

$$B = \frac{1}{2}\left(\widetilde{V}_L - Z_0\widetilde{I}_L\right) \tag{7.117}$$

The constants A and B are thus expressed in terms of the boundary conditions at the load. The solution for $\widetilde{V}$:

$$\boxed{\widetilde{V} = \widetilde{V}(d) = \frac{1}{2}\left(\widetilde{V}_L + Z_0\widetilde{I}_L\right)e^{+\gamma d} + \frac{1}{2}\left(\widetilde{V}_L - Z_0\widetilde{I}_L\right)e^{-\gamma d}} \tag{7.118}$$

Current Particular Solution: The phasor solution of the T-line current is obtained by starting with the general current solution of Eq. (7.113), inserting $\dfrac{\gamma}{j\omega L'} = \dfrac{1}{Z_0} = Y_0$, inserting the values of A and B, and manipulating the result into the following form:

$$\boxed{\widetilde{I} = \widetilde{I}(d) = \frac{1}{2}\left(\widetilde{I}_L + Y_0\widetilde{V}_L\right)e^{+\gamma d} + \frac{1}{2}\left(\widetilde{I}_L - Y_0\widetilde{V}_L\right)e^{-\gamma d}} \tag{7.119}$$

Frequency is also an independent variable in $\widetilde{V}$ and $\widetilde{I}$, which is incorporated notation-wise in Eq. (7.38).

Appendix 7.C   Alternate Development of Reflection Coefficient vs. Position

Consider the ratio of a reflected wave to an incident wave at *any* position d from the load, which is the reflection coefficient at position d. Utilizing the phasor expressions for the incident and reflected traveling voltage waves:

$$\Gamma = \frac{\widetilde{V}^-}{\widetilde{V}^+} = \frac{\frac{1}{2}\left(\widetilde{V}_L - Z_0\widetilde{I}_L\right)e^{-\gamma d}}{\frac{1}{2}\left(\widetilde{V}_L + Z_0\widetilde{I}_L\right)e^{+\gamma d}} \cdot \frac{\frac{1}{\widetilde{I}_L}}{\frac{1}{\widetilde{I}_L}} = \frac{\widetilde{Z}_L - Z_0}{\widetilde{Z}_L + Z_0}\, e^{-2\gamma d} \tag{7.120}$$

One sees that the load reflection coefficient is embedded in this general reflection coefficient at position d:

$$\boxed{\Gamma = \Gamma(f, d) = \left(\frac{\widetilde{Z}_L - Z_0}{\widetilde{Z}_L + Z_0}\right)e^{-2\gamma d} = \Gamma_L\, e^{-2\gamma d}} \tag{7.121}$$

Notice that Γ at any position d along the T-line can be found in terms of the Γ at the load, Γ_L. The $e^{-2\gamma d}$ factor expresses how Γ changes as one moves away from the load position. On a lossless T-line, recall that the magnitude of the reflection coefficient is constant as a function of position (and is equal to the load reflection coefficient magnitude), and only the phase changes with position. As d increases on a lossy line (move farther from the load), $|\Gamma|$ decreases due to the attenuation constant. Insertion of transmission loss is one practical technique to reduce reflection levels in a circuit or system. The reflection coefficient is also frequency dependent.

An equivalent expression to Eq. (7.121) is to use the impedance at position d:

$$\boxed{\Gamma = \Gamma(f,d) = \frac{\widetilde{Z}(f,d) - Z_0}{\widetilde{Z}(f,d) + Z_0}} \tag{7.122}$$

One could algebraically manipulate this reflection coefficient equation to solve for impedance:

$$\widetilde{Z} = \widetilde{Z}(f,d) = Z_0 \frac{1 + \Gamma_L e^{-2\gamma d}}{1 - \Gamma_L e^{-2\gamma d}} = Z_0 \frac{1 + \Gamma(f,d)}{1 - \Gamma(f,d)} \tag{7.123}$$

Thus, the reflection coefficient and the impedance are directly related not only at the load but also at every position on the T-line. The expression in Eq. (7.123) is actually the mathematical basis for the Smith chart:

$$z(f,d) = \frac{\widetilde{Z}(f,d)}{Z_0} = \frac{1 + \Gamma_L e^{-2\beta d}}{1 - \Gamma_L e^{-2\beta d}} \tag{7.124}$$

where $z(f, d)$ is the normalized impedance at position d.

7.10 Homework

Note: Use $c = 2.998 \times 10^8$ m/s when c is required in calculations.

1. In the lossless T-line *model* for AC sinusoidal signals,
   a) How do the inductance and capacitance parameters differ from those used for discrete (lumped) inductors and capacitors?
   b) Why is L' in series in the model?
   c) Why is C' in parallel in the model?
   d) Why is the model over a differential length of T-line instead of a finite length?

2. In the *solution* of the phasor differential wave equation for a T-line,
   a) Why must the T-line be <u>uniform</u> and <u>fixed</u> (L' and C' are constant with respect to position and time)?
   b) Mathematically, why are there two terms in the solution?
   c) The arbitrary constants in the general solution are in terms of what quantities in the particular solution?

3. In the *interpretations* of the traveling wave solutions of the phasor differential wave equation for a T-line,
   a) Physically why are there two terms in the T-line voltage (and current) solutions?
   b) Why must the T-line be <u>linear</u>? Hint: consider the two terms in the solution.
   c) What is independent variable d?
   d) What is the physical significance of $\gamma = j\beta$?
   e) Which wave, incident or reflected, corresponds to the phasor term $\widetilde{V}_L^- e^{-\gamma d}$? Justify in the time domain.

4. a) What is *characteristic impedance*?
   b) How does characteristic impedance relate to T-line parameters?
   c) How is characteristic impedance developed theoretically for a given T-line?

5. Explain how to determine which term in the time domain lossy T-line voltage solution is the incident wave term and which term is the reflected wave term by reasoning from
   a) The exponential amplitude factors, and
   b) The arguments of the sinusoidal functions.

6. a) What are the physical loss mechanisms on a T-line?
   b) Mathematically, why is the propagation constant imaginary on a lossless T-line yet complex on a lossy T-line?
   c) What is the mathematical difference between the lossy and lossless T-line solutions?

7. a) What is the mathematical expression for voltage as a function of position on a *lossless* T-line if $\widetilde{Z}_L = Z_0$? *Briefly* explain what the expression implies for $\widetilde{Z}_L = Z_0$.
   b) Repeat part a) if the T-line is *lossy*. A sketch of the magnitude may help.

8. a) If $d = 0$ is inserted into the lossless T-line impedance result, what should $\widetilde{Z}$ equal? Why?
   b) If $d \rightarrow \infty$, what should $\widetilde{Z}$ equal? Why?
   c) Show both results mathematically.

9. For an RG-6 coaxial cable with an inner radius of 0.370 mm, an outer radius of 2.35 mm, and a dielectric insulator having properties $\mu_r = 1$ and $\varepsilon_r = 2.2$, determine:
   a) The characteristic impedance, and
   b) Speed of propagation on the cable.

   Compare these values to a datasheet for RG-6. How do the calculated results compare?

10. Repeat Problem 9 if the dielectric is air. Compare the impedance and velocity results.

11. A lossless T-line is 23.0 m long, has a characteristic impedance of 100 Ω, the dielectric has $\varepsilon_r = 2.1$, and the load impedance is $123 + j121$ Ω at 10.0 MHz.
    a) Calculate $\widetilde{Z}_{in}$ using the most appropriate equation.
    b) Calculate $\widetilde{Z}_L$ from $\widetilde{Z}_{in}$ using the most appropriate equation. Your result should match the original $\widetilde{Z}_L$.

12. For the T-line circuit below, calculate
    a) input impedance,
    b) load reflection coefficient,
    c) input reflection coefficient,
    d) load *VSWR*, and
    e) input *VSWR*.
    f) Conclude on the *VSWR* and reflection coefficient magnitude along a lossless T-line.

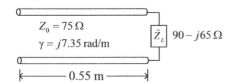

13. Verify the results of Example 7.5.1 using
   a) manual calculations (try without looking at Example 7.5.1; show intermediate results; then check against Example 7.5.1), and
   b) a mathematical software program.

14. Verify the results of Example 7.5.2 using
   a) manual calculations for the most appropriate equation for the impedance of a *lossless* T-line (try without looking at Example 7.5.2; show intermediate results; then check against Example 7.5.2), and
   b) a mathematical software program.
   c) Did the input impedance change relative to the lossy T-line? Why or why not? (use physical, not mathematical, reasoning).

15. For the T-lines in Problems 13, calculate
   a) load reflection coefficient,
   b) input reflection coefficient,
   c) load *VSWR*, and
   d) input *VSWR*.

16. For the T-lines in Problems 14, calculate
   a) load reflection coefficient,
   b) input reflection coefficient,
   c) load *VSWR*, and
   d) input *VSWR*.
   e) Conclude on the *VSWR* and the reflection coefficient magnitude on a *lossy* T-line relative to a lossless T-line.

17. A T-line has a characteristic impedance of approximately $50\,\Omega$ and an attenuation constant of 103 dB/100 ft. The dielectric has $\varepsilon_r = 2.2$, and the load impedance is $22 + j74\,\Omega$. Calculate
   a) the load reflection coefficient,
   b) the *VSWR* at the load,
   c) the return loss of the load, and
   d) the input impedance for both the lossy and a lossless T-lines at 0.154 m if $f = 8\,\text{GHz}$.
   e) Calculate $\widetilde{Z}_L$ from $\widetilde{Z}_{\text{in}}$ for both T-lines in d) using the most appropriate equations. Your result should match the original $\widetilde{Z}_L$.

18. A lossless T-line has a characteristic impedance of $50\,\Omega$ and a dielectric with $\varepsilon_r = 2.2$. The load impedance is $22 + j74\,\Omega$. Use a Smith chart to determine
   a) the load reflection coefficient,
   b) the *VSWR* at the load,
   c) the return loss of the load, and
   d) the input impedance at 0.154 m if $f = 8\,\text{GHz}$.
   e) Compare the Smith chart results with the analytic results (Problem 17). If different, explain why.
   f) Determine the load impedance from the input impedance using a Smith chart.

19. A T-line is 23.0 m long, has a characteristic impedance of $100\,\Omega$, the dielectric has $\varepsilon_r = 2.1$, and the load impedance is $123 + j121\,\Omega$ at 10.0 MHz.
   a) Determine the input impedance using a Smith chart. Compare the Smith chart result to the analytic result (Problem 11).
   b) Determine the load impedance from the input impedance using a Smith chart.

20. For the T-line circuit in Problem 12, use a Smith chart to determine
    a) input impedance,
    b) load reflection coefficient,
    c) input reflection coefficient,
    d) load *VSWR*,
    e) input *VSWR*,
    f) the minimum impedance along the T-line, and its electrical distance from the load,
    g) the maximum impedance along the T-line, and its electrical distance from the load, and
    h) the electrical distance between the minimum and maximum impedance positions.
    i) Conclude on the *VSWR* and reflection coefficient magnitude along a lossless T-line. Compare the Smith chart refection coefficient and *VSWR* results to the analytic results (Problem 12).

21. Describe
    a) What a Smith chart is,
    b) Why the Smith chart is useful when modern computational tools are available, and
    c) What is meant by the term *electrical distance*?
    d) How can electrical distance be changed?

22. a) What is the mathematical expression for *impedance* as a function of position on a lossless T-line if $\widetilde{Z}_L = Z_0$?
    b) Where is the impedance locus for $\widetilde{Z}_L = Z_0$ located on the Smith chart as βd increases?
    c) What is the mathematical expression for *impedance* as a function of position on a lossless T-line if $\widetilde{Z}_L = 0$?
    d) Where is the impedance locus for $\widetilde{Z}_L = 0$ located on the Smith chart as βd increases?

23. As the position on a lossy T-line is moved toward the load,
    a) does the radius on the Smith chart increase, decrease, or remain the same? Justify your answer.
    b) Is the impedance closer or further from being matched? Justify your answer.
    c) Does the magnitude of the reflection coefficient stay the same, increase, or decrease? Justify your answer.

24. Explain the general difference(s) between the loci on a Smith chart of moving position on a lossless T-line from a given general load (has both resistance and reactance) at a fixed frequency versus of a fixed position from a given load with changing frequency (being linearly swept, for example).

25. In microwave engineering, it is common to use a length of T-line to realize a desired reactance in a circuit. Assume an inductive reactance is needed in a circuit using a length of T-line with a short-circuited load.
    a) Will the T-line length need to be greater than or less than a quarter wavelength at the operating frequency?
    b) Justify your answer to part a) using an appropriate equation.
    c) Justify your answer to part b) using the Smith chart.

26. Repeat Problem 25 for capacitive reactance.

Development-type homework problems

27. In the *development* of the phasor differential wave equation for a T-line, physically interpret each of the telegrapher's equations:
    a) $\dfrac{d\widetilde{V}}{dz} = -j\omega L' \widetilde{I}$ and

    b) $\dfrac{d\widetilde{I}}{dz} = -j\omega C'\widetilde{V}.$

    c) Why are the two telegrapher's equations combined to form $\dfrac{d^2\widetilde{V}}{dz^2} = -\omega^2 L'C'\widetilde{V}$?

    d) What mathematical function is in the solution to this differential equation? Why?

28. Show that the phasor voltage particular solution satisfies the lossless T-line wave equation. Note that $z = \ell - d$, so $dz = -d(d)$ and $d^2z = +d^2(d)$ in the wave equation.

29. a) Insert the expressions for L' and C' of coax into $1/\sqrt{L'C'}$ and simplify.

    b) Identify the result of a). Thus, what does $1/\sqrt{L'C'}$ equal for a T-line?

    c) Show that velocity $= \omega/\beta$ using $\beta = \omega\sqrt{L'C'}$.

30. Perform a units balance analysis on $\beta = \omega\sqrt{L'C'}$. Advice: Use the time-domain circuit relationships for inductors and capacitors to determine the units of a henry and the units of a farad in terms of volts and amperes.

31. As suggested in Section 7.4, the lossy T-line solution can be developed by incorporating a distributed resistance R' in series with L' and a distributed conductance G' in parallel with C' into the T-line model. Draw and label the lossy T-line model. Develop the following phasor general solution from it:

$$\widetilde{V} = \widetilde{V}(z) = A\,e^{-\gamma z} + B e^{+\gamma z} \text{ where } \gamma = \pm\sqrt{(R' + j\omega L')(G' + j\omega C')}.$$

32. a) Determine the expression for the input impedance of a lossless T-line that is $\lambda/4$ long using the most appropriate equation. Hint: What happens to the $\tan(\beta d)$ when $\beta d = \lambda/4$? How does this result simplify the $\widetilde{Z}_{\text{in}}$ expression?

    b) If the load is pure capacitance, what is the nature of the input impedance? Justify the result with a Smith chart argument.

    c) Repeat b) for a load that is a pure inductance.

    d) Repeat b) for a load that is a pure resistance.

Application note: This circuit is called the *quarter-wave transformer* and it functions as an *impedance inverter*. It is commonly used in narrowband high-frequency circuit designs, such as filters and impedance matching circuits.

33. Show how Eqs. (7.65) through (7.70) are obtained from the general lossless T-line solutions in Eqs. (7.61), (7.62), and (7.64).

34. a) Rearrange the $\widetilde{Z}_{\text{in}} = \widetilde{Z}(f, \ell)$ result to solve for $\widetilde{Z}_L$ in the lossless T-line case.

    b) Repeat a) for the lossy T-line case.

8

Antennas and Links

Electromagnetic waves guided by transmission lines (T-lines) have been emphasized in the last two chapters. Now we transition to electromagnetic waves that propagate in "free space," that is, propagating in vacuum or air with no T-line to guide them, and the interface from a T-line to/from free space, which is the antenna! These antennas and *unguided* electromagnetic waves are the basis for wireless systems, such as cell phones, Wi-Fi, GPS, Bluetooth, and so forth, which we are surrounded by on a daily basis. This chapter will utilize the electromagnetics and transmission line concepts covered so far to introduce the fundamentals of the antennas and links that are critical in modern wireless communication technology.

An intuitive transition from the T-line to the antenna is initially presented along with major antenna concepts (8.1). The electromagnetic (EM) plane wave is then examined with a direct analogy to the T-line in order to determine the power received by an antenna from an incident plane wave (8.2). Key antenna parameters are surveyed (8.3). Then the characteristics of antennas are utilized to determine the power received in a link by utilizing the well-known Friis Transmission equation (8.4).

Electromagnetics and Transmission Lines: Essentials for Electrical Engineering, Second Edition.
Robert A. Strangeway, Steven S. Holland, and James E. Richie.
© 2023 John Wiley & Sons, Inc. Published 2023 by John Wiley & Sons, Inc.
Companion website: www.wiley.com/go/Strangeway/ElectromagneticsandTransmissionLines

8.1 Introduction to Antennas

8.1.1 An Intuitive Transition from a Transmission Line to an Antenna

What is electromagnetic **radiation**?[1] It is the generation of an electromagnetic wave in free space due to charge motion on conductors. When is radiation significant? Start with a T-line with an open-circuit load as shown in Figure 8.1a. What completes the current path for time-varying signals? Displacement current! (See Section 6.1.1.) Now begin to flare the ends of the T-line, as shown in Figure 8.1b. Displacement current still completes the current path, but the fields "bulge out" from the open end of the T-line. Finally, flare one-quarter wavelength ($\lambda/4$) at a right angle to the T-line, as shown in Figure 8.1c, to form a **dipole** antenna. On the T-line, the currents are in *opposite* directions (antiparallel), whereas in the flared end, the currents are in the *same* direction. Insignificant radiation occurs when the currents are in opposite directions and the spacing between the conductors is much less than one wavelength (a condition assumed throughout Chapters 6 and 7) because the radiated field intensities of the two currents subtract and nearly cancel at a distance far from the T-line. In contrast, significant radiation can occur[2] when the currents are in the same direction because the radiated field intensities of the two currents add constructively at a distance far from the antenna. By the Poynting vector $\overline{S} = \overline{E} \times \overline{H}$, electromagnetic power radiates.

A brief, qualitative explanation of the radiated electromagnetic waves is given here. Assume the T-line in Figure 8.1c is excited with a steady-state sinusoidal source such that the currents and fields on the T-line and the antenna are periodically changing direction (sinusoidally). The time snapshot in this figure shows radiated wavefronts that have already detached from the antenna and are comprised of alternating directions of $\overline{E}$ and similarly for $\overline{H}$. The Poynting vector and RHR can be used to show that these field directions are consistent with waves

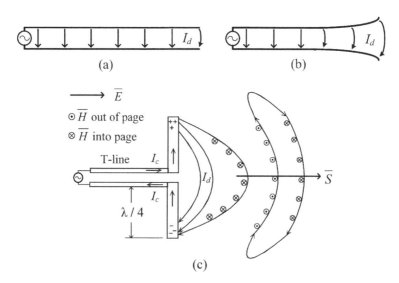

Figure 8.1 Transition from a T-line to a half-wave dipole antenna (I_c = conduction current, I_d = displacement current). (a) Displacement current on a T-line. (b) Flaring the load end of the T-line. (c) Dipole with currents and a cross-section of $\overline{H}$ shown.

---

1 "Radiate," "radiating," and "radiation" here refer to electromagnetic transmission from or reception by an antenna, not atomic radioactivity.

2 Significant radiation is also possible from transmission lines with oppositely directed currents but with a separation between the conductors that is an appreciable fraction of a wavelength since the radiated fields from the currents will no longer cancel at a distance away from the transmission line. Thus it is usually important to keep the separation between transmission line conductors as small as possible to avoid unintentional radiation!

propagating to the right. The magnetic field forms closed loops surrounding the dipole antenna (solenoidal magnetic flux) at an ever-increasing radius as the wavefront propagates away from the antenna. Notice also that the displacement currents (electric fields) form closed loops to complete the electric flux path between wavefronts, since there are no charges for the electric field to terminate on in free space. The time snapshot also shows the electric fields "bulging out" from the dipole arms, and when the current changes direction a half-cycle later, these fields will detach to form another wavefront. As time increases, this process repeats, producing sinusoidally varying, *unguided* electromagnetic wavefronts that propagate away and spread out spherically from the antenna. In this manner, an antenna *transitions a guided electromagnetic wave on a T-line to an unguided electromagnetic wave in free space* (and vice versa).

8.1.2 Antenna Concepts

What is an **antenna**? As discussed in the previous section, it is a structure to transition between a guided EM wave on a T-line and an EM wave in free space. An EM wave transitions from a T-line to free space in a *transmitting antenna*. Likewise, an EM wave transitions from free space to a T-line in a *receiving antenna*. Antennas are used in wireless links (to be addressed in Section 8.4) to accommodate a moving transmitter and/or receiver, when distances are too far for installing a T-line (practical and economic reasons[3]), and to accomplish broadcasting from one transmitter to several receivers. Antennas are used to direct *unguided* electromagnetic waves in free space into generally desired directions. What function does an antenna fundamentally perform? An antenna converts an electrical signal in a circuit to/from an electromagnetic wave in free space. The schematic symbol for an antenna is shown in Figure 8.2.

A principle from physics is that the *primary source of radiation is accelerated charges*: Qdv/dt, where v is the *velocity* of the charges (not voltage) and dv/dt is the acceleration of the charges. Consider the following visualization of electromagnetic radiation when charge is accelerated.[4,5,6] A charge moving with a constant velocity has a "constant" electromagnetic field pattern around it. See Figure 8.3a. Any *change* in the electromagnetic field can only propagate at the speed of light.

Now examine Figure 8.3b, where the charge has been accelerated. The electromagnetic field from some earlier time before the acceleration of the charge (shown as the electric flux lines outside of the dotted circle) will not match radially to the electric field lines when the charge is accelerated (shown as the electric flux lines inside of the dotted circle). A component of the electric field pattern that is perpendicular to the radial direction (the "kink") is created to complete the flux path. This electric field component, along with its associated magnetic field component, propagates outward. This EM wave is referred to as electromagnetic radiation.

Recall from Section 6.2.3 that the Poynting vector $\overline{S}$ gives electromagnetic wave power density (W/m^2) and direction:

$$\overline{S} = \overline{E} \times \overline{H} \tag{8.1}$$

Assume that the charge is accelerated in the horizontal direction. Take the cross product of the nonradial component of the electric field (that resulted due to acceleration of the charge) and the magnetic field "that goes around the current" as sketched in Figure 8.3b. The result is a Poynting vector with a component in the radial direction which represents electromagnetic power density flowing away from the accelerated charge. This component represents radiation of the electromagnetic field.

Charges accelerate in a time-changing current. Hence, *a time-changing current can radiate*. The sinusoidal case is fundamentally important because signals can be constructed

Figure 8.2 Schematic symbol for an antenna.

---

3 Aside from the physical constraints that may preclude connecting two locations with a T-line such as coax or a fiber-optic cable, we will see in Section 8.4 that beyond a certain distance, a wireless link can also have lower overall loss than a transmission line.

4 Ohanian, H.C. (1988). *Classical Electrodynamics*, 411–412. Massachusetts: Allyn and Bacon.

5 Scott, W.T. (1966). *The Physics of Electricity and Magnetism*, 2e, 570–571. Wiley.

6 Weidner, R.T. and Sells, R.L. (1965). *Elementary Classical Physics*, vol. **2**, 1024–1028. Massachusetts: Allyn and Bacon.

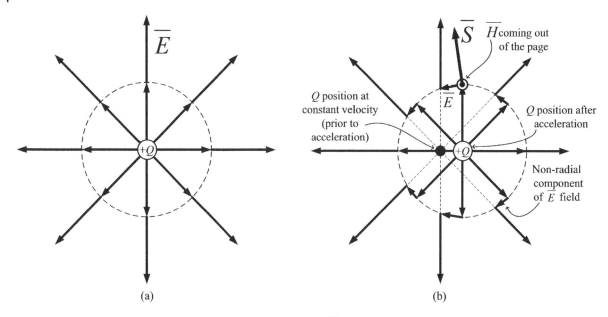

Figure 8.3 Visualization of radiation from an accelerated charge. (a) $\bar{E}$ field of Q with constant velocity just before acceleration. (b) $\bar{E}$ field of accelerated Q.

from sinusoids (recall the application of Fourier series and transforms to represent signals in terms of sinusoids of different frequencies). The question is when is radiation significant? Frequency and antenna size are examined in this context.

You may have noticed that antennas and wireless links are most commonly used at higher frequencies (MHz and GHz). A major practical factor[7] that motivates operating a wireless link at higher frequencies is the required size of the antenna. Re-examine Figure 8.1c. Notice that the total length of this dipole antenna is $\lambda/2$. In order to efficiently radiate (or receive) electromagnetic waves, an antenna must be an appreciable fraction of a wavelength ($\lambda/2$ is common). High frequencies have shorter wavelengths, and thus the physical size of the antenna is smaller. Low-frequency communication links exist, but the physical antenna size required often limits the practicality of these systems.

8.2 Uniform Plane Waves

Imagine an antenna is radiating in free space. The unguided EM wave travels away from the antenna at a constant velocity radially, which means the radiated waves have spherical wavefronts. See Figure 8.4. At a large distance from the transmit antenna, in what is called the **far field**,[8] the spherical wavefront is approximately flat, or planar, in a small cross-sectional area that represents the receiving antenna. Thus, the radiated spherical waves of the transmit antenna can be approximated as "plane waves" at distances far from the transmitting antenna (in the far field). For the introductory discussion in this chapter, we will assume we are always in the far field of any transmitting antenna and thus can utilize this plane wave approximation when analyzing wireless links. The uniform plane wave model will be examined next.

---

7 Not considered in this discussion is the fact that the information capacity of a communication system generally increases with operating frequency and the associated increased system bandwidth, which adds an additional practical limitation to operating a communication link at low frequencies. See for example: Haykin, S. and Moher, M. (2009). *Communication Systems*, 5e. New Jersey: Wiley.

8 A common estimate of the far field distance is $R_{\text{far field}} > \dfrac{2D^2}{\lambda}$, where D is the largest dimension of the antenna.

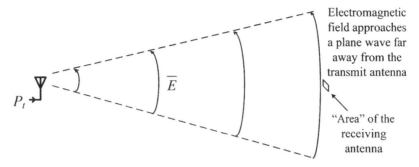

Figure 8.4 Spherical wavefront approaching a plane wave far from the transmit antenna.

8.2.1 Comparison of Uniform Plane Wave and Transmission Line Solutions

The **uniform plane wave** is an electromagnetic wave that has constant electric and magnetic field intensities in all dimensions normal to the direction of propagation, with infinitely large planar wavefronts (to simplify the mathematical description). See Figure 8.5. Assume the plane wave is propagating in the $+z$ direction, the electric field intensity $\overline{E}$ is in the x-direction, and the magnetic field intensity $\overline{H}$ is in the y-direction. The electric field intensity will be uniform in the *entire x–y plane*, that is, the strength and direction of $\overline{E}$ will be the same at any position within any given x–y plane. Similarly, $\overline{H}$ will be uniform in any given x–y plane.

Note: In this text, only *uniform* plane waves are considered. Any discussion pertaining to a "plane wave" is assumed to be a uniform plane wave.

The uniform plane wave field equations will be compared to the corresponding T-line voltage and current equations. The electric field intensity solution to Maxwell's equations (technically the electromagnetic wave equation) for a plane wave is:[9]

$$\overline{E}(z,t) = E_o \cos{(\omega t - \beta z + \theta)}\, \overline{a}_x \tag{8.2}$$

where E_o is the peak magnitude of the sinusoidal electric field intensity and θ is a phase shift with respect to a $t = 0$ and $z = 0$ reference. Note the similarity to the incident voltage wave solution for a lossless T-line:

$$v^+(z,t) = V_L^+ \cos{(\omega t - \beta z + \theta^+)} \tag{8.3}$$

Thus, the propagation behavior of the electric field vector in a plane wave and the voltage wave on a T-line are the same.

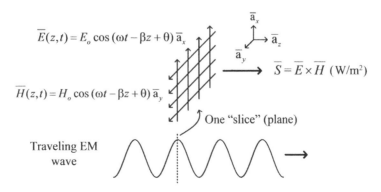

Figure 8.5 Visualization of one "slice" of a plane wave.

---

9 A lossless, homogeneous, isotropic, time-invariant medium is assumed.

The magnetic field intensity solution is also obtained by inserting the electric field solution into Faraday's law (differential form; see appendices to Chapter 5) and the result is:

$$\overline{H}(z,t) = H_o \cos\left(\omega t - \beta z + \theta\right)\overline{a}_y \tag{8.4}$$

Note the similarity to the incident current wave solution for a lossless T-line:

$$i^+(z,t) = I_L^+ \cos\left(\omega t - \beta z + \theta^+\right) = \frac{V_L^+}{Z_0} \cos\left(\omega t - \beta z + \theta^+\right) \tag{8.5}$$

A characteristic impedance-type quantity for plane waves, the **intrinsic impedance**, η (Greek letter lowercase eta; units Ω), results:

$$\overline{H} = \frac{E_o}{\eta} \cos\left(\omega t - \beta z + \theta\right)\overline{a}_y \tag{8.6}$$

The analogies between plane waves and waves on a T-line are listed in Table 8.1.

Several observations can be made on the plane wave solution:

- Per Maxwell's equations, a sinusoidal electric field "generates" a sinusoidal magnetic field, and a sinusoidal magnetic field "generates" a sinusoidal electric field. Propagation of the electromagnetic wave results.
- The electric and magnetic fields propagate together as a single electromagnetic wave. A conductor is not necessary for propagation.
- The electric field intensity and the magnetic field intensity vector orientations are orthogonal, and both are orthogonal to the direction of propagation.
- The intrinsic impedance is the ratio of electric field intensity to magnetic field intensity for a *single propagating wave* and represents power transfer through space (just like characteristic impedance does on a T-line). Further, note the analogous forms of the equations for intrinsic impedance and characteristic impedance.
- The phase constant and wave velocity are identical for a T-line and a plane wave.

8.2.2 The Poynting Vector and Electromagnetic Wave Power

How is the power determined in an electromagnetic wave in terms of the electric and magnetic fields? Recall instantaneous power for signals in circuit theory:

$$p(t) = v(t)i(t) \tag{8.7}$$

Table 8.1 Comparison of the T-line and uniform plane wave solutions (assume *z increases* as the load is approached).

T-line solution	Uniform plane wave
$v^+(z,t) = V^+ \cos(\omega t - \beta z)$	$\overline{E}^+(z,t) = E_o^+ \cos(\omega t - \beta z)\overline{a}_x$
$i^+(z,t) = I^+ \cos(\omega t - \beta z) = \dfrac{V^+}{Z_0} \cos(\omega t - \beta z)$	$\overline{H}^+ = H_o^+ \cos(\omega t - \beta z)\overline{a}_y = \dfrac{E_o^+}{\eta} \cos(\omega t - \beta z)\overline{a}_y$
$Z_0 = \dfrac{V^+}{I^+} = \sqrt{\dfrac{L}{C}}$ Characteristic impedance	$\eta = \dfrac{E_o^+}{H_o^+} = \sqrt{\dfrac{\mu}{\varepsilon}}$ Intrinsic impedance
$\beta = \omega\sqrt{\mu\varepsilon} = \omega\sqrt{LC}$	$\beta = \omega\sqrt{\mu\varepsilon}$
$v = \lambda f = \dfrac{1}{\sqrt{\mu\varepsilon}} = \dfrac{1}{\sqrt{LC}}$	$v = \lambda f = \dfrac{1}{\sqrt{\mu\varepsilon}}$
$P_{\text{ave}} = \dfrac{V_p I_p}{2} = \dfrac{V_p^2}{2Z_0} = \dfrac{I_p^2 Z_0}{2}$ Power in a wave on a T-line (W)	$S_{\text{ave}} = \dfrac{E_o H_o}{2} = \dfrac{E_o^2}{2\eta} = \dfrac{H_o^2 \eta}{2}$ Magnitude of the Poynting vector $\overline{S}$ (W/m^2)

Similarly, the power *density* flow in an electromagnetic wave is calculated from the previously introduced Poynting vector $\overline{S}$:

$$\overline{S} = \overline{E} \times \overline{H} \quad \text{units}: \quad (\text{V/m}) \cdot (\text{A/m}) = \left(\text{W/m}^2\right) \tag{8.8}$$

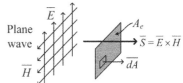

which gives both the magnitude and the direction of the electromagnetic power density flow. Recall that for the plane wave in Section 8.2.1, $\overline{E}$ was in the x-direction and $\overline{H}$ was in the y-direction. The direction of wave propagation and of $\overline{S}$ is in the z-direction for this particular plane wave. The reason why $\overline{E} \perp \overline{H}$ in the plane wave should now be apparent from the cross product. Since the plane wavefront is infinite in extent, it doesn't make sense to talk about the "total" power of a plane wave (which is infinite and thus not physical). Instead, we consider the power of the plane wave passing through a particular area, which is determined by integrating the average power density $\overline{S}_{\text{ave}}$ (Poynting vector) over the cross-sectional area of concern that the plane wave crosses:[10]

Figure 8.6 Set-up for received power from a plane wave using $\overline{S}$.

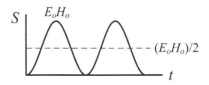

Figure 8.7 S plot vs. time.

$$P_{\text{ave}} = \int \overline{S}_{\text{ave}} \cdot \overline{dA} \tag{8.9}$$

where the differential surface vector $\overline{dA}$ points outward from the surface through which the electromagnetic wave crosses, as shown in Figure 8.6. Conceptually, one can consider Eq. (8.9) to represent the plane wave power that is absorbed or "captured" over some cross-sectional area by a receiving antenna. For plane waves, this relation often reduces to $\left|\overline{S}_{\text{ave}}\right| A_e$ because the power density is uniform in a plane wave, where A_e is the "effective area" of the receiving antenna (A_e will be formally described in Section 8.3), and this area is assumed to be in a plane parallel to the wavefront (in which case $\overline{S}_{\text{ave}} \parallel \overline{dA}$).

Thus, the significance of the *Poynting vector* is that it gives the *power density flow* in an electromagnetic wave. The capability to determine both the direction of the power flow and the total power through an area carried in an electromagnetic plane wave is both important and useful for wireless link analysis.

The expression for the AC Poynting vector is now developed. Start with the time-domain AC electric and magnetic field intensities:

$$\overline{E} = E_o \cos\left(\omega t - \beta z + \theta\right) \overline{a}_x \, (\text{V/m}) \quad \text{and} \quad \overline{H} = H_o \cos\left(\omega t - \beta z + \theta\right) \overline{a}_y \, (\text{A/m}) \tag{8.10}$$

Insert these field expressions into the Poynting vector and simplify:

$$\overline{S} = \overline{E} \times \overline{H} = E_o \cos\left(\omega t - \beta z + \theta\right) \overline{a}_x \times H_o \cos\left(\omega t - \beta z + \theta\right) \overline{a}_y \tag{8.11}$$

$$\overline{S} = \overline{E} \times \overline{H} = E_o H_o \cos^2\left(\omega t - \beta z + \theta\right) \overline{a}_z \, \left(\text{W/m}^2\right) \tag{8.12}$$

Note that $\overline{S}$ pulsates, as graphed in Figure 8.7, just as AC power in an electric circuit does. The peak power density is $E_o H_o$. The average power density is *half of the peak amplitude*, as sketched in the figure. (The circuit analogy is the average power in an AC signal, $V_{\text{peak}} I_{\text{peak}}/2$.) Thus, the peak and average electromagnetic power densities are

$$S_{\text{peak}} = E_o H_o \left(\text{W/m}^2\right) \quad \text{and} \quad S_{\text{ave}} = \frac{S_{\text{peak}}}{2} = \frac{E_o H_o}{2} \left(\text{W/m}^2\right) \tag{8.13}$$

---

10 This equation is one term from *Poynting's Theorem*, as we are only concerned here with calculating real power flow in a plane wave over a small, open surface.

Recall how electric flux density D was brought outside of the integral in Gauss's law. A similar approach is taken with the time-average power that crosses a defined area A_e:

$$P = \int \overline{S} \cdot \overline{dA} = \underbrace{\int \frac{1}{2} E_o H_o \, \overline{a}_z \cdot dA \overline{a}_z}_{\overline{S}_{\text{ave}} \, \| \, \overline{dA}} = \underbrace{\int \frac{1}{2} E_o H_o \, dA}_{S_{\text{ave}} \text{ const}} = \frac{1}{2} E_o H_o \underbrace{\int dA}_{A_e} = \frac{1}{2} E_o H_o A_e = S_{\text{ave}} A_e \text{ (W)} \tag{8.14}$$

Thus, the time-average power density in a uniform plane wave is (recalling $\eta = E_o/H_o$):

$$\boxed{S_{\text{ave}} = \frac{1}{2} E_o H_o = \frac{1}{2\eta} E_o^2 = \frac{1}{2} \eta H_o^2} \tag{8.15}$$

as indicated in Table 8.1 (compare Eq. (8.15) to the analogous circuits expressions for power). The average power received by an antenna of area A_e due to an incident uniform plane wave is:

$$\boxed{P_{\text{ave}} = S_{\text{ave}} A_e} \tag{8.16}$$

The next example illustrates practical determination of these quantities.

Example 8.2.1

Given a uniform plane wave of peak amplitude 100 μV/m at 2.3 GHz in a Teflon medium, determine

a) the intrinsic impedance
b) the phase constant
c) the velocity of the wave
d) the peak magnetic field intensity
e) the peak and average Poynting vector magnitudes
f) the average power that crosses a 4 m x 4 m cross-sectional area

Strategy

Given: $E_o = 100 \, \mu\text{V/m}$ in a uniform plane wave
$\qquad f = 2.3 \, \text{GHz}$
$\qquad \varepsilon_r = 2.1$ for Teflon

Desired : $\eta, \beta, v, S_{\text{peak}}, S_{\text{ave}}, P_{\text{ave}}$

$$\eta = \sqrt{\frac{\mu}{\varepsilon}}, \quad \beta = \omega\sqrt{\mu\varepsilon}, \quad v = \frac{1}{\sqrt{\mu\varepsilon}}, \quad H_o = \frac{E_o}{\eta}, \quad S_{\text{peak}} = E_o H_o, \quad P_{\text{ave}} = S_{\text{ave}} A_e$$

Solution

$$\eta = \sqrt{\frac{\mu}{\varepsilon}} = \sqrt{\frac{4\pi \times 10^{-7}}{(2.1)(8.854 \times 10^{-12})}} = 260 \, \Omega$$

$$\beta = \omega\sqrt{\mu\varepsilon} = \underbrace{2\pi(2.3 \times 10^9)}_{14.451 \text{ G rad/s}} \sqrt{(4\pi \times 10^{-7})(2.1)(8.854 \times 10^{-12})} = 69.9 \, \text{rad/m}$$

$$v = \frac{1}{\sqrt{\mu\varepsilon}} = \frac{1}{\sqrt{(4\pi \times 10^{-7})(2.1)(8.854 \times 10^{-12})}} = 2.07 \times 10^8 \, \text{m/s}$$

$$H_o = \frac{E_o}{\eta} = \frac{100\,\mu\text{V/m}}{260\,\Omega} = 0.384\,\mu\text{A/m}$$

$$S_{\text{peak}} = E_o H_o = (100\,\mu\text{V/m})(0.384\,\mu\text{A/m}) = 38.4\,\text{pW/m}^2$$

$$S_{\text{ave}} = \frac{S_{\text{peak}}}{2} = 19.2\,\text{pW/m}^2$$

$$P_{\text{ave}} = S_{\text{ave}}A_e = (19.2)\underbrace{[(4)(4)]}_{A_e} = 307\,\text{pW}$$

8.2.3 Polarization

Polarization is important in antenna system design and operation. **Polarization** is defined as the spatial orientation of the electric field vector as an electromagnetic wave propagates. The electric field vector is in a plane perpendicular to the direction of plane wave propagation. Two common types of polarization that are used with antennas are:

- *Linear polarization*: As the EM wave propagates, $\overline{E}$ points in the same direction relative to the physical environment, such as the surface of the earth. Two common types of linear polarization, vertical and horizontal, are shown in Figure 8.8a and b, respectively. Note that the polarization of the EM wave from the dipole in Figure 8.1 is vertical (assuming the dipole is vertical with respect to the earth).
- *Circular polarization*: As the EM wave propagates, the direction of $\overline{E}$ (and $\overline{H}$) rotates as viewed from any fixed point in space. The tip of $\overline{E}$ rotates in a circle. There are two types of circular polarization: left-hand circular polarization (LHCP) and right-hand circular polarization (RHCP) – see Figure 8.8c for LHCP. One can determine the "sense" of the polarization, right-handed or left-handed, by pointing your right-hand thumb in the direction of the plane wave propagation. The E field vector rotates in the direction of your curled right-hand fingers for RHCP and opposite that direction for LHCP, again as viewed from a fixed position.

The polarization concept of waves also applies to antennas, as every antenna responds strongly to a particular wave polarization (or transmits a particular wave polarization). Thus, the polarization of an antenna is an important specification. The usefulness of polarization arises from the need to match the transmitting and receiving antenna polarizations. For example, the vertical dipole antenna in Figure 8.1c transmits and receives vertically polarized waves. Now imagine a transmit dipole and a receive dipole. If the polarizations of the transmitting

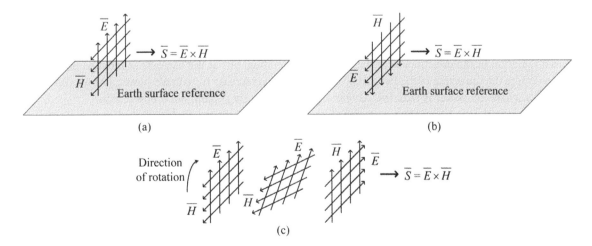

Figure 8.8 Polarization of plane waves. (a) Vertical linear polarization. (b) Horizontal linear polarization. (c) Circular polarization (LHCP shown). In part (c), three instances in time, $\omega t = 0°$, $45°$, and $90°$, are shown for a fixed position.

and receiving antennas are mismatched, a reduced signal is received.[11] If the polarizations are perpendicular (orthogonal), zero signal is received.

The isolation between orthogonally polarized waves can be practically utilized: two channels of information can be communicated over the same link. If an antenna can receive one polarization and reject the other, then the same frequencies can be *reused* in the other polarization. Thus, double the number of channels can be utilized in the same bandwidth! The same antenna can be used to receive both polarizations, but it must have some method to separate the polarizations. An important practical implication is that the polarizations of the transmitting antenna and the receiving antenna must be properly aligned.

It is generally more complicated to design antennas to transmit/receive circularly polarized waves. However, circular polarization is useful in systems where the receive and transmit antenna orientations are impractical to remain aligned, because only the sense of polarization (LHCP or RHCP) needs to be the same to receive a signal. Example applications include mobile devices and satellite-based systems, such as the Global Positioning System (GPS). In GPS, satellites may spin relative to a stationary place on earth and the atmosphere (specifically, the ionosphere) of the earth causes polarization orientation to change as a wave propagates between the earth and space (called Faraday rotation), making linear polarization alignment impractical. Thus, circular polarization is utilized in GPS.

8.3 Antenna Parameters

8.3.1 Antenna Gain

One of the fundamental parameters for an antenna is antenna gain. In order to understand this parameter, it is helpful to answer the question: What is the **total power radiated** P_{rad} by an antenna relative to the input power? The antenna is usually constructed with conductors that cannot increase power – it is a passive component. For a lossless, matched antenna, the radiated power equals the input power to the antenna: $P_{\text{rad}} = P_{\text{in}}$. It is emphasized that **antenna gain** G_{ant} is *not* power gain in the sense of an amplifier: $P_{\text{rad}} \neq G_{\text{ant}}P_{\text{in}}$!

Antenna gain represents how much the antenna concentrates the radiated (or received) power at the expense of other directions. For example, imagine a light bulb in the center of a spherical room. The light from the bulb would equally illuminate all portions of the wall. However, if the light bulb is placed at the focal point of a parabolic reflector, such as in a flashlight, a smaller area of the wall on axis from the flashlight would be greatly illuminated, but at the expense of significantly lower illumination in other directions. The parabolic reflector similarly focuses the EM wave and thus has an antenna gain (the basis of common dish antennas). Another analogy that is used to envision antenna gain is a balloon. Imagine you start with a spherical balloon, which represents the distribution of the radiated power equally in all directions. If one were to then squeeze one side of the balloon, the balloon would extend in other directions, mostly in the directions away from where you are squeezing it. This scenario represents increased radiation in the directions that the balloon is extending, but at the expense of radiated power in the directions where the squeezing of the balloon occurs. There is no increase in radiated power, just a redistribution (focusing) of where the power is radiated. Antenna gain is a "focusing" gain, not a power gain.

How is this description of antenna gain quantified? The concept of the **isotropic radiator** is needed. An isotropic radiator is an ideal antenna that radiates (and receives) electromagnetic power density equally in all directions. Antenna gain is equal in transmission and in reception, that is, antennas are *reciprocal*.[12] Consider power received from an antenna ($P_{\text{received - antenna}}$) versus power received from an isotropic radiator ($P_{\text{received - isotropic}}$). See Figure 8.9. The following equation is one way to intuitively quantify antenna gain.

---

11 The reduction in signal due to polarization misalignment is described by the *polarization loss factor*. See for example: Balanis, C.A. (2016). *Antenna Theory: Analysis and Design*, 4e. New Jersey: Wiley.

12 For passive antennas, which are passive components (reciprocity was discussed with *S*-parameters in Chapter 6).

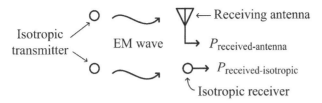

Figure 8.9 Illustration of the antenna gain concept.

$$G_{\text{ant}} = \frac{P_{\text{received-antenna}}}{P_{\text{received-isotropic}}} \tag{8.17}$$

For an antenna receiving a signal from a particular direction, the antenna gain represents the ratio of power received by an antenna relative to what would be received if the antenna were replaced by an isotropic radiator. In this sense, $G_{\text{ant}} > 1$ represents a receive signal increase, and $G_{\text{ant}} < 1$ represents a receive signal reduction, compared with an isotropic radiator. The gain can be expressed in decibel form as $G_{\text{ant}}(\text{dBi}) = 10\log(G_{\text{ant}})$, where "dBi" indicates "dB relative to an isotropic radiator." While antenna gain is a function of angle (see the discussion of radiation patterns in Section 8.3.2), often antenna datasheets report the peak antenna gain in dBi. A general derivation[13] results in the following key result for the peak antenna gain:

$$antenna\ gain \qquad \boxed{G_{ant} = \frac{4\pi A_e}{\lambda^2}} \tag{8.18}$$

where A_e is the **effective area** of the antenna (also called *effective aperture area*), which is the area of an incident plane wave that is "scooped out" by the antenna. This area does not usually equal the physical area of the antenna. For example, a wire antenna has an A_e significantly larger than the cross-sectional area of the wire. On the other hand, a dish antenna has an A_e somewhat smaller than the physical area of the dish.[14]

8.3.2 Radiation Patterns

A radiation pattern is commonly plotted as antenna gain versus direction. Why are spherical coordinates utilized in radiation patterns? See Figure 8.10a. Direction is specified solely from two angles, θ and ϕ, and polar plots of antenna gain vs. these two angles allow a three-dimensional visualization of the radiation pattern. Consider two commonly used radiation patterns (assuming the antenna is aligned with the z-axis):

- Vertical radiation pattern: antenna gain vs. θ, as shown in Figure 8.10b
- Horizontal radiation pattern: antenna gain vs. ϕ, as shown in Figure 8.10c

For example one can visualize the three-dimensional radiation pattern for a dipole antenna: a doughnut-shaped pattern, shown in Figure 8.10a. This three-dimensional visualization can be related to the vertical and horizontal patterns in (b) and (c).

Radiation pattern measurements are usually performed in one of two environments that minimize reflections near the antenna to isolate the radiation pattern of the antenna itself:

- *Range* – outdoors, open space, away from interfering EM sources

---

13 See, for example, Balanis, C.A. (2016). *Antenna Theory: Analysis and Design*, 4e. New Jersey: Wiley.
14 The relationship between the effective area and the physical area depends on the specific antenna. See Kraus, J.D. and Marhefka, R.J. (2001). *Antennas: For All Applications*, 3e. New York: McGraw Hill, for example.

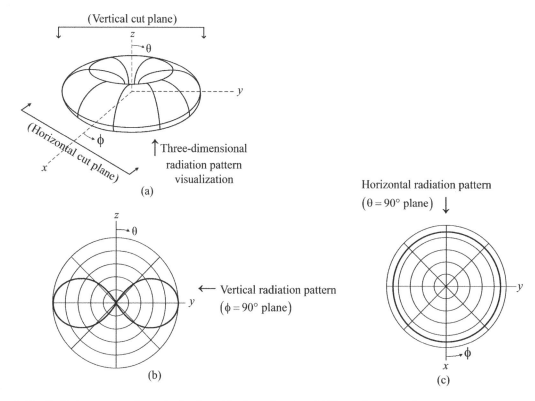

Figure 8.10 Radiation patterns for a dipole aligned in the *z*-direction. (a) Three-dimensional radiation pattern visualization. (b) Vertical radiation pattern ($\phi = 90°$ plane). (c) Horizontal radiation pattern ($\theta = 90°$ plane).

- *Anechoic chamber* – a room where all surfaces are covered with electromagnetic absorbing material

  Other radiation pattern-related quantities include half-power beamwidth, "lobes," and "zones":

- *Half-power beamwidth*: θ_{HP} and ϕ_{HP} are the angular spans in the θ and ϕ "planes" of the spherical coordinate system, shown in Figures 8.11a, b respectively, within which the antenna gain is one-half or greater relative to the on-axis maximum antenna gain (on a dB scale, this corresponds to gain within 3 dB of the maximum gain).
- *Side lobes*: Other lobes in the beam pattern apart from the main lobe for a directional antenna, as sketched in Figure 8.11c, where antenna gain is present but often undesired.

It should be noted that radiation patterns are normally plotted in the region far from the antenna where the fields are propagating away and the radiation patterns are well defined. Fields in this region and the region itself are referred to as **far field** (also referred to as *far zone*). All radiation patterns in this text are in the far field. Also, radiation patterns are the same for transmission and reception for passive antennas due to *reciprocity*.

There is also a **near field** (also referred to as *near zone*) where, in addition to the radiating fields, there are *reactive* fields. Reactive fields do not propagate but instead store electrical energy in a manner similar to an inductor or a capacitor.

8.3.3 Radiation Resistance and *VSWR*

If the antenna size dimension is on the order of a λ (one tenth of a wavelength ($\lambda/10$) and larger), then it is observed that the losses in the antenna circuit are much greater than the I^2R losses of the conductors. Why?

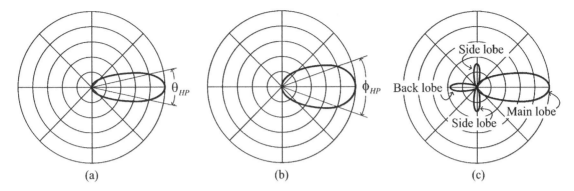

Figure 8.11 Half power beamwidths, side lobes, back lobe (radial division is 3 dB). (a) Vertical half-power beamwidth. (b) Horizontal half-power beamwidth. (c) Side lobes and back lobe.

Because power is leaving the circuit through radiation! This effect is characterized by **radiation resistance** R_{rad}. The antenna (in part) looks like a resistance to the circuit because real power is "lost" through radiation. It is analogous to characteristic impedance and intrinsic impedance, in that radiation resistance represents a *transfer of power* from a guided wave to a radiated wave, not dissipation.

The input impedance of an antenna consists of a radiation resistance, a resistance due to ohmic losses in the antenna, and a reactance due to the reactive fields (non-radiating portion of the EM fields in the near field) around the antenna. Thus, an antenna may be (and often is) mismatched to the characteristic impedance of the T-line and has an associated *VSWR*. As a reminder:

$$VSWR = \frac{|\widetilde{V}_{max}|}{|\widetilde{V}_{min}|}\Bigg|_{\substack{\text{standing} \\ \text{wave} \\ \text{pattern}}} = \frac{1 + |\Gamma|}{1 - |\Gamma|} \tag{8.19}$$

Recall that the *VSWR* ranges from one for a matched load to infinity for a short, open, or purely reactive load. The load here is the input impedance of the antenna. A low *VSWR* is important for antennas because if the *VSWR* is too high, (1) on transmit, too much reflected power from the antenna can damage or destroy the transmitter ($P_{refl} = |\Gamma|^2 P_{inc}$) as well as result in less power that is actually transmitted, and (2) on receive, reflecting too much of the "collected" power from the incoming plane wave means less power available at the antenna terminals and the receiver circuitry. A common specification is for an antenna to have *VSWR* < 2 over the operating frequency band. However, different applications have different requirements for *VSWR*. Antennas connected to high-power transmitters, for example those used in AM or FM broadcast, may require extremely well-matched antennas with *VSWR* ≪ close to one. On the other hand, antennas used in mobile devices, such as smartphones, often have more relaxed requirements of *VSWR* < 3, where sacrificing impedance match to increase the antenna bandwidth is common.[15]

---

15 For a T-line circuit, there is an inherent tradeoff between low reflection coefficient and the frequency bandwidth over which a device is impedance matched, called the Bode–Fano criterion. See: Pozar, D.M. (2012). *Microwave Engineering*, 4e. New Jersey: Wiley.

8.4 Links

A **link** is defined here to be from the input of the transmitting antenna to the output of the receiving antenna. See Figure 8.12a. The purpose of this section is to determine the relationship between the transmitted power, P_t, and the received power, P_r, which is the link loss. These relationships provide the tools necessary for first-order analysis of line-of-sight links, such as estimating the power received in a system (link budget analysis).

The **link loss**, L, is the total loss of a link from the input of the transmitting antenna to the output of the receiving antenna. Major factors in link loss include (1) free-space loss L_{fs}, (2) antenna gains G, (3) atmospheric losses L_a, and other losses L_o, all unitless ratios. The free-space loss is examined in the next section.

8.4.1 Free-Space Loss

Recall that *loss* is fundamentally a unitless ratio of input power to output power (Section 6.4.1). **Free-space loss** (also called *spreading loss* or *path loss*) is the loss between two ideal isotropic radiators. See Figure 8.12b. Why consider the free-space loss using ideal isotropic radiators? This approach distinguishes the spreading out of the wave as it propagates from the focusing (antenna gain) of the wave accomplished by the actual antennas.

Start with a transmitting isotropic radiator that radiates an equal power density S_{ave} over the surface area of a sphere of radius d ($d = $ distance, *not* diameter):

$$S_{ave} = \frac{P_t}{4\pi d^2} \tag{8.20}$$

At a distance d from the transmit antenna in the far field, S_{ave} represents the power density (W/m^2) of a plane wave incident on the receiving antenna. The receiving isotropic radiator, which has an effective area of A_{e-iso} (m^2), collects a portion of the transmitted power according to:

$$P_r = S_{ave}A_{e-iso} = \frac{P_t}{4\pi d^2}\, A_{e-iso} \tag{8.21}$$

where: P_r is the power received by the receiving isotropic radiator (W), A_{e-iso} is the effective area of the receiving isotropic radiator (m^2), d is the distance between the two antennas (m), and P_t is the power transmitted by the transmitting isotropic radiator (W).

The free space loss then is : $L_{fs} = \dfrac{P_t}{P_r} = \dfrac{4\pi d^2}{A_{e-iso}} \tag{8.22}$

Recall the general expression for the gain of an antenna: $G_{ant} = 4\pi A_e/\lambda^2$. The effective area of an isotropic radiator is $A_{e-iso} = \lambda^2/4\pi$ because G_{ant} is one (unity). Substitute this result into Eq. (8.22):

Free $-$ space loss : $\boxed{L_{fs} = \dfrac{P_t}{P_r} = \left(\dfrac{4\pi d}{\lambda}\right)^2} \tag{8.23}$

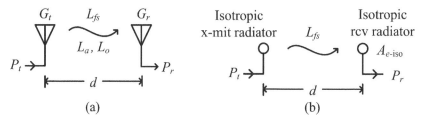

(a)   (b)

Figure 8.12 Link loss and free-space loss. (a) Link loss between the actual antennas. (b) Free-space loss between two isotropic radiators.

The P_t and P_r in Eq. (8.23) are for *isotropic radiators*. In dB form:

$$L_{fs}(\text{dB}) = P_t(\text{dBW}) - P_r(\text{dBW}) = 20 \log \left(\frac{4\pi d}{\lambda} \right) \tag{8.24}$$

where $P(\text{dBW})$ is the power in dBW (see Section 6.4.1), which is commonly used in links as opposed to the dBm. The free-space loss can be very large. For example, a typical value of free-space loss for geostationary satellites (synchronous orbit around the Earth) at 4 GHz is 200 dB!

While the free-space loss can be quite large, it has an interesting property: for every doubling of distance, the *free-space loss increases by only* 6 dB! For short distances, the free-space loss may be large compared with a similar length T-line. For large distances, $L_{fs}(\text{dB})$ is often significantly *lower* than a similar length T-line (which has a loss directly proportional to length). This is an additional benefit of wireless links over wired T-line connections in a system (aside from the many practical benefits of having no wires). Further, this "6 dB increase for every doubling of distance" property keeps $L_{fs}(\text{dB})$ manageable even for very large distance links, such as geosynchronous satellites links and the interplanetary links of NASA's deep space network!

Note that L_{fs} increases as frequency increases. One can think of increasing electrical distance (d/λ) between the antennas as frequency increases. Free-space loss is due only to the spreading of the wave as it propagates. Other losses and antenna gains must be incorporated to determine link loss, the loss between the two actual antennas, which is examined next.

8.4.2 Friis Transmission Equation for Link Loss

The link between the antennas under consideration is shown in Figure 8.12a. Antennas spatially concentrate the radiation pattern so as to minimize "waste" of the radiated power in undesired directions. Antenna gain *decreases* the total link loss. Other losses increase the total link loss:

L_a = signal loss due to *atmospheric* absorption
L_o = *other* signal losses, which include:

- T-line and connector losses,
- Mismatch of the antenna impedance to the connecting T-line Z_0,
- Antenna pointing loss (antenna orientations for maximum gain are misaligned), and
- Polarization loss (antenna polarizations are misaligned).

If the transmitting antenna has gain G_t, the radiated power density is increased (in the direction of peak antenna gain) by a factor equal to the transmitting antenna gain. Let the effective area of the receiving antenna be A_{er}.

$$\text{Then the power collected by the receiving antenna is}: P_r = \left(\frac{G_t P_t}{4\pi d^2} \right) A_{er} \tag{8.25}$$

$$\text{The effective area of the receiving antenna with gain } G_r \text{ is}: A_{er} = \frac{\lambda^2}{4\pi} G_r \tag{8.26}$$

$$\text{Substitute Eq.(8.26) into Eq.(8.25) and rearrange to obtain}: P_r = P_t G_t G_r \left(\frac{\lambda}{4\pi d} \right)^2 \tag{8.27}$$

Together with L_a and L_o, the total link loss can be solved from Eq. (8.27):

$$L = \frac{P_t}{P_r} = \frac{1}{G_t G_r} \left(\frac{4\pi d}{\lambda} \right)^2 L_a L_o \tag{8.28}$$

Friis transmission equation $\quad \boxed{L = \frac{P_t}{P_r} = \frac{L_{fs} L_a L_o}{G_t G_r}} \tag{8.29}$

$$L(\text{dB}) = P_t(\text{dBW}) - P_r(\text{dBW}) = L_{fs}(\text{dB}) + L_a(\text{dB}) + L_o(\text{dB}) - G_t(\text{dBi}) - G_r(\text{dBi}) \tag{8.30}$$

where P_t and P_r are the transmitted and received powers from the actual antennas, respectively. Note that link loss is frequency dependent. For a fixed physical distance, the free-space loss clearly increases as frequency increases. Atmospheric loss also depends on frequency. In general, antenna gain also depends on frequency. For example, a larger antenna typically has a larger effective area, and hence a higher antenna gain. Increasing the gain of the receive and transmit antennas, when possible, decreases the link loss. This motivates the need for very high gain antennas in long distance links, such as the aforementioned geosynchronous satellite links or NASA's deep space network that supports interplanetary communication.

Often, P_t and G_t are combined as the **effective isotropic radiated power** (*EIRP*) to quantify the transmitted power density for regulation, safety, and interference purposes. The *EIRP* is the equivalent power that an isotropic radiator would have to radiate in order to equal the power received from the actual transmitting antenna.

$$EIRP = G_t P_t \tag{8.31}$$

$$P_r = \frac{P_t}{L} = \frac{(P_t G_t) G_r}{L_{fs} L_a L_o} = \frac{(EIRP) G_r}{L_{fs} L_a L_o} \tag{8.32}$$

Application: There are standard maps that indicate the *EIRP* of a satellite vs. geographic location by showing constant *EIRP* contours. These patterns are often called "footprints."

Example 8.4.1

A satellite link is 23 500 miles long and has an atmospheric loss of 20 dB. The uplink power output from a 60 dBi gain antenna is 150 W at 27.5 GHz. The receiving antenna is a parabolic dish with a gain of 48 dBi. Determine:

a) the free-space loss to the nearest dB,
b) the *EIRP* in dBW,
c) the diameter of the receiving dish antenna in m (simplistically assume A_e equals the physical area of the dish), and
d) the received power to the nearest dBW.

Solution

a) $L_{fs} = 10 \log \left(\frac{4 \pi d}{\lambda} \right)^2 = 20 \log \left[\dfrac{4 \pi (23500 \text{ mi}) \left(\dfrac{1609 \text{ m}}{\text{mi}} \right)}{\left(\dfrac{2.998 \times 10^8 \dfrac{\text{m}}{\text{s}}}{27.5 \times 10^9 \dfrac{1}{\text{s}}} \right)} \right] = 213 \text{ dB}$

b) $EIRP(\text{dBW}) = 10 \log (G_t P_t) = 10 \log (G_t) + 10 \log (P_t) = 60 \text{ dBi} + 10 \log (150) = 82 \text{ dBW}$

c) This calculation assumes that A_e equals the physical area of the dish antenna:

$$G_{ant} = \frac{4 \pi A_e}{\lambda^2} = \frac{4 \pi \frac{\pi}{4} (dia)^2}{\lambda^2} \quad \rightarrow \quad dia = \sqrt{\frac{G_{ant} \lambda^2}{\pi^2}} = \sqrt{\frac{10^{48/10}}{\pi^2} \left(\frac{2.998 \times 10^8}{27.5 \times 10^9} \right)^2} = 0.872 \text{ m}$$

d) $P_r(\text{dBW}) = P_t(\text{dBW}) - L(\text{dB}) = P_t(\text{dBW}) - [L_{fs}(\text{dB}) + L_a(\text{dB}) + L_o(\text{dB}) - G_t(\text{dBi}) - G_r(\text{dBi})]$ Assume the "other losses" $L_o = 0$ dB.

$$P_r(\text{dBW}) = 10 \log (150) - [213 + 20 + 0 - 60 - 48] = -103 \text{ dBW}$$

8.5 Summary of Important Equations

$$\beta = \omega\sqrt{\mu\varepsilon} \qquad v = \lambda f = \frac{1}{\sqrt{\mu\varepsilon}} \qquad \eta = \frac{E_o^+}{H_o^+} = \sqrt{\frac{\mu}{\varepsilon}} \qquad \eta_o = 377\,\Omega$$

$$\bar{S} = \bar{E} \times \bar{H} \qquad S_{ave} = \frac{E_o H_o}{2} = \frac{E_o^2}{2\eta} = \frac{H_o^2 \eta}{2} \qquad P_{ave} = \int \bar{S}_{ave} \cdot \overline{dA} \qquad P_{ave} = S_{ave} A_e$$

$$G_{ant} = \frac{4\pi A_e}{\lambda^2} \qquad L_{fs} = \frac{P_t}{P_r} = \left(\frac{4\pi d}{\lambda}\right)^2 \qquad L = \frac{P_t}{P_r} = \frac{L_{fs} L_a L_o}{G_t G_r} \qquad EIRP = G_t P_t$$

8.6 Homework

1. a) Show that the intrinsic impedance of free space is 377 Ω.
    b) What is the intrinsic impedance of polyethylene?

2. If the electric field intensity of an AC plane wave is 10 μV/m (peak) and is oriented in the $\bar{a}_y$ direction, and the magnetic field intensity is oriented in the $\bar{a}_x$ direction.
    a) In what direction does the plane wave travel? Include the sign of the unit vector.
    b) What is the peak magnetic field intensity?

3. Determine the radiated power in dBW and W for the system shown below. Explain your answer.

4. a) Does radiated power generally increase or decrease as frequency increases? What is a practical advantageous consequence of this fact?
    b) Does free-space loss increase or decrease as frequency increases? Explain why.
    c) Given a satellite link with 200 dB of free-space loss, identify two primary ways to increase the received power level. Assume the frequency cannot be changed.
    d) What is the effect on the Friis Transmission equation if the polarization of the transmitting and receiving antennas do not match? Identify the relevant variable.

5. An electric field intensity of 120 μV/m (peak) is present at the receiving antenna of a 4 GHz satellite downlink. What is the power received in (a) W, (b) dBW, and (c) dBm if the receiving antenna has an effective area of 3.142 m^2?

6. Repeat Problem 5 if the peak magnetic field intensity is instead given as 1 μA/m.

7. If 1 kW were present over a 10 cm by 10 cm area of a plane wave, determine (a) the peak electric field intensity, (b) the peak magnetic field intensity, and (c) the average power density. (d) Are these dangerous levels? Why or why not?

8. Imagine radiation at 60 Hz. Assume a plane wave with a peak electric field intensity of 1 V/m. (a) What area is required to receive enough power to light a 60 W bulb (assuming 100% efficiency)? (b) Is this area practical?

9. A link consists of a transmitting antenna with a gain of 25 dBi and a receiving antenna with a gain of 23 dBi. The antennas are separated by 30 km. The transmitter has an output of 2 W at 5 GHz.

    a) Assuming atmospheric and other losses are negligible, determine the power received in W at the output of the receiving antenna.

    b) Considering the gain of the transmitting antenna, what is the power transmitted from the transmitting antenna?

10. For a satellite downlink at 4 GHz, shown below, with a transmitter power of 10 W, a transmitting antenna gain of 55 dBi, and a receiving antenna gain of 40 dBi, determine:

    a) the satellite *EIRP* in dBW, and

    b) the power received in W if the link distance is 24 000 miles (not m).

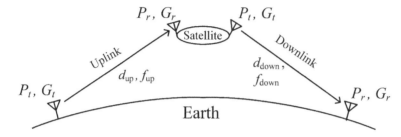

    Figure for problems 10 and 11 (frequencies, antenna gains, and power levels usually differ between an uplink and a downlink)

11. Given a satellite link with the following parameters:

    For the uplink: $P_t = 1\,\text{kW}$, $G_r = 20\,\text{dBi}$, $G_t = 55\,\text{dBi}$, $f = 6\,\text{GHz}$, $d_{up} = 24\,576$ miles

    For the downlink: $P_t = 5\,\text{W}$, $G_r = 30\,\text{dBi}$, $G_t = 17\,\text{dBi}$, $f = 4\,\text{GHz}$, $d_{down} = 24\,576$ miles

    Determine:

    a) the uplink link loss,

    b) the downlink link loss,

    c) the *EIRP* of the transmitting Earth station,

    d) P_r at the satellite, and

    e) P_r at the receiving Earth station.

12. You are designing a communication system operating at 900 MHz with a distance of 1 km between the transmitter and receiver. In order to minimize the transmitter power requirements, you must determine whether a wireless link or a coaxial cable offers the lowest overall loss. The coaxial cable has a low loss of 4 dB/100 ft at 900 MHz.

    a) Obtain an idea of the loss levels for each option by calculating the total coaxial cable loss and calculating the total link loss for a wireless system having transmit and receive antenna gains of 2 dBi (dipole antennas). Perform both calculations twice: one at a distance of 500 m and the second at 1 km.

    b) Is a wireless link or coaxial cable connection the best choice for the 1 km system?

    c) Compare the changes in link loss and in cable loss if the distance is halved.

9

Signal Integrity

CHAPTER MENU

This chapter is a brief introduction to the highly relevant subject within electrical engineering known as signal integrity. It will build upon and show additional applications for the electromagnetics topics covered in Chapters 2 through 8. It is an informational chapter with the purpose to increase your awareness of some basic signal integrity concepts and issues.[1] It is by no means an exhaustive examination of this subject. The main categories of signal integrity are examined in Sections 9.2 through 9.5.

9.1 Introduction to Signal Integrity

Modern signals, especially high-speed digital signals, generally have very high-frequency components. Higher clock speeds increase the bandwidth requirements. But the frequency content of the rising and falling edges of a digital signal usually increases the significant frequency content far beyond that due to the nominal clock frequency content. Recall the Fourier series analysis of a pulse train. As the pulse width narrows, the significant frequency content of the signal increases even if the fundamental frequency remains constant. More frequencies are required to "build" rapid changes in the signal. Rise and fall times of high-speed digital signals, often called *edge rates*, must become shorter as the clock speed increases. Hence, more frequency components are present in the signal, and transmission line and other effects become potential problems. A common rule of thumb to estimate the bandwidth required for a signal based on the rise (or fall) time is

---

1 Text references include Bogatin, E. (2018). Signal Integrity – Simplified, 3e. New York: Pearson; Schmitt, R. (2002). *Electromagnetics Explained – A Handbook for Wireless/RF, EMC, and High-Speed Electronics*. United Kingdom: Newness-Elsevier Science; Paul, C.R., Scully, R.C., and Setffka, M.A. (2022). *Introduction to Electromagnetic Compatibility*, 3e. New Jersey: Wiley; Ott, H.W. (2009). *Electromagnetic Compatibility Engineering*. New Jersey: Wiley. Also consult the vendor literature for up-to-date application notes.

Electromagnetics and Transmission Lines: Essentials for Electrical Engineering, Second Edition.
Robert A. Strangeway, Steven S. Holland, and James E. Richie.
© 2023 John Wiley & Sons, Inc. Published 2023 by John Wiley & Sons, Inc.
Companion website: www.wiley.com/go/Strangeway/ElectromagneticsandTransmissionLines

$$BW_{\text{rise time}} \approx \frac{0.35}{t_{\text{rise}}} \text{ to } \frac{0.5}{t_{\text{rise}}}$$

$$(9.1)$$

This equation is usually stated for the 10–90 rise time. For example, a 2 GHz clock signal has a half period of 0.25 ns. If the rise and fall time in each half period is 0.05 ns, then a signal bandwidth of at least 7 GHz is required! Hence, the signal in this particular example is subject to interference anywhere within the 7 GHz bandwidth.

Modern high-speed digital signals are subject to a multitude of interfering signals and noise, all in a shrinking physical environment where circuit traces are extremely close together (think of the density of the lead connections to a modern computer chip). Hence, the "opportunities" to corrupt a signal are better than ever, and the problems become worse as clock speeds increase and circuit density grows. There is a need to determine the "quality" of a signal both in the design cycle as well as with the finished product.

Signal integrity is the degree to which a signal performs its designed function. One scheme uses four general classes within the topic of signal integrity: transmission line effects (especially reflections), crosstalk (electric and magnetic field coupling), electromagnetic interference (EMI), and switching noise in the power/ground circuitry. For example, one must take an analog viewpoint of a digital channel in order to design and test for good signal integrity. Each effect is briefly described in this chapter. Sometimes, timing issues, such as bus contention and setup and hold violations, are considered another class under signal integrity, but will not be discussed in this chapter. Modern circuit layout software contains signal integrity simulation – it is an integral part of the modern circuit design process.

There are product and user safety standards that must also be specified and met, such as grounding, flammability, and electrostatic discharge (ESD). These topics are not covered in this chapter, but there are numerous commercial vendor application notes available.

9.2 Transmission Line Effects

Transmission line effects can harm the integrity of a signal. Reflections due to impedance mismatches (Chapters 6 and 7) are the most troublesome cause. Impedance mismatches may be due to:

- Mismatched loads and sources (recall the bounce diagram)
- Changes in the transmission line itself, such as a change in the trace width on a printed circuit board (PCB), which changes the characteristic impedance Z_0
- **Vias** that are used to connect traces between layers on a multilayer PCB
- Lead connections to packaged components
- Gaps in the return plane on a PCB (such as gaps for routing a trace)
- Coupling from other nearby transmission lines (which changes Z_0)

These impedance mismatches can also create troublesome resonances and electromagnetic radiation.

Dispersion can also occur in some transmission media, where the propagation velocity is a function of frequency. How does dispersion affect the integrity of a signal? The frequency-dependent velocity on the transmission line means that different frequencies within the signal arrive at the load at different times. Signal distortion results. Transmission line loss generally causes dispersion. Dispersion is inherent in waveguide and optical fiber media. It can also occur in microstrip and other transmission lines. The effects of dispersion are increasingly prevalent as the bandwidth of the signal increases.

Care must be given when measuring transmission line effects because the probes that are used to measure the effects can actually cause reflections! Refer to the extensive body of vendor application note literature that covers this topic.

9.3 Crosstalk

9.3.1 Electric and Magnetic Field Coupling

Closely spaced parallel conductors often occur in PCBs, such as data and address lines. The clock, trigger signals, and analog signal lines also must be routed. Board space is limited, forcing many lines to be parallel to one another for at least a portion of their lengths. Whenever two circuits are in the vicinity of one another, they might "see" each other through the mechanism of the electric and/or magnetic field coupling from one T-line to another. Thus, we generally have the potential problem of *coupled transmission lines*.

Crosstalk is the undesired coupling of a signal from one transmission line to another transmission line. The coupling mechanisms are both the electric and magnetic fields, although one may dominate under certain conditions. The sketches from Chapter 5 are reproduced in Figure 9.1a and b. Electric field coupling is equivalent to capacitive coupling, and these capacitances are called **mutual capacitances**. Mutual capacitance is commonly called *parasitic capacitance* or *stray capacitance* because it is present due to proximity of conductors but is normally not of intentional design. Magnetic field coupling is equivalent to inductive coupling, and these inductances are called **mutual inductances**. Not only does crosstalk introduce an undesired signal and noise from one transmission line to another but coupling also affects the characteristic impedance of both lines, potentially resulting in undesired transmission line effects as discussed in the previous section.

One subset of the coupled T-line scenario involves the frequencies present. If the frequencies are "low," the T-line lengths of concern are much less than one wavelength and the voltages and currents, and electric and magnetic fields, are uniform along the T-lines as a function of position. This coupling is called **quasi-static**[2] because the field is fairly uniform along the transmission line lengths. As a result, the coupling between the T-lines can be expressed in terms of electrostatic and magnetostatic fields, and one could model the coupling with a lumped model of mutual capacitance and mutual inductance. Note that the "static" part of these terms does not mean that the fields are DC. Instead it means that the frequencies are low enough for the T-line lengths involved such that distributed circuit effects are not significant.

The quasi-static magnetic field coupling between two transmission lines can be considered from the mutual inductance relationship in circuits:

$$v_{coupled} = M \frac{di_{driven}}{dt} \tag{9.2}$$

where M is the mutual inductance between the transmission lines, i_{driven} is the current in the driven line, and $v_{coupled}$ is the induced voltage in the coupled line. In a similar manner, quasi-static electric field coupling between the transmission lines can be considered from the mutual capacitance i–v relationship from Chapter 5:

$$i_{coupled} = C_m \frac{dv_{driven}}{dt} \tag{9.3}$$

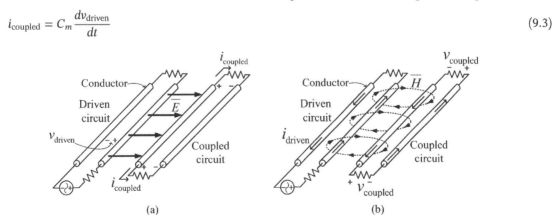

(a)   (b)

Figure 9.1 Generic sketches of coupling between T-lines. (a) Electric field coupling. (b) Magnetic field coupling.

---

2 Quasi-static is used here to indicate that the field distributions are approximately the same as electrostatic and magnetostatic fields when the T-line length is much less than one wavelength.

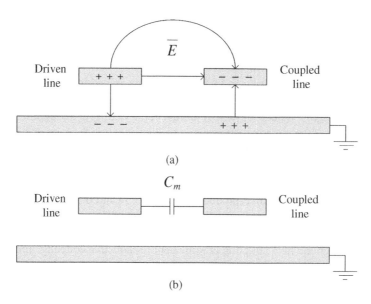

Figure 9.2 Electrostatic field coupling between transmission lines. (a) End view of microstrip T-lines and the $\overline{E}$ field. (b) Mutual capacitance between the T-lines.

where v_{driven} is the voltage on the driven line and i_{coupled} is the induced current in the coupled line. Note that the induced waveform in both cases will change relative to the waveform in the driven line because of the derivative relationships. Fast edge rates couple as impulse-type "spikes."

How can one visualize electrostatic coupling of transmission lines? Recall the discussion of displacement current and mutual capacitance in Chapter 5. Refer to Figure 9.2a. Assume that the electric field is time-changing. Note that the time-changing electric field on the driven T-line has electric flux lines that couple to the conductors on the second T-line, often called the coupled line. Consequently, charges will move in the coupled line circuit, that is, current will flow and a voltage will develop on the coupled line. Hence, a simplified model is a mutual capacitance C_m that represents the electric field coupling from the driven line to the coupled line, as shown in Figure 9.2b.

How can one visualize magnetostatic coupling of transmission lines? Refer to Figure 9.3a. Recall the operational principles of the mutual inductor from Chapter 5. The time-changing magnetic field of the driven line will induce a voltage into the conductors of the coupled line. A simplified model is mutual inductance M that represents the magnetic field coupling from the driven line to the coupled line, as shown in Figure 9.3b.

If the frequencies are "high," and the T-line lengths of concern are not much less than one wavelength, then there is a spatial variation of the voltages and currents, and electric and magnetic fields, along the T-lines. Coupling becomes distributed, that is, a function of position at higher frequencies, just as voltage and current vary as a function of position on T-lines at higher frequencies. Coupling between the T-lines would need to incorporate position as a primary variable, as it does in T-line theory. This case is more common (and complicated) because of the high-frequency content in the rising and falling edges of modern high-speed digital signals.

9.3.2 Shielding

Shielding, in the sense of crosstalk, is the arrangement of conductors, whether ground (return lines) or additional conductors placed within the transmission media, to reduce coupling between the transmission lines. For example, if two PCB traces are parallel to one another, placing the ground line of one transmission line adjacent to the hot line of the other transmission line generally reduces capacitive coupling relative to the case of adjacent hot lines of both transmission lines. The use of a coax shield can also reduce coupling. Often, shielding for one type of coupling does not significantly affect or can even worsen the other type of coupling.

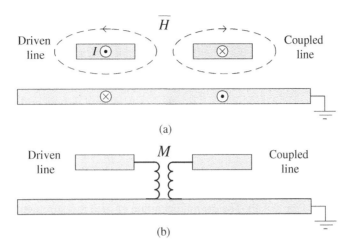

Figure 9.3 Magnetostatic field coupling between transmission lines. (a) End view of microstrip T-lines and the $\overline{H}$ field. (b) Mutual inductance between the T-lines.

9.4 Electromagnetic Interference

9.4.1 Overview

EMI is the undesired effect on one circuit, the *victim*, due to another circuit(s), the *aggressor*, through **conducted interference** via the power, data, or other lines that connect to the victim circuit, or through **radiated interference** (coupled and/or radiated electromagnetic fields) from the aggressor circuit(s) to the victim circuit. An older term for EMI was radio frequency interference (RFI).

The goal of the designer is to limit EMI to acceptable levels such that the circuit functions properly (signal integrity) and EMI levels meet regulatory standards. The electromagnetics foundation developed throughout this text provides an engineer with the necessary concepts to understand basic EMI considerations. The EMI discussion here is intended to help develop awareness of EMI issues. For example, we learned in Chapters 6 through 8 that the separation between the transmission line conductors must be very small compared with wavelength to ensure negligible radiation. The foundational electromagnetic and T-line concepts allow one to proceed into more advanced EMI-conscious design strategies and best practices, such as PCB layout rules.

A product may malfunction due to EMI because of interference from other products or interference from itself. EMI requires a (an):

- Interference source
- Channel (coupling mechanism)
  - Conduction paths
  - Radiation paths
- Receptor (antenna, case, cable, ...)

If the product is insufficient with regards to EMI, then troubleshooting is used to determine the root cause(s) of EMI problems. A sampling of EMI terminology from the EMI literature follows:

- **Emissions**: interference from one product that interferes with the operation of other products
- **Electromagnetic Immunity** (also called *Susceptibility*): the ability of the product to operate with externally generated EMI present
- **Electromagnetic Compatibility** (EMC): the ability of the product to operate with EMI present and to not overly cause EMI
- **Device under test** (DUT) or **equipment under tes**t EUT

EMI is routinely regulated in standards and laws and is specified and measured in products. It is required that most products meet designated EMI standards. Many organizations set standards, especially the *American National Standards Institute*, **ANSI**, and the *Comité International Spécial des Perturbations Radioélectriques* (International Special Committee on Radio Interference), **CISPR**. It is recommended that you determine the standards that are pertinent to your engineering projects so that the appropriate EMI requirements can be considered in the design and planning phases, not later where it is often costly to troubleshoot and remedy poor EMI performance that violates standards.

9.4.2 EMI Measurements

There are two major EMI measurement setups, one for conducted interference, shown in Figure 9.4a, and one for radiated interference, shown in Figure 9.4b. The overall operation of each measurement will be examined in this section.

An instrument that is common to both measurements is the **EMI analyzer**. The EMI analyzer is a specialized RF/microwave spectrum analyzer that accepts input frequencies from the kHz into the GHz. It does not calculate the frequency spectrum from the time domain signal, as is usually done in oscilloscopes because scopes cannot accommodate the large amplitude range (dynamic range) of signals. Instead, it can be viewed as a tunable bandpass filter[3] (BPF). See Figure 9.5 for a block diagram. As the center frequency of the filter is tuned (swept), if frequency components of the input signal are present within the bandwidth of the filter at a given center frequency, they are passed by the filter, detected, and displayed. The bandwidth of the BPF is called the **resolution bandwidth** (RBW) because it sets the ability to distinguish (resolve) two closely spaced signals. If two signals are separated by less than the RBW, then they cannot be resolved by the BPF. Thus, the EMI analyzer effectively measures the frequencies within the RBW. The RBW can be changed by the user. For EMI tests, the RBW is specified in regulation standards. Resolution bandwidth is an important parameter in EMI specifications and measurements.

The EMI analyzer has additional specialized EMI hardware and software. An EMI analyzer has three different types of detectors (peak, average, and quasi-peak) and has equipment correction factors and emission test limits programmed into the instrument. EMI standards will specify one or more detector types in a measurement because EMI has, in general, time-dependent fluctuations. The three detectors measure EMI basically as follows:

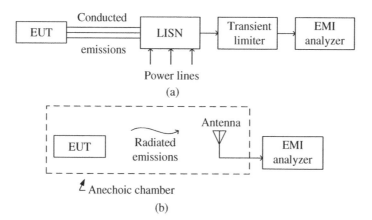

Figure 9.4 Basic EMI measurement setups. (a) Conducted emissions measurements. (b) Radiated emissions measurements (an anechoic chamber is indicated in (b), but it could be situated in an open air test site, OATS).

---

3 RF/microwave spectrum analyzers usually use a superheterodyne architecture with a tunable local oscillator. Consult instrumentation literature and/or application notes for more information on this topic.

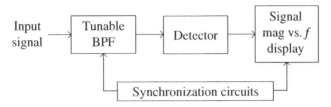

Figure 9.5 A functional view of an RF/microwave spectrum analyzer.

- *Peak detector*: detects the highest level of EMI fluctuations that are present versus frequency. If the EUT passes with a peak detector, then the other detectors are unnecessary.
- *Average detector*: averages the time-dependent EMI fluctuations that are present from the peak detector.
- *Quasi-peak detector*: detects the interference with a specified time constant circuit after the peak detector. Essentially, the faster the EMI fluctuations that are present, the higher the output from this detector because the time constant circuit cannot fully discharge. The quasi-peak detector output is usually between that of the peak and average detector outputs. The time constants are specified in EMI standards.

The main test equipment and basic procedures for conducted and radiated EMI measurements follow.

Conducted emissions measurements: See Figure 9.4a.

- EMI analyzer
- Line impedance stabilization network (LISN)
  - Isolates interference from power mains to the EUT and vice versa
  - Routes conducted emissions from EUT to the EMI analyzer
- Transient limiter: protects the EMI analyzer from power line spikes and transients

The LISN in a conducted emissions test has filter networks to isolate interference from the power mains to the EUT and vice versa. It also has a circuit to isolate the 60 Hz power from the EMI analyzer and to route the EMI signal from the EUT to the EMI analyzer via a transient limiter. *The transient limiter must be present between the LISN and the EMI analyzer.* Otherwise, a power line transient, surge, or spike can destroy the sensitive input circuits in the EMI analyzer, resulting in a costly repair bill. The test is performed in a screen room, which is basically a Faraday cage. The room is completely enclosed by conductors so that no interfering signal enters or leaves the room. Input power lines are heavily filtered and output signals are shielded.

In a conducted emissions measurement, correction factors (basically transfer functions) for the LISN and transient limiter, usually stored within the EMI analyzer, are enabled. The EMI analyzer is set to a standards-defined frequency range, such as 150 kHz–30 MHz. The EUT is operated at "full load" so that worst-case emissions are present. The received voltage is measured on the EMI analyzer, usually in dBμV. A voltage expressed in dBμV is:[4]

$$V(\text{dB}\mu\text{V}) = 20\log\left[V(\text{RMS}) \times 10^6\right] \tag{9.4}$$

The conducted emissions are measured versus frequency and compared to standards-defined limits to determine if the EUT passes or fails the conducted emissions EMI test.

Radiated emissions measurements: See Figure 9.4b.

- EMI analyzer
- Different receiving antennas for frequency ranges, examples:
  - Biconical: 30–300 MHz
  - Log periodic: 200 MHz–1 GHz
  - Horn: above 1 GHz

---

4 The voltage in dBμV is related to the power in dBm in a 50 Ω system by $V(\text{dB}\mu\text{V}) = P(\text{dBm}) + 107$.

- Suitable environment for radiated EMI measurements (discussed below)
  - Open air test site (OATS)
  - Anechoic chamber
- Means to move the antenna and the EUT with respect to each other
  - Turntable
  - Relative height adjustments
  - Relative polarization adjustments

Radiated tests are performed in an environment that is free from reflections except for the ground or the floor below the EUT. The open air test site, **OATS**, is geographically located far from buildings and other reflecting obstacles. Usually, it is far from urban areas so that it is also relatively free from interfering signals. The **EMI anechoic chamber** has electromagnetic absorbing material on the walls and ceiling but not the floor. The floor is conductive to emulate the Earth (technically, this type of anechoic chamber is often classified as a *semi-anechoic* chamber). This room prevents interfering radiated signals from entering and affecting the measurement. It also suppresses radiated EMI from the EUT from reflecting off of the walls except the floor and affecting the measurement. The antenna in a radiated emissions test is selected for the specified frequency range in the pertinent EMI standards. It is situated at a prescribed distance, such as 3 m, from the EUT and a prescribed height above the ground or floor.

In a radiated emissions measurement, an antenna factor and other correction factors (such as the loss of the cable that connects the antenna to the EMI analyzer) are enabled. The electric field antenna factor[5] $AF_{electric}$ relates the incident electric field intensity to the received (rcv) voltage at the antenna terminals (connector):

$$AF_{electric} = \frac{E_{inc}}{V_{rcv}} \tag{9.5}$$

The radiated emission measurements and standards are often expressed in dB form:

$$E(dB\mu V/m) = 20\log\left[E(V/m) \times 10^6\right] \tag{9.6}$$

Then the incident electric field intensity in dBμV/m is determined from the measurement of the received voltage by

$$E_{inc}(dB\mu V/m) = V_{rcv}(dB\mu V) + AF_{electric}(dB/m) + \text{other correction factors} \tag{9.7}$$

Normally the correction factors and these calculations are incorporated into the specialized software that is loaded into an EMI analyzer.

The antenna and the EUT are relatively positioned and oriented (polarization-wise) at the prescribed distance from each other to result in the *largest* interfering signal, which must pass the EMI standards. As with conducted emissions, the EUT is operated at "full load" so that worst-case emissions are present. Once the largest radiated emissions are identified, it is measured versus frequency and compared to standards-defined limits to determine if the EUT passes or fails the radiated emissions EMI test.

These EMI tests are time consuming and expensive, with one to five days being typical. Products that intentionally radiate take the longest time to test. Companies that are accredited (they passed EMI measurement facility standards) must perform and document the final tests. The measurement results from both conducted and radiated EMI tests should be repeatable in different EMI accredited facilities. Companies often perform **precompliance tests** in product development stages where they test with their own EMI equipment in unaccredited facilities. If the EUT passes a precompliance test, then it is more likely to pass the official EMI test. Some larger companies also have EMI accredited facilities for cost-effectiveness. In any event, you should check within your company for product testing standards and procedures long before the project nears completion.

---

5 There are also antenna factors for magnetic field intensity.

9.5 Power/Ground Switching Noise[6]

Switching noise is the additional noise generated in one signal path due to transients, which are usually due to switching of signals, in another signal path via the voltage changes in the power grid and/or the ground plane. The reference voltage (ground) of chips can change due to voltage drops across the finite impedance (resistance and inductive reactance) of the ground plane. This effect can especially occur in ground paths, where multiple return currents can flow simultaneously. **Simultaneous switching noise** (SSN) occurs when multiple drivers are switching digital states. It is often called **ground bounce** because the ground reference voltage is "bouncing" around instead of staying at zero volts. It is also called *rail collapse.*

An important conclusion in this area is to not treat ground as a zero resistance ideal path for currents to complete their circuits. Instead one must adopt a "return path" mentality when designing ground planes, where the circuit imperfections of the ground plane (again, finite resistance and significant inductive reactance) are incorporated into the analysis and design. Simulation of circuit design layouts is critical to operational success on the first pass.

9.6 Summary of Important Equations

$$BW_{\text{rise time}} \approx \frac{0.35}{t_{\text{rise}}} \text{ to } \frac{0.5}{t_{\text{rise}}} \qquad v_{\text{coupled}} = M\frac{di_{\text{driven}}}{dt} \qquad i_{\text{coupled}} = C_m\frac{dv_{\text{driven}}}{dt}$$

$$\text{Radiated EMI measurements}: \qquad AF_{\text{electric}} = \frac{E_{\text{inc}}}{V_{\text{rcv}}} \qquad E(\text{dB}\mu\text{V/m}) = 20\log\left[E(\text{V/m}) \times 10^6\right]$$

$$E_{\text{inc}}(\text{dB}\mu\text{V/m}) = V_{\text{rcv}}(\text{dB}\mu\text{V}) + AF_{\text{electric}}(\text{dB/m}) + \text{other correction factors}$$

9.7 Homework

The following homework questions relate to the informational nature of this chapter.

1. Name and briefly describe four main categories of signal integrity.

2. a) Why does a change in the characteristic impedance on a T-line cause reflections?
   b) Would a change in the width of a PCB trace cause reflections? Why or why not?

3. a) What is a **via** on a PCB?
   b) Why is it an issue in signal integrity?

4. a) Can both major types of **crosstalk** exist simultaneously? Why or why not?
   b) What primarily determines the dominant type of crosstalk? Hint: impedance levels

5. State and generally describe the two broad types of EMI measurements, including the test equipment that is utilized in each type of measurement.

6. a) What is EMI **precompliance**?
   b) Why is it used in industry?

7. a) What is the purpose of the **antenna factor** (*AF*) in a radiated EMI measurement?
   b) Would you expect that the *AF* is related to antenna gain? Provide reasoning.

---

6 Caution: The terminology in this category of signal integrity varies widely.

8. a) What is a **screen room**? In which EMI measurement is it utilized?

  b) What is an **anechoic chamber**? In which EMI measurement is it utilized?

  c) What is an **OATS**? In which EMI measurement is it utilized?

9. Develop $V(\text{dB}\mu\text{V}) = P(\text{dBm}) + 107$. Assume a 50 Ω system.

10. a) For what reasons are a **LISN** used in conducted EMI measurements?

  b) Why is it essential to use a **transient limiter** in conducted EMI measurements?

11. Why is **ground bounce** a problem in digital circuits?

Appendix A

Alphabetical Characters, Names, and Units

The following alphabetical characters and abbreviations ("symbols") are named for their most common use in this textbook. Some symbols have multiple uses, such as b, c, d, r, v, x, y, and z.

Symbol	Name	Units
$\overline{a}$	Unit vector	(unitless)
A	Area	meter^2 (m^2)
A	Attenuation (power ratio)	(unitless)
A_e	Effective area (of an antenna)	meter^2 (m^2)
$\overline{A}$	General vector (magnitude is A)	(depends on context)
a, b, c, d	Dimensions	meter (m)
ACL	Ampere's circuital law	—
b	Normalized susceptance	(unitless)
$\overline{B}$	Magnetic flux density	tesla (T) = weber/meter^2 (Wb/m^2)
BSL	Biot–Savart law	—
BW	Bandwidth	hertz (Hz) or radian/second (rad/s)
c	Speed of light in vacuum	meter/second (m/s)
C	Capacitance	farad (F)
C	Coupling factor	(unitless)
C'	Capacitance per unit length	farad/meter (F/m)
CL	Coulomb's law	—
d	Distance from load on a T-line	meter (m)
$\overline{D}$	Electric flux density	coulomb/meter^2 (C/m^2)
dB	Decibel	(unitless)
dce	Differential current element	ampere-meter (A-m)
DUT	Device under test	—
div	Divergence (vector operator)	meter^{-1} (m^{-1})
e	Base of the natural logarithm	(unitless)

Electromagnetics and Transmission Lines: Essentials for Electrical Engineering, Second Edition.
Robert A. Strangeway, Steven S. Holland, and James E. Richie.
© 2023 John Wiley & Sons, Inc. Published 2023 by John Wiley & Sons, Inc.
Companion website: www.wiley.com/go/Strangeway/ElectromagneticsandTransmissionLines

Symbol	Name	Units
$\overline{E}$	Electric field intensity	volt/meter (V/m)
EIRP	Effective isotropic radiated power	watt (W)
EM	Electromagnetic	—
emf	Electromotive force	volt (V)
EMI	Electromagnetic interference	—
EUT	Equipment under test	—
f	Frequency (cyclic frequency)	hertz (Hz)
$\overline{F}$	Force	newton (N)
$\mathfrak{F}$	Magnetomotive force (script F)	ampere (A)
g	Normalized conductance	(unitless)
G	Conductance	siemens (S)
G	Power gain (power ratio)	(unitless)
GS	Gaussian surface	—
h	Height	meter (m)
$\overline{H}$	Magnetic field intensity	ampere/meter (A/m)
I, i	Electric current	ampere (A)
$\widetilde{I}$	Phasor (complex) current	ampere (A)
IL	Insertion loss (power ratio)	(unitless)
j	Imaginary part ($j = e^{j\pi/2} = 1\angle 90^{\circ}$)	(unitless)
$\overline{J}$	Volume current density	ampere/meter^2 (A/m^2)
K, k	Constant, often proportionality constant	(need context)
$\overline{K}$	Surface current density	ampere/meter (A/m)
ℓ	Length	meter (m)
L	Loss (power ratio); link loss	(unitless)
L	Inductance	henry (H)
L'	Inductance per unit length	henry/meter (H/m)
LISN	Line impedance stabilization network	—
mmf	Magnetomotive force	ampere (A)
M	Mutual inductance	henry (H)
$\overline{M}$	Magnetization	ampere/meter (A/m)
n	Variable integer number	—
n	Turns ratio (transformers)	(unitless)
$(\ldots)_n$	Normal component of vector (…)	(need context)
N	Fixed integer number; number of turns	—
P, p	Power	watt (W)
P	Observation point	—
$\overline{P}$	Polarization	coulomb/meter^2 (C/m^2)
PCB	Printed circuit board	—
Q	Electric charge	coulomb (C)

Symbol	Name	Units
$\bar{r}$	Position vector	meter (m)
r	Radius in spherical coordinates	meter (m)
r	Normalized resistance	(unitless)
$\bar{R}$	Distance vector	meter (m)
R	Resistance	ohm (Ω)
$\mathfrak{R}$	Reluctance (script R)	henry^{-1} (H^{-1}) (also A/Wb)
RBW	Resolution bandwidth	hertz (Hz)
RL	Return loss	(unitless)
S_{ii}, S_{ij}	Scattering parameter	(unitless)
$\bar{S}$	Poynting vector	watt/meter^2 (W/m^2)
t	Time	second (s)
$(...)_t$	Tangential component of vector (...)	(need context)
T	Period	second (s)
V, v	Voltage, electric potential	volt (V)
V	Volume	meter^3 (m^3)
$\widetilde{V}$	Phasor (complex) voltage	volt (V)
$VSWR$	Voltage standing wave ratio	(unitless)
$v, \bar{v}, vel.$	Velocity	meter/second (m/s)
w	Energy density	joule/meter^3 (J/m^3)
W	Energy, work	joule (J)
x, y, z	Cartesian (rectangular) coordinates	meter (m)
x	Normalized reactance	(unitless)
X	Reactance	ohm (Ω)
y	Normalized admittance	(unitless)
Y_0	Characteristic admittance	siemens (S)
$\widetilde{Y}$	Complex admittance	siemens (S)
z	Distance from source on a T-line	meter (m)
z	Normalized impedance	(unitless)
Z_0	Characteristic impedance	ohm (Ω)
$\widetilde{Z}$	Complex impedance	ohm (Ω)

Appendix B

Greek Letters, Names, and Units

Greek Letter		Name	Units
α	Lowercase alpha	Attenuation constant	neper/meter (Np/m)
β	Lowercase beta	Phase constant	radian/meter (rad/m)
γ	Lowercase gamma	Propagation constant	meter^{-1} (m^{-1})
Γ	Uppercase gamma	Reflection coefficient	(unitless)
Δ	Uppercase delta	Small increment of a quantity	(depends on context)
ε	Lowercase epsilon	Permittivity or dielectric constant	farad/meter (F/m)
η	Lowercase eta	Intrinsic impedance	ohm (Ω)
θ	Lowercase theta	Angle in spherical coordinate system; or AC phase shift	(degree or radian)
λ	Lowercase lambda	Wavelength	meter (m)
Λ	Uppercase lambda	Linked magnetic flux	weber (Wb)
μ	Lowercase mu	Permeability	henry/meter (H/m)
π	Lowercase pi	3.14159...	—
ρ	Lowercase rho	Radius in cylindrical coordinate system	meter (m)
ρ	Lowercase rho	Resistivity	ohm-meter (Ω-m)
ρ_L	Lowercase rho	Line charge density	coulomb/meter (C/m)
ρ_S	Lowercase rho	Surface charge density	coulomb/meter^2 (C/m^2)
ρ_V	Lowercase rho	Volume charge density	coulomb/meter^3 (C/m^3)
σ	Lowercase sigma	Conductivity	siemens/meter (S/m)
Σ	Uppercase sigma	Summation symbol	—
τ_d	Lowercase tau	Time delay (indicated by subscript d)	second (s)
ϕ	Lowercase phi	Angle in cylindrical and spherical coordinate systems; or AC phase shift	(degree or radian)
Φ	Uppercase phi	Magnetic flux	weber (Wb)
χ	Lowercase chi	Susceptibility	(unitless)
Ψ	Uppercase psi	Electric flux	coulomb (C)
ω	Lowercase omega	Radian (or angular) frequency	radian/second (rad/s)

Electromagnetics and Transmission Lines: Essentials for Electrical Engineering, Second Edition.
Robert A. Strangeway, Steven S. Holland, and James E. Richie.
© 2023 John Wiley & Sons, Inc. Published 2023 by John Wiley & Sons, Inc.
Companion website: www.wiley.com/go/Strangeway/ElectromagneticsandTransmissionLines

Appendix C

A Short List of Physical Constants

Symbol	Name	Approximate value
c	Speed of light in vacuum	2.998×10^{8} m/s
C	Coulomb – unit of charge	6.24×10^{18} charges
e	Base of the natural logarithm	2.7183
e^{-}	Charge of one electron	-1.602×10^{-19} C
ε_{o}	Permittivity of vacuum	8.8542×10^{-12} F/m
η_{o}	Intrinsic impedance of free space	$377\,\Omega$
μ_{o}	Free-space permeability	$4\pi \times 10^{-7}$ H/m
π	Pi	3.14159

Electromagnetics and Transmission Lines: Essentials for Electrical Engineering, Second Edition.
Robert A. Strangeway, Steven S. Holland, and James E. Richie.
© 2023 John Wiley & Sons, Inc. Published 2023 by John Wiley & Sons, Inc.
Companion website: www.wiley.com/go/Strangeway/ElectromagneticsandTransmissionLines

Appendix D

A Short List of Common Material Electrical Properties

The values for the following material electrical properties are approximate and are given to communicate a "feel for the numbers." These values can vary significantly based on manufacturing processes, purity, frequency, operating level of the electric or magnetic field, temperature, and so forth. Consult specific vendor literature for precise values under specified conditions. Do not use these values for design or applications. The decimal places carry no significant figure information.

Conductor	Conductivity σ (S/m)
Aluminium	3.5×10^7
Copper	6×10^7
Gold	4×10^7
Silver	6.2×10^7
Dielectric	**Relative permittivity ε_r**
Mica	5
Mylar	3
Polyethylene (PE)	2.3
Teflon	2.1
Vacuum	1
Material	**Relative permeability μ_r**
Ferrites	10–20 000
Iron/alloys/steel	100–200 000
Nickel	100–500
Most dielectrics	1
Nonmagnetic conductors	1
Vacuum	1

Electromagnetics and Transmission Lines: Essentials for Electrical Engineering, Second Edition.
Robert A. Strangeway, Steven S. Holland, and James E. Richie.
© 2023 John Wiley & Sons, Inc. Published 2023 by John Wiley & Sons, Inc.
Companion website: www.wiley.com/go/Strangeway/ElectromagneticsandTransmissionLines

Appendix E

Summary of Important Equations

Chapter 1 Vectors, Vector Algebra, and Coordinate Systems

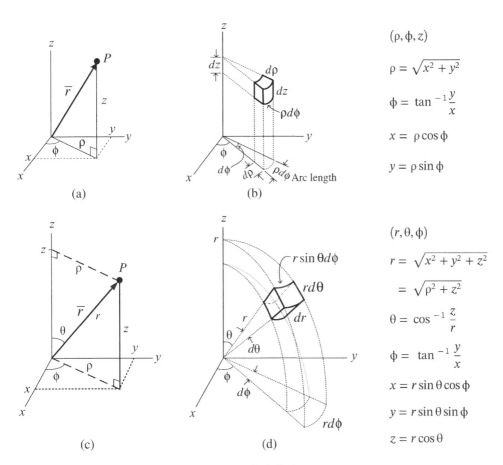

Appendix E Figure 1 (a) Cylindrical coordinates. (b) Cylindrical differential length elements. (c) Spherical coordinates. (d) Spherical differential length elements.

(ρ, ϕ, z)

$\rho = \sqrt{x^2 + y^2}$

$\phi = \tan^{-1}\dfrac{y}{x}$

$x = \rho \cos\phi$

$y = \rho \sin\phi$

(r, θ, ϕ)

$r = \sqrt{x^2 + y^2 + z^2}$

$\;\; = \sqrt{\rho^2 + z^2}$

$\theta = \cos^{-1}\dfrac{z}{r}$

$\phi = \tan^{-1}\dfrac{y}{x}$

$x = r \sin\theta \cos\phi$

$y = r \sin\theta \sin\phi$

$z = r \cos\theta$

Electromagnetics and Transmission Lines: Essentials for Electrical Engineering, Second Edition.
Robert A. Strangeway, Steven S. Holland, and James E. Richie.
© 2023 John Wiley & Sons, Inc. Published 2023 by John Wiley & Sons, Inc.
Companion website: www.wiley.com/go/Strangeway/ElectromagneticsandTransmissionLines

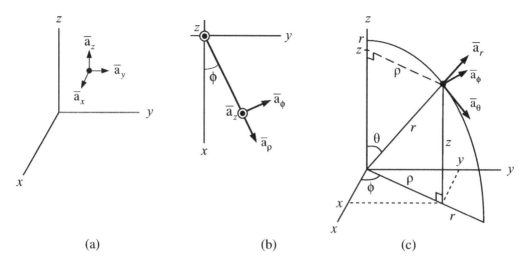

Appendix E Figure 2   (a) Cartesian unit vectors. (b) Cylindrical unit vectors. (c) Spherical unit vectors.

$$\bar{r} = x\bar{a}_x + y\bar{a}_y + z\bar{a}_z \qquad\qquad \bar{r} = \rho\bar{a}_\rho + z\bar{a}_z \qquad\qquad \bar{r} = r\bar{a}_r$$

$$\bar{R} = R_x\bar{a}_x + R_y\bar{a}_y + R_z\bar{a}_z \qquad \bar{R} = R_\rho\bar{a}_\rho + R_\phi\bar{a}_\phi + R_z\bar{a}_z \qquad \bar{R} = R_r\bar{a}_r + R_\theta\bar{a}_\theta + R_\phi\bar{a}_\phi \quad \bar{R} = R\bar{a}_R$$

$$\bar{A} = A_x\bar{a}_x + A_y\bar{a}_y + A_z\bar{a}_z \qquad \bar{A} = A_\rho\bar{a}_\rho + A_\phi\bar{a}_\phi + A_z\bar{a}_z \qquad \bar{A} = A_r\bar{a}_r + A_\theta\bar{a}_\theta + A_\phi\bar{a}_\phi \quad \bar{A} = A\bar{a}_A$$

$$A = |\bar{A}| = \sqrt{A_x^2 + A_y^2 + A_z^2} \quad A = |\bar{A}| = \sqrt{A_\rho^2 + A_\phi^2 + A_z^2} \quad A = |\bar{A}| = \sqrt{A_r^2 + A_\theta^2 + A_\phi^2}$$

$$\bar{R}_{QP} = \bar{r}_P - \bar{r}_Q \qquad\qquad \bar{a}_r = \frac{\bar{r}}{r} \qquad\qquad \bar{a}_R = \frac{\bar{R}}{R} \qquad\qquad \bar{a}_A = \frac{\bar{A}}{A}$$

$$\bar{A} \cdot \bar{B} = |\bar{A}||\bar{B}|\cos\theta = AB\cos\theta = A_xB_x + A_yB_y + A_zB_z$$

$$\bar{A} \times \bar{B} = AB\sin\theta\,\bar{a}_n = \left(+ A_yB_z - A_zB_y \right)\bar{a}_x + \left(+ A_zB_x - A_xB_z \right)\bar{a}_y + \left(+ A_xB_y - A_yB_x \right)\bar{a}_z = |\bar{A} \times \bar{B}|\,\bar{a}_n$$

$$\bar{a}_x \times \bar{a}_y = \bar{a}_z \qquad \bar{a}_\rho \times \bar{a}_\phi = \bar{a}_z \qquad \bar{a}_r \times \bar{a}_\theta = \bar{a}_\phi$$

$\bar{a}_{(\ )} \cdot \bar{a}_{(\ )}$	$\bar{a}_\rho$	$\bar{a}_\phi$	$\bar{a}_r$	$\bar{a}_\theta$
$\bar{a}_x$	$\cos\phi$	$-\sin\phi$	$\sin\theta\cos\phi$	$\cos\theta\cos\phi$
$\bar{a}_y$	$\sin\phi$	$\cos\phi$	$\sin\theta\sin\phi$	$\cos\theta\sin\phi$
$\bar{a}_z$	0	0	$\cos\theta$	$-\sin\theta$

Chapter 2 The Superposition Laws of Electric and Magnetic Fields

$$\bar{E} = \sum_{i=1}^{N} \bar{E_i} = \sum_{i=1}^{N} \frac{Q_i}{4\pi\varepsilon_o R_i^2}\,\bar{a}_{R_i} = \sum_{i=1}^{N} \frac{Q_i\bar{R}_i}{4\pi\varepsilon_o R_i^3} \;\; = \sum_{i=1}^{N} \frac{Q_i(\bar{r} - \bar{r}_i')}{4\pi\varepsilon_o|\bar{r} - \bar{r}_i'|^3}$$

$$\overline{E} = \int d\overline{E} = \int \frac{dQ}{4\pi\varepsilon_o R^2}\overline{a}_R = \int \frac{dQ\overline{R}}{4\pi\varepsilon_o R^3} = \int \frac{dQ(\overline{r}-\overline{r}')}{4\pi\varepsilon_o |\overline{r}-\overline{r}'|^3} \qquad \varepsilon_o = 8.8542\times10^{-12}\ \mathrm{F/m}$$

$dQ = \rho_L d\ell$ where ρ_L = line charge density $(\mathrm{C/m})$

$dQ = \rho_S dA$ where ρ_S = surface charge density $(\mathrm{C/m^2})$

$dQ = \rho_V dV$ where ρ_V = volume charge density $(\mathrm{C/m^3})$

$$\overline{E} = \frac{\rho_L}{2\pi\varepsilon_o\rho}\overline{a}_\rho \ \text{(infinitely long line charge)} \quad \overline{E} = \frac{\rho_S}{2\varepsilon_o}\overline{a}_n \ \text{(infinite plane of surface charge)}$$

$$\overline{F} = Q\left(\overline{E}+\overline{v}\times\overline{B}\right) \qquad \overline{F} = \int I\overline{d\ell}\times\overline{B}$$

$$\overline{B} = \mu\overline{H} \qquad\qquad \mu = \mu_o\mu_r \qquad\qquad \mu_o = 4\pi\times10^{-7}\ \mathrm{H/m}$$

$$I = \int \overline{J}\cdot\overline{dA} \qquad\qquad I = \int \overline{K}\cdot\overline{dw}$$

$$\overline{H} = \int d\overline{H} = \int \frac{Id\ell\overline{a}_I\times\overline{R}}{4\pi R^3} \ \text{(line currents)} \quad \overline{H} = \frac{I}{2\pi\rho}\overline{a}_\phi \ \text{(infinitely long straight line current)}$$

Chapter 3 The Flux Laws of Electric and Magnetic Fields

$$\Psi_{\mathrm{net}} = \oint \overline{D}\cdot\overline{dA} = Q_{\mathrm{encl}} \qquad \overline{D} = \varepsilon_o\overline{E} \ \text{(free space)} \qquad \overline{D}\cdot\overline{a}_n = D_n = \rho_S \ \text{(perfect conductor)}$$

$$\Phi_{\mathrm{net}} = \oint \overline{B}\cdot\overline{dA} = 0 \qquad\qquad \Phi = \int \overline{B}\cdot\overline{dA} \qquad\qquad \overline{B} = \mu\overline{H} = \mu_o\mu_r\overline{H} \qquad B_{n1} = B_{n2}$$

Chapter 4 The Path Laws and Circuit Principles

$$V_{BA} = \frac{W}{Q} = -\int_A^B \overline{E}\cdot\overline{d\ell} = V_B - V_A \qquad\qquad \oint\overline{E}\cdot\overline{d\ell} = 0 \quad (\text{static } \overline{E}) \qquad\qquad \overline{D} = \varepsilon\overline{E} = \varepsilon_o\varepsilon_r\overline{E}$$

Perfect conductor BCs : $\qquad\qquad E_t = 0 \quad D_n = \rho_S \qquad\qquad\qquad \varepsilon_o = 8.8542\times10^{-12}\ (\mathrm{F/m})$

$$C = \frac{Q}{V} \qquad\qquad C_{\mathrm{parallel\ plate}} = \frac{\varepsilon A}{d} \qquad\qquad C'_{\mathrm{coax}} = \frac{2\pi\varepsilon}{\ln\left(\dfrac{b}{a}\right)} \ (\mathrm{F/m}) \qquad\qquad W = \frac{1}{2}\varepsilon\int E^2 dV$$

$$\overline{J} = \sigma\overline{E} \qquad R = \frac{\rho\ell}{A} \qquad\qquad P = \int \sigma E^2 dV \qquad\qquad \oint\overline{H}\cdot\overline{d\ell} = I_{\mathrm{encl}} \qquad\qquad \overline{H} = \frac{I}{2\pi\rho}\overline{a}_\phi$$

$$L = \frac{\Lambda}{I} = \frac{N\Phi}{I} \qquad\qquad L_{\mathrm{coil}} = \frac{\mu N^2 A}{\ell} \qquad\qquad L'_{\mathrm{coax}} = \frac{\mu}{2\pi}\ln\left(\frac{b}{a}\right) \ (\mathrm{H/m})$$

$$\overline{B} = \mu\overline{H} = \mu_o\mu_r\overline{H} \qquad\qquad \mu_o = 4\pi\times10^{-7}\ (\mathrm{H/m}) \qquad\qquad W = \frac{1}{2}\mu\int H^2 dV$$

Magnetic BCs : $\qquad\qquad B_{n1} = B_{n2} \qquad H_{t1} = H_{t2}(K = 0) \qquad\qquad \overline{H} = \overline{K}\times\overline{a}_n \text{(perfect conductor)}$

Chapter 5 Maxwell's Equations

$$\text{FL:} \oint \overline{E} \cdot \overline{d\ell} = -\frac{d}{dt}\int \overline{B} \cdot \overline{dA} = -\int \frac{\partial \overline{B}}{\partial t} \cdot \overline{dA} + \oint (\overline{v} \times \overline{B}) \cdot \overline{d\ell} \qquad \text{GL:} \oint \overline{D} \cdot \overline{dA} = Q_{\text{encl}}$$

$$\text{ACL:} \oint \overline{H} \cdot \overline{d\ell} = I_c + \frac{d}{dt}\int \overline{D} \cdot \overline{dA} \qquad \text{solenoidal } \Phi: \oint \overline{B} \cdot \overline{dA} = 0$$

Chapter 6 Transmission Lines: Waves and Reflections

$$Z_0 \equiv \left.\frac{V}{I}\right|_{\substack{\text{one wave}\\ \text{(one direction)}}} \qquad \text{DC: } \Gamma_L = \frac{V_{\text{reflected}}}{V_{\text{incident}}} = \frac{R_L - Z_0}{R_L + Z_0} \qquad \text{AC: } \Gamma_L = \frac{\widetilde{V}_L^-}{\widetilde{V}_L^+} = \frac{\widetilde{Z}_L - Z_0}{\widetilde{Z}_L + Z_0}$$

$$\text{vel.} = \frac{\ell}{\tau_d} = \frac{1}{\sqrt{\mu\varepsilon}} = \frac{1}{\sqrt{\mu_o\varepsilon_o\mu_r\varepsilon_r}} = \frac{c}{\sqrt{\mu_r\varepsilon_r}} \qquad \text{vel.} = \lambda f \qquad \beta = \frac{2\pi}{\lambda} \qquad \overline{S} = \overline{E} \times \overline{H}$$

$$VSWR \equiv \frac{|\widetilde{V}_{\text{max}}|}{|\widetilde{V}_{\text{min}}|} = \frac{1 + |\Gamma|}{1 - |\Gamma|} \qquad RL(\text{dB}) = -20\log|\Gamma| \qquad \text{normalization: } z_L = \frac{\widetilde{Z}_L}{Z_0}$$

$$A(\text{dB}) = 10\log\left(\frac{P_{\text{in}}}{P_{\text{out}}}\right) = -G\,(\text{dB}) \qquad P(\text{dBm}) \equiv 10\log\left[\frac{P(\text{W})}{1\,\text{mW}}\right]$$

$$S_{ii} = \Gamma_i = \left.\frac{\widetilde{V}_i^-}{\widetilde{V}_i^+}\right|_{\text{port } j \text{ terminated in } Z_0} \qquad S_{12} = \left.\frac{\widetilde{V}_1^-}{\widetilde{V}_2^+}\right|_{\text{port 1 terminated in } Z_0} \qquad S_{21} = \left.\frac{\widetilde{V}_2^-}{\widetilde{V}_1^+}\right|_{\text{port 2 terminated in } Z_0}$$

$$S_{ij} = S_{ji} \ (i \neq j) \text{ for reciprocal devices}$$

$$|S_{ii}|(\text{dB}) = 20\log(|S_{ii}|) = 20\log(|\Gamma_i|) = -RL_i(\text{dB}) \qquad |S_{21}|(\text{dB}) = 20\log|S_{21}| = G(\text{dB}) = -IL(\text{dB})$$

Chapter 7 Transmission Lines: Theory and Applications

Telegrapher's equations: $\dfrac{d\widetilde{V}}{dz} = -j\omega L'\widetilde{I}$ and $\dfrac{d\widetilde{I}}{dz} = -j\omega C'\widetilde{V}$ T-line wave equation: $\dfrac{d^2\widetilde{V}}{dz^2} = -\omega^2 L'C'\widetilde{V}$

$z = $ distance from the source toward the load $d = $ distance from the load toward the source

$$\widetilde{V} = \widetilde{V}(f,d) = \frac{1}{2}\left(\widetilde{V}_L + Z_0\widetilde{I}_L\right)e^{+\gamma d} + \frac{1}{2}\left(\widetilde{V}_L - Z_0\widetilde{I}_L\right)e^{-\gamma d} \qquad \gamma = j\omega\sqrt{L'C'} = j\beta \text{ (lossless T-line)}$$

$$v(t,d) = V_L^+\cos\left(\omega t + \beta d + \theta_L^+\right) + V_L^-\cos\left(\omega t - \beta d + \theta_L^-\right) \qquad Z_0 \equiv \sqrt{\frac{L'}{C'}} \qquad Z_0 = \frac{\widetilde{V}^+}{\widetilde{I}^+} = \frac{\widetilde{V}^-}{\left(-\widetilde{I}^-\right)}$$

$$\beta = \frac{2\pi}{\lambda} = \frac{\omega}{vel.} = \frac{2\pi\sqrt{\mu_r\varepsilon_r}}{c}f \qquad \gamma = \alpha + j\beta = \alpha\,(\text{Np/m}) + j\beta(\text{rad/m}) \text{ (lossy T-line)} \quad 1\,\text{Np} = 8.686\,\text{dB}$$

$$\widetilde{Z} = \widetilde{Z}(f,d) = Z_0\frac{\left(\widetilde{Z}_L + Z_0\right)e^{+\gamma d} + \left(\widetilde{Z}_L - Z_0\right)e^{-\gamma d}}{\left(\widetilde{Z}_L + Z_0\right)e^{+\gamma d} - \left(\widetilde{Z}_L - Z_0\right)e^{-\gamma d}} \qquad \text{(general)}$$

$$\widetilde{Z} = \widetilde{Z}(f,d) = Z_0\frac{\widetilde{Z}_L + jZ_0\tan\beta d}{Z_0 + j\widetilde{Z}_L\tan\beta d} \qquad \text{(lossless T-line only)}$$

$$\widetilde{Z}_L = Z_0 \frac{\left(\widetilde{Z}_{\text{in}} + Z_0\right) e^{-\gamma\ell} + \left(\widetilde{Z}_{\text{in}} - Z_0\right) e^{+\gamma\ell}}{\left(\widetilde{Z}_{\text{in}} + Z_0\right) e^{-\gamma\ell} - \left(\widetilde{Z}_{\text{in}} - Z_0\right) e^{+\gamma\ell}} \qquad \widetilde{Z}_L = Z_0 \frac{\widetilde{Z}_{\text{in}} - jZ_0 \tan \beta\ell}{Z_0 - j\widetilde{Z}_{\text{in}} \tan \beta\ell}$$

$$\Gamma_{\text{in}} = \Gamma(f, \ell) = \frac{\widetilde{Z}_{\text{in}} - Z_0}{\widetilde{Z}_{\text{in}} + Z_0} \qquad \Gamma = \Gamma(f, d) = \frac{\widetilde{Z}(f, d) - Z_0}{\widetilde{Z}(f, d) + Z_0} \qquad \Gamma_L = \Gamma_L(f) = \frac{\widetilde{V}_L^-}{\widetilde{V}_L^+} = \frac{-\widetilde{I}_L^-}{\widetilde{I}_L^+} = \frac{\widetilde{Z}_L - Z_0}{\widetilde{Z}_L + Z_0}$$

$$\beta d[\lambda] = \beta d(\text{rad}) \left(\frac{1[\lambda]}{2\pi \, (\text{rad})}\right) \qquad \beta d[\lambda] = d(\text{m}) \frac{1[\lambda]}{\lambda(\text{m})}$$

Chapter 8 Antennas and Links

$$v = \lambda f = \frac{1}{\sqrt{\mu\varepsilon}} \qquad \beta = \omega\sqrt{\mu\varepsilon} \qquad \eta = \frac{E_o^+}{H_o^+} = \sqrt{\frac{\mu}{\varepsilon}} \qquad \eta_o = 377 \, \Omega$$

$$\overline{S} = \overline{E} \times \overline{H} \qquad S_{\text{ave}} = \frac{E_o H_o}{2} = \frac{E_o^2}{2\eta} = \frac{H_o^2\eta}{2} \qquad P_{\text{ave}} = \int \overline{S}_{\text{ave}} \cdot \overline{dA} \qquad P_{\text{ave}} = S_{\text{ave}} A_e$$

$$G_{\text{ant}} = \frac{4\pi A_e}{\lambda^2} \qquad L_{fs} = \frac{P_t}{P_r} = \left(\frac{4\pi d}{\lambda}\right)^2 \qquad L = \frac{P_t}{P_r} = \frac{L_{fs} L_a L_o}{G_t G_r} \qquad EIRP = G_t P_t$$

Chapter 9 Signal Integrity

$$BW_{\text{rise time}} \approx \frac{0.35}{t_{\text{rise}}} \text{ to } \frac{0.5}{t_{\text{rise}}} \qquad v_{\text{coupled}} = M \frac{di_{\text{driven}}}{dt} \qquad i_{\text{coupled}} = C_m \frac{dv_{\text{driven}}}{dt}$$

Radiated EMI measurements: $AF_{\text{electric}} = \dfrac{E_{\text{inc}}}{V_{\text{rcv}}} \qquad E(\text{dB}\mu\text{V/m}) = 20\log\left[E(\text{V/m}) \times 10^6\right]$

$$E_{\text{inc}}\left(\frac{\text{dB}\mu\text{V}}{\text{m}}\right) = V_{\text{rcv}}(\text{dB}\mu\text{V}) + AF_{\text{electric}}\left(\frac{\text{dB}}{\text{m}}\right) + \text{other correction factors}$$

Bibliography

This bibliography is not intended to be encyclopedic but does reference many works, new and old, that readers may find of interest in their studies of electromagnetics and transmission lines, antennas, and signal integrity.

 1. Anderson, E.M. (1985). *Electric Transmission Lines Fundamentals*. Reston.
 2. Balanis, C.A. (2016). *Antenna Theory: Analysis and Design*, 4e. New Jersey: Wiley.
 3. Bogatin, E. (2018). *Signal Integrity – Simplified*, 3e. New York: Pearson.
 4. Brown, R.G., Sharpe, R.A., Hughes, W.L., and Post, R.E. (1973). *Lines, Waves, and Antennas – The Transmission of Electric Energy*, 2e. New York: Wiley.
 5. Cheng, D.K. (1989). *Field and Wave Electromagnetics*, 2e. Massachusetts: Addison-Wesley.
 6. Ghannouchi, F.M. and Mohammadi, A. (2009). *The Six-Port Technique with Microwave and Wireless Applications*. Massachusetts: Artech House.
 7. Haykin, S. and Moher, M. (2009). *Communication Systems*, 5e. New Jersey: Wiley.
 8. Hayt, W.H. and Buck, J.A. (2019). *Engineering Electromagnetics*, 9e. New York: McGraw-Hill.
 9. Ishii, T.K. (1995). *Microwave Engineering*, 2e. Oxford.
10. Johnson, H. and Graham, M. (2003). *High-Speed Signal Propagation: Advanced Black Magic*. New York: Pearson.
11. Jordan, E.C. and Balmain, K.G. (1968). *Electromagnetic Waves and Radiating Systems*, 2e. Prentice-Hall.
12. Kraus, J.D. and Marhefka, R.J. (2001). *Antennas: For All Applications*, 3e. New York: McGraw Hill.
13. Lance, A.L. (1964). *Introduction to Microwave Theory and Measurements*. New York: McGraw-Hill.
14. Laverghetta, T.S. (2005). *Microwave and Wireless Simplified*, 2e. Massachusetts: Artech House.
15. Ludwig, R. and Bretchko, P. (2000). *RF Circuit Design – Theory and Applications*. New York: Pearson.
16. Morrison, R. (2016). *Grounding and Shielding: Circuits and Interference*, 6e. New Jersey: Wiley.
17. Ohanian, H.C. (1988). *Classical Electrodynamics*. Massachusetts: Allyn and Bacon.
18. Ott, H.W. (2009). *Electromagnetic Compatibility Engineering*. New Jersey: Wiley.
19. Paul, C.R., Scully, R.C., and Setffka, M.A. (2022). *Introduction to Electromagnetic Compatibility*, 3e. New Jersey: Wiley.
20. Pozar, D.M. (2012). *Microwave Engineering*, 4e. New Jersey: Wiley.
21. Ramo, S., Whinnery, J.R., and Van Duzer, T. (1994). *Fields and Waves in Communication Electronics*, 3e. New York: Wiley.
22. Rizzi, P.R. (1988). *Microwave Engineering – Passive Circuits*. New York: Pearson.
23. Sadiku, M.N.O. (2018). *Elements of Electromagnetics*, 7e. Oxford.
24. Schmitt, R. (2002). *Electromagnetics Explained—A Handbook for Wireless/RF, EMC, and High-Speed Electronics*. United Kingdom: Newness-Elsevier Science.

Electromagnetics and Transmission Lines: Essentials for Electrical Engineering, Second Edition.
Robert A. Strangeway, Steven S. Holland, and James E. Richie.
© 2023 John Wiley & Sons, Inc. Published 2023 by John Wiley & Sons, Inc.
Companion website: www.wiley.com/go/Strangeway/ElectromagneticsandTransmissionLines

25. Scott, W.T. (1966). *The Physics of Electricity and Magnetism*, 2e. New York: Wiley.

26. Skilling, H.H. (1948). *Fundamentals of Electric Waves*, 2e. Wiley reprinted by Krieger, 1974.

27. Smith, P.H. (1969). *Electronic Applications of the Smith Chart*. New York: McGraw-Hill.

28. Strangeway, R.A., Petersen, O.G., Gassert, J.D., and Lokken, R.J. (2019). *Electric Circuits*. Wisconsin: RacademicS.

29. Weidner, R.T. and Sells, R.L. (1965). *Elementary Classical Physics*, vol. 2. Massachusetts: Allyn and Bacon.

30. Wentworth, S.M. (2005). *Fundamentals of Electromagnetics with Engineering Applications*. New Jersey: Wiley.

Select Answers to Homework Problems

All symbolic answers without specific values for constants are in SI units unless stated otherwise.

Chapter 1

1. a) $7\bar{a}_x + 5\bar{a}_y - 2\bar{a}_z$
   c) $+58.4°$
   d) $2.40\bar{a}_x + 0\bar{a}_y + 4.80\bar{a}_z$
   e) $-0.512\bar{a}_x + 0.820\bar{a}_y + 0.256\bar{a}_z$

2. a) $3\bar{a}_x - 1\bar{a}_y + 9\bar{a}_z$
   c) $+62.1°$
   d) $4.18\bar{a}_x - 0.465\bar{a}_y + 0.930\bar{a}_z$
   e) $-0.093\bar{a}_x - 0.993\bar{a}_y - 0.079\bar{a}_z$

3. a) $-9\bar{a}_x - 2\bar{a}_y - 7\bar{a}_z$
   b) $-0.941\bar{a}_x + 3.765\bar{a}_y$
   c) $-0.587\bar{a}_x - 0.147\bar{a}_y + 0.796\bar{a}_z$

4. a) $+9\bar{a}_x + 4\bar{a}_y - 2\bar{a}_z$
   b) $+75.3°$
   c) 1.94
   d) $-0.433\bar{a}_x + 0.882\bar{a}_y - 0.186\bar{a}_z$

5. $\pi(b^2 - a^2)h$

6. $\pi(b^2 - a^2) + 2h(b - a) + \pi h(b + a)$

7. $\dfrac{4}{3}\pi(b^3 - a^3)$

8. $\pi(b^2 - a^2) + 2\pi(b^2 + a^2)$

13. $7.06\bar{a}_\rho + 17.6\bar{a}_\phi + 18.0\bar{a}_z$

14. $-10.1\bar{a}_x + 16.1\bar{a}_y - 7.0\bar{a}_z$

15. $6.40\bar{a}_x - 5.20\bar{a}_y + 5.00\bar{a}_z$

16. $1.79\bar{a}_x + 8.05\bar{a}_y + 5.00\bar{a}_z$

17. $632\bar{a}_\rho - 652\bar{a}_\phi - 368\bar{a}_z$

18. $-779\bar{a}_\rho + 467\bar{a}_\phi - 368\bar{a}_z$

19. $16.0\bar{a}_x - 11.5\bar{a}_y + 23.3\bar{a}_z$

20. $6.50\bar{a}_x + 17.9\bar{a}_y + 23.9\bar{a}_z$

21. $-438\bar{a}_r + 233\bar{a}_\theta - 845\bar{a}_\phi$

22. $891\bar{a}_r - 4\bar{a}_\theta - 407\bar{a}_\phi$

23. $\bar{A} = 12.0\bar{a}_x + 16.0\bar{a}_y - 13.6\bar{a}_z$
   $\bar{A} = 19.8\bar{a}_\rho + 2.8\bar{a}_\phi - 13.6\bar{a}_z$
   $\bar{A} = 8.3\bar{a}_r + 22.5\bar{a}_\theta + 2.8\bar{a}_\phi$

24. $\bar{A} = 36.0\bar{a}_x + 16.0\bar{a}_y + 5.50\bar{a}_z$
   $\bar{A} = -26.6\bar{a}_\rho + 29.1\bar{a}_\phi + 5.50\bar{a}_z$
   $\bar{A} = -27.0\bar{a}_r + 2.70\bar{a}_\theta + 29.1\bar{a}_\phi$

27. $\bar{A} - \dfrac{\bar{A} \cdot \bar{B}}{B}$

28. $\dfrac{1}{2}\left|\overline{MP} \times \overline{MQ}\right|$

Chapter 2

3. a) $\bar{E} = +189.5\bar{a}_x + 252.7\bar{a}_y - 94.8\bar{a}_z$ V/m
   b) $\bar{E} = +189.5\bar{a}_x - 166.6\bar{a}_y - 24.9\bar{a}_z$ V/m

Electromagnetics and Transmission Lines: Essentials for Electrical Engineering, Second Edition.
Robert A. Strangeway, Steven S. Holland, and James E. Richie.
© 2023 John Wiley & Sons, Inc. Published 2023 by John Wiley & Sons, Inc.
Companion website: www.wiley.com/go/Strangeway/ElectromagneticsandTransmissionLines

4. a) $\overline{E} = +308\overline{a}_x + 184.8\overline{a}_y + 154.0\overline{a}_z \text{ V/m}$

    b) $\overline{E} = +516.2\overline{a}_x + 155.0\overline{a}_y + 198.6\overline{a}_z \text{ V/m}$

5. a) 226 C

    b) 452 C

6. a) 679 C

    b) 151 C

8. b) $\overline{E} \approx \dfrac{\rho_L a}{2\varepsilon_o h^2}(-\overline{a}_z)$

    c) $\overline{E} \approx 0$

10. $\overline{E} = \dfrac{\rho_L}{4\pi\varepsilon}\left(\dfrac{1}{z_1-h} - \dfrac{1}{z_1+h}\right)\overline{a}_z = \cdots(\text{algebra})\cdots$

    $= \dfrac{\rho_L h}{2\pi\varepsilon\left(z_1^2 - h^2\right)}\overline{a}_z$

11. a) $2\rho\overline{a}_y$

    b) $2\rho(-\overline{a}_x)$

12. a) $2r\overline{a}_\rho$

    b) $2r(-\overline{a}_z)$

    c) $r\overline{a}_y - r\overline{a}_z$

13. $\overline{E} = \dfrac{\rho_L}{2\pi\varepsilon a}\overline{a}_x$

14. a) $\overline{E} = \dfrac{\rho_S}{2\varepsilon}\left[1 - \dfrac{a}{\sqrt{a^2 + h^2}}\right](-\overline{a}_z)$

    b) $\overline{E} \approx 0$

15. a) $\overline{E} = \dfrac{\rho_S h}{2\pi\varepsilon\sqrt{a^2 + h^2}}(+\overline{a}_x)$

    $+ \dfrac{\rho_S}{4\varepsilon}\left[1 - \dfrac{a}{\sqrt{a^2 + h^2}}\right](-\overline{a}_z)$

    b) $\overline{E} \approx \dfrac{\rho_S h}{2\pi\varepsilon_o a}\overline{a}_x$

19. 10.6 mA

20. 66.6 mA

21. $\overline{H} = \dfrac{I}{4\pi a}(-\overline{a}_x)$

22. $\overline{H} = \dfrac{I a^2}{8\left(a^2 + h^2\right)^{3/2}}\overline{a}_z$

Chapter 3

3. $\overline{D} = \dfrac{\rho_\ell}{2\pi\rho}\overline{a}_\rho$

4. $\overline{D} = \dfrac{\rho_s a^2}{r^2}\overline{a}_r$

5. $\overline{D} = \rho_V \dfrac{a^2}{2\rho}\overline{a}_\rho$

6. $\overline{D} = \pm\dfrac{\rho_S}{2}\overline{a}_z$

8. a) $\overline{E} = \dfrac{\rho_S a}{\varepsilon_o \rho}\overline{a}_\rho$

9. b) $\overline{E} = 0$

    c) $\overline{D} = \dfrac{\rho_v\left(b^2 - a^2\right)}{2\rho}\overline{a}_\rho$

    d) $\overline{D} = \dfrac{\rho_v\left(\rho^2 - a^2\right)}{2\rho}\overline{a}_\rho$

10. b) $\overline{E} = 0$

    c) $\overline{D} = \dfrac{\rho_v\left(b^3 - a^3\right)}{3r^2}\overline{a}_r$

    d) $\overline{D} = \dfrac{\rho_v\left(r^3 - a^3\right)}{3r^2}\overline{a}_r$

11. b) $\overline{E} = 0$

    c) $\overline{D} = \dfrac{K\left(b^3 - a^3\right)}{3\rho}\overline{a}_\rho$

    d) $\overline{D} = \dfrac{K\left(\rho^3 - a^3\right)}{3\rho}\overline{a}_\rho$

12. b) $\overline{E} = 0$

    c) $\overline{D} = \dfrac{K\left(b^5 - a^5\right)}{5r^2}\overline{a}_r$

    d) $\overline{D} = \dfrac{K\left(r^5 - a^5\right)}{5r^2}\overline{a}_r$

13. $Q_{\text{encl}} = \dfrac{K_1 2\pi\ell}{a} + K_2\pi a^2\ell^2$

14. $Q_{\text{encl}} = \dfrac{4\pi\varepsilon K}{a}$

17. $\Phi = \dfrac{\mu I\ell}{2\pi}\ln\left(\dfrac{b}{a}\right)$

18. $\Phi = \dfrac{2\pi\mu_o H_o}{k}\left(e^{-ka} - e^{-kb}\right)$

19. $\Phi = \dfrac{\mu NIA}{\ell}$

20. a) $H_{\text{air}} = 796 \text{ kA/m}, \quad H_{\text{core}} = 796 \text{ A/m}$

Chapter 4

1. -6.85 V

2. -62.3 V

3. -11.0 V

4. 2.71 MV

7. V same, E same, Q decreases, D decreases, Ψ decreases, C decreases

8. Q same, D same, E decreases, V decreases, Ψ same, C increases

14. b) increase

17. a) $\overline{H} = \dfrac{J_o b^5}{5\rho}\overline{a}_\phi$

    b) $\overline{H} = \dfrac{J_o \rho^4}{5}\overline{a}_\phi$

18. a) 0

    b) $\overline{H} = \dfrac{J_o\left(b^5 - a^5\right)}{5\rho}\overline{a}_\phi$

    c) $J_o\Big|_{\text{hollow}} = \dfrac{b^5}{b^5 - a^5}J_o\Big|_{\text{solid}}$

20. a) $\overline{H} = 0$

21. $\overline{H} = \dfrac{I}{2\pi\,\rho}\overline{a}_\phi$

23. a) No

25. d) Yes

27. $L = \dfrac{\mu N^2 A}{2\pi\rho_{\text{mean}}}$

28. $L = \dfrac{\mu N^2 h}{2\pi}\ln\left(\dfrac{b}{a}\right)$

31. b) inside: $+\overline{a}_z$; outside: $-\overline{a}_z$

32. $-\overline{a}_y$

33. a) 4.78 mA

Chapter 5

1. a) no
   b) yes

2. yes

4. b) yes

5. a) no
   b) yes

6. for 60 Hz, $|V| = 1.2\,\mu V$
   for 1 GHz, $|V| = 20 \text{ V}$
   square wave

13. a) 8.96 A
    b) 46.7 mA

14. a) 46.7 mA
    b) 8.96 A

16. a) yes
    b) yes

Chapter 6

1. 6.13 mm; 0.613 mm

2. 0.106 m

3. 0.126 µs (using $\varepsilon_r = 2.3$)

4. 1034 m

7. V_S: 50 V step at $t = 0$,
   step to 16.7 V at $t = 8\,\mu s$;
   DC steady state $= 16.7$ V
   V_L: 16.7 V step at $t = 4\,\mu s$;
   DC steady state $= 16.7$ V

8. a) V_S: 50 V step at $t = 0$,
  step to 66.7 V at $t = 8$ µs;
  DC steady state = 66.7 V
  V_L: 66.7 V step at $t = 4$ µs;
  DC steady state = 66.7 V

  b) V_S: 50 V step at $t = 0$,
  step to 0 V at $t = 8$ µs;
  DC steady state = 0 V
  V_L: remains at 0 V;
  DC steady state = 0 V

  c) V_S: 50 V step at $t = 0$,
  step to 100 V at $t = 8$ µs;
  DC steady state = 100 V
  V_L: 100 V step at $t = 4$ µs;
  DC steady state = 100 V

13. $VSWR = 1.98$, $RL = 9.68$ dB
  cannot determine $\angle\Gamma$ from
  $VSWR$ or RL

14. $1.756 + j0.525$

17. 14.1 dBm, 25.7 mW

18. 13.0 dB

19. +0.9 dBm, −11.8 dBm

20. a) 49.8 mW = 17.0 dBm
  b) 89.7 mW = 19.5 dBm

21. a) $|s_{11}| = |s_{22}| = 0.0562$
  $|s_{12}| = |s_{21}| = 0.316$

22. a) $|s_{11}| = 0.0476$, $|s_{12}| = 0.0316$
  $|s_{21}| = 10.0$, $|s_{22}| = 0.333$

Chapter 7

7. b) $\widetilde{V}(d) = \widetilde{V}_L\, e^{+\gamma d}$

9. a) 75 Ω
  b) 2.02×10^8 m/s

10. a) 111 Ω
  b) 2.998×10^8 m/s

11. $195 - j123$ Ω

12. a) $34.9 - j11.1$ Ω
  b) $0.376 \angle -55.5°$
  c) $0.376 \angle -158.7°$
  d) and e) 2.21

14. $13.8 + j11.8$ Ω

15. c) 5.565
  d) 4.396

16. c) 5.565
  d) 5.565

17. a) $0.766 \angle +64.9°$
  b) 7.56
  c) 2.31 dB
  d) $255.5 - j40.9$ Ω (lossy T-line)
    $357.8 - j84.1$ Ω (lossless T-line)

18. a) $0.766 \angle +65°$
  b) 7.5
  c) 2.3 dB
  d) $360 - j80$ Ω

20. a)–e) see Problem 12
  f) 34 Ω, 0.173 [λ]
  g) 165 Ω, 0.423 [λ]
  h) 0.250 λ

23. a) increase
  b) further
  c) increase

25. a) less than quarter wavelength

26. a) more than quarter wavelength

Chapter 8

1. b) 248 Ω

2. a) $-\overline{a}_z$
  b) 26.5 nA/m

3. 100 W = 20 dBW

5. a) 60 pW
  b) −102 dBW
  c) −72 dBm

6.   a)   592 pW

      b)   −92 dBW

7.   a)   8.68 kV/m

      b)   23.0 A/m

      c)   100 kW/m^2

8.   a)   45 000 m^2

9.   a)   3.19 nW

      b)   2 W

10.   a)   65 dBW

      b)   754 pW = −91 dBW

11.   a)   125 dB

      b)   149.5 dB

      c)   85 dBW

      d)   −95 dBW

      e)   −142.5 dBW

12.   a)   for 500 m: 81.5 dB (link)

             65.6 dB (cable)

             for 1 km: 87.5 dB (link)

             131.2 dB (cable)

Index

a

ACL. *See* Ampere's circuital law (ACL)

AC sinusoidal waves on T-lines   153–158

American National Standards Institute (ANSI)   238

Ampere's circuital law (ACL)   96

   application to boundary condition   103–105

   determination of magnetic field intensity   97–100

   and displacement current   140

   intuitive development of   96–97

Anechoic chamber   226, 240

Antenna(s)   215

   charge acceleration   217–218

   description   217

   dipole   216

   effective area of   225

   factor   240

   far field   218, 226

   free-space loss   228–229

   gain   224–225

   isotropic radiator   224

   in links   228–230

   near field   226

   parameters   224–227

   polarization of   223–224

   radiation pattern   20, 225–226

   radiation resistance   226–227

   symbol   217

   transition from T-line to antenna   216–217

   uses of   217

   VSWR   227

Attenuation   167

Attenuation constant   192

Average detector   239

b

Bandwidth

   resolution   238

   rise time   233–234

Biot–Savart law (BSL)   33, 50–55

Bounce diagram   152–153

Boundary conditions   71

   dielectric–conductor electric field   70–71, 86–87

   dielectric–dielectric electric field   106–108

   magnetic field   103–105

Bound charge   91, 108–109

c

Capacitance   87

   approach to determination of   88

Electromagnetics and Transmission Lines: Essentials for Electrical Engineering, Second Edition.
Robert A. Strangeway, Steven S. Holland, and James E. Richie.
© 2023 John Wiley & Sons, Inc. Published 2023 by John Wiley & Sons, Inc.
Companion website: www.wiley.com/go/Strangeway/ElectromagneticsandTransmissionLines

Printed and bound by CPI Group (UK) Ltd, Croydon, CR0 4YY

16/04/2025

14658353-0005